Physical Security and Environmental Protection

Physical Security and Environmental Protection

John Perdikaris

CRC Press
Taylor & Francis Group
Boca Raton   London   New York

CRC Press is an imprint of the
Taylor & Francis Group, an **informa** business

CRC Press
Taylor & Francis Group
6000 Broken Sound Parkway NW, Suite 300
Boca Raton, FL 33487-2742

First issued in paperback 2017

© 2014 by Taylor & Francis Group, LLC
CRC Press is an imprint of Taylor & Francis Group, an Informa business

No claim to original U.S. Government works

Version Date: 20140317

ISBN 13: 978-1-4822-1194-8 (hbk)
ISBN 13: 978-1-138-07493-4 (pbk)

Library of Congress Cataloging-in-Publication Data

Perdikaris, John.
   Physical security and environmental protection / author, John Perdikaris.
      pages cm
   Includes bibliographical references and index.
   ISBN 978-1-4822-1194-8 (hardback)
   1. Emergency management. 2. Disaster relief. 3. Environmental protection. I. Title.

HV551.2.P39 2014
363.34'8--dc23   2014001343

Visit the Taylor & Francis Web site at
http://www.taylorandfrancis.com

and the CRC Press Web site at
http://www.crcpress.com

Contents

Preface...ix

Author xi

Chapter 1   Emergency Management..1

    1.1   Introduction to Disasters and Emergency Management............1
    1.2   Types of Disasters..2
    1.3   Emergency Planning...4
    1.4   All-Hazards Approach to Disaster Management 12
    1.5   Decision Making and Problem Solving................................. 15
    1.6   Leadership and Influence 18
    1.7   Effective Communication Skills 22
    1.8   Developing and Managing Volunteers 24

Chapter 2   Threat Ensemble, Vulnerability, and Risk Assessments....................29

    2.1   Introduction 29
    2.2   Threat Assessment.. 31
    2.3   Vulnerability Assessment 39
    2.4   Risk Assessment 45

Chapter 3   Critical Infrastructure Protection...47

    3.1   Introduction 47
    3.2   Building Security..48
            3.2.1   Unauthorized Entry (Forced and Covert)................... 67
            3.2.2   Insider Threats..68
            3.2.3   Explosive Threats: Stationary and Moving
                       Vehicle-Delivered Bombs, Mail Bombs,
                       and Package Bombs...68
            3.2.4   Ballistic Threats 69
            3.2.5   WMD: CBR...70
            3.2.6   Cyber and Information Security Threats 71
            3.2.7   Development and Training on Occupant
                       Emergency Plans 71
    3.3   Water Supply Systems Security... 71
    3.4   Security for Energy Facilities..87
    3.5   Food and Agricultural Security..93
    3.6   Aviation Security 95
    3.7   Maritime Security and Asset Protection 99

3.8   Land Transportation Security Systems 103
3.9   Cybersecurity 107

Chapter 4   Targeted Violence and Violent Behavior.. 113

4.1   Introduction 113
4.2   Methods and the Mind-Set of a Terrorist 114
4.3   Terrorism 116
4.4   Organized Crime 120
4.5   Maritime Piracy.. 123
4.6   Incidents and Indicators... 126
4.7   Suspicious Activity 129
4.8   Avoiding an Attack.. 133

Chapter 5   Protection Strategies.. 137

5.1   Physical Security 137
5.2   Improvised Explosive Devices Awareness 143
5.3   Surveillance and Countersurveillance.................................. 145
5.4   Conducting a Site Security Survey....................................... 154
5.5   Geospatial Intelligence 157
5.6   Technical Surveillance and Countermeasures 159
5.7   Protection against Explosives and Blast Effects.................. 162
        5.7.1   Stationary Vehicle along Secured Perimeter Line ... 165
        5.7.2   Stationary Vehicle in a Parking Garage or
                 Loading Dock... 165
        5.7.3   Moving Vehicle Attack... 165
        5.7.4   Hand-Carried Weapon Placed against the
                 Exterior Envelope 166
5.8   Responding to Cyberattacks... 175
5.9   Executive and Close Personal Protection 178
5.10   Travel Security.. 194

Chapter 6   Management Strategies 201

6.1   Crisis and Incident Management... 201
6.2   Bomb Threat Management.. 206
6.3   Managing a Technical Surveillance and
        Countermeasures Detail 210
6.4   Managing a Close Protection Detail 217
6.5   Managing Disasters in the Data Center................................ 219
        6.5.1   Section 1—Document Introduction 221
        6.5.2   Section 2—Crisis Scenarios/Situations.................... 221
        6.5.3   Section 3—Crisis Considerations 221

6.5.4 Section 4—Crisis Management Team 222
6.5.5 Section 5—Crisis Management Facility 222
6.5.6 Section 6—Notification Procedures........................222
6.5.7 Section 7—Action Procedures 222
6.5.8 Section 8—Postcrisis Analysis 222
6.5.9 Section 9—Plan Exercising....................................223
6.5.10 Section 10—Appendix 223
6.6 Hostage Survival and Crisis Negotiations.............................225
6.7 Managing Violent Behavior in the Workplace 241
6.7.1 Type I Attacker—Criminal Intent...........................242
6.7.2 Type II Attacker—Customer and/or Client.............242
6.7.3 Type III Attacker—Worker-on-Worker 243
6.7.4 Type IV Attacker—Personal Relationship...............243
6.8 Event and Crowd Management...247

Chapter 7 Contingency Plans...253
7.1 Developing a Force Protection Plan 253
7.1.1 Estimate the Threat 253
7.1.2 Assess Vulnerabilities 254
7.1.3 Develop Protective Measures 254
7.1.4 Conduct Routine Security Operations.....................254
7.1.5 Conduct Contingency Operations 254
7.2 Developing a Security Plan 255
7.2.1 Approvals 255
7.2.2 Executive Summary 255
7.2.3 Communications and Consultations........................255
7.2.4 Context 256
7.2.5 Security Risk Assessment 257
7.2.6 Security Risk Treatment Process 257
7.2.7 Implementation...257
7.3 Developing a Business Continuity Plan................................258
7.3.1 Analysis Phase..258
7.3.2 Solution Design Phase 259
7.3.3 Implementation Phase 260
7.3.4 Testing Phase..260
7.3.5 Maintenance Phase..261

Chapter 8 Response and Recovery Operations...263
8.1 Responding to Natural and Human-Induced Disasters.........263
8.1.1 Natural Disasters...263
8.1.2 Human-Induced Disasters 265

8.2   Responding to Chemical, Biological, Radiological,
      Nuclear, Explosive, and Incendiary Events 265
      8.2.1   Biological Agents 266
              8.2.1.1   Wet or Dry Agent from a Point Source 266
              8.2.1.2   Threat of Dry Agent Placed in HVAC
                        or Package with No Physical Evidence.....266
              8.2.1.3   Confirmed Agent Placed in HVAC
                        System (Visible Fogger, Sprayer, or
                        Aerosol Device) 267
      8.2.2   Nuclear or Radiological Agents 267
      8.2.3   Incendiary Devices..268
      8.2.4   Chemical Agents 268
      8.2.5   Explosives...270
              8.2.5.1   Unexploded Device and Preblast
                        Operations...270
              8.2.5.2   Explosive Device Preblast.........................270
              8.2.5.3   Explosive Device Postblast 271
              8.2.5.4   Agency-Related Actions, Fire
                        Department 272
              8.2.5.5   Emergency Medical Services 273
              8.2.5.6   Law Enforcement.......................................274
              8.2.5.7   HazMat Group 275
8.3   Responding to a Terrorist Event 276
8.4   Disaster Recovery Operations 281
8.5   Special Response Teams 283
8.6   Stress Management after a Disaster 285

Appendix A: Select Emergency Management Organizations..........................289

Appendix B: Top 10 Global Disasters Since 1900 297

Appendix C: Select Global Special Operations Teams 299

Appendix D: Select Global Terrorist Organizations 303

Bibliography 307

Index...323

Preface

A disaster is a natural or man-made hazard resulting in an event of substantial magnitude causing significant physical damage or destruction, loss of life, or drastic change to the environment. It can be defined as any tragic event stemming from events such as earthquakes, floods, catastrophic accidents, fires, or explosions. It is a phenomenon that can cause damage to life and property and destroy the economic, social, and cultural life of people.

When disaster strikes a vulnerable population, the incident evokes a combination of horror and empathy among a country's population. Similarly, unless a response is carefully planned and successfully carried out, a government can appear impotent or nonresponsive to the dangers facing its citizens. Often disasters result from mismanagement of the risks involved. The risks involved are a product of both hazards and vulnerability.

Hazards that strike areas with low vulnerability never become disasters. Similarly, developing countries suffer the most when it comes to disasters, since they lack the tools and coping capacity to effectively deal with a disaster. However, industrialized nations are more resilient and capable in dealing with disasters.

In this book I have attempted to assemble a guide through the disaster management process including prevention, mitigation, preparedness, response, and recovery for individuals interested in this field of study. The handbook guides the reader through the various phases of disaster management. Chapter 1 is an introduction to the emergency management process, which includes sections on hazard analysis, emergency planning, effective communication, leadership, and management of volunteers. Chapter 2 discusses threats assessment including an all-hazard approach to threat assessments, vulnerability assessments, and risk analysis. Chapter 3 provides an overview of critical infrastructure protection, particularly those sectors that are vulnerable to the threats identified in Chapter 2 including facility and/or building security, water security, energy security, food security, transportation security, and cybersecurity.

Chapter 4 provides information on violence and violent behavior, specifically looking at targeted violence through criminal behavior such as organized crime, terrorism, and maritime piracy. It also provides an overview of protection strategies on how to identify suspicious behavior and how to avoid a potential attack. Chapters 5 and 6 examine protection strategies and how to properly manage those strategies. Chapter 7 provides an overview or outline on how to develop force protection plans, security plans, and business continuity plans. The plans themselves include the information gained in the previous chapters. Finally, Chapter 8 examines response and recovery operations and postincident stress management, the focus of which is responding to a chemical, biological, radiological, nuclear, and explosive (CBRNE) event.

It must be borne in mind, however, that protection and response strategies and standard operating procedures will differ greatly from country to country based upon the form of government, culture, religion, and other factors. For example, in some countries certain features of society including race, sex, or caste might be considered more valuable than others. Even in democracies where everyone is considered to be equal, a more influential group or population group will garner a more significant response from the authorities and the media. Cultural influences may also play an important role in how an incident transpires. The issue of investment prompts further distinctions in practice. Developing countries tend to be the hardest hit when a disaster strikes, since they have limited resources, whereas industrialized nations tend to cope with disasters much better. However, there is more investment in developing nations' infrastructure and in building up their resilience and coping capacity to disasters, whereas in industrialized nations there is more of a focus on protection strategies and response and recovery operations.

The more authoritarian the government, the more likely there will be pressure on response agencies to act quickly to end an incident or respond to a disaster to prevent the government from appearing weak. More democratic governments, on the other hand, will be under pressure to place great emphasis on the lives of the population at risk. As a result, the strategies and procedures listed in this book are by no means exhaustive. I do believe that the information contained in this book though gives the reader an understanding of the principles and steps necessary to manage a hazard or threat effectively and prevent it from becoming a disaster. Form the initial threat assessment to response and recovery operations, the handbook on physical security and environmental protection allows the reader to progress through the various stages of emergency planning and management.

John Perdikaris

Author

Dr. John Perdikaris is a registered professional engineer in the province of Ontario, Canada. He has 15 years of varied engineering and emergency management experience on a variety of projects within the province of Ontario, including acting as project manager on various engineering and emergency preparedness projects. He holds a master's and a PhD degree in engineering from the University of Guelph, Ontario. His fields of expertise include force protection, critical infrastructure protection, emergency management, water resources management, forecasting and warning systems, and modeling and simulation.

1 Emergency Management

1.1 INTRODUCTION TO DISASTERS AND EMERGENCY MANAGEMENT

A disaster is a state in which a population, population group, or an individual is unable to cope with or overcome the adverse effects of an extreme event without outside help. The impact of an extreme event may include significant physical damage or destruction, loss of life, or drastic change to the environment. It is a phenomenon that can cause damage to life and property and destroy the economic, social, and cultural life of people.

The above definition perceives disasters as the consequence of inappropriately managed risk. For example, when looking at an extreme event such as a flood, although the primary cause for a flood is extreme rainfall, snowmelt, or a combination of both, the impact or magnitude of a flood is determined by human influences. The risks associated with a disaster are a function of both the hazards and the vulnerability of the affected group. Hazards that strike in areas with low vulnerability will never become disasters, as is the case with uninhabited regions. Developing countries suffer the greatest costs when a disaster hits; more than 95% of all deaths caused by disasters occur in developing countries, and losses due to natural disasters are 20 times greater [as a percentage of gross domestic product (GDP)] in developing countries than in industrialized countries.

Disaster management is defined as the organization and management of resources and responsibilities for dealing with all humanitarian aspects of emergencies, in particular preparedness, mitigation, response, and recovery in order to lessen the impact of disasters. It deals specifically with the processes used to protect populations and/or organizations from the consequences of disasters. However, it does not necessarily extend to the prevention or elimination of the threats themselves, although the study and prediction of threats is an imminent part of disaster management. International organizations focus on community-based disaster preparedness, which assists communities to reduce their vulnerability to disasters and strengthen their coping capacities.

When the capacity of a community or country to respond to and recover from a disaster is exceeded, outside help is necessary. This assistance comes from different sources including government and nongovernment organizations. The definition of disaster as a state where people at risk can no longer help themselves conforms to the modern view of a disaster as a social event, where the people at risk are vulnerable to the effects of an extreme event because of their social conditions. According to this view, disaster management is not only a technical task, but also a social task. Therefore, community-based disaster preparedness becomes an integral part of disaster management. By reducing a community's vulnerability to a disaster and by building upon its coping capacities, skills, and resources, these same communities are better able to meet future crises.

The first people to respond to a disaster are those living in the local community, and they are also the first to start rescue and relief operations. They know what their needs are, have an intimate knowledge of the area, and may have experienced similar events in the past. Therefore, community members need to be consulted and involved in the response and recovery operations, including assessment, planning, and implementation. Consultation can take place through community leaders, representatives of women's or other community associations, beneficiaries, and other groups.

1.2 TYPES OF DISASTERS

Disasters can be classified into two subcategories: natural hazards and technological or man-made hazards. Natural hazards are naturally occurring physical phenomena caused by either rapid or slow onset events that can be geophysical, hydrological, climatological, meteorological, and/or biological in nature. Geophysical disasters include earthquakes, landslides, tsunamis, and volcanic activity; hydrological disasters include hazards such as avalanches and floods; climatological disasters include hazards such as extreme temperatures, droughts, and wildfires; meteorological disasters include cyclones, hurricanes, storms, and/or wave surges; and biological disasters include disease epidemics and infestations such as insect and/or animal plagues. Some natural disasters can result from a combination of different hazards, for example, floods can be the result of tsunamis, storm surges, hurricanes, or cyclones or a combination of all four. However, after a flood, epidemics such as cholera, malaria, and dengue fever begin to emerge.

Technological or man-made hazards are events that are caused by humans and occur in or close proximity to human settlements. This can include environmental degradation, pollution, complex emergencies and/or conflicts, cyberattacks, famine, displaced populations, industrial accidents, and transport accidents. Workplace fires are more common and can cause significant property damage and loss of life. Communities are also vulnerable to threats posed by extremist groups who use violence against both people and property. High-risk targets include military and civilian government facilities, international airports, large cities, and high-profile landmarks. Cyberterrorism involves attacks against computers and networks done to intimidate or coerce a government or its people for political or social objectives.

Technological disasters are complex by their very nature and could include a combination of both natural and man-made hazards. In addition, there are a range of challenges such as climate change, unplanned urbanization, underdevelopment and poverty, as well as the threat of pandemics that will shape disaster management in the future. These aggravating factors will result in the increased frequency, complexity, and severity of disasters.

Climate change ranks among the greatest global problems of the twenty-first century, and the scientific evidence on climate change is stronger than ever. For example, the Intergovernmental Panel on Climate Change (IPCC) released its Fourth Assessment Report in early 2007, saying that climate change is now unequivocal. It confirms that extremes are on the rise and that the most vulnerable people, particularly in developing countries, face the brunt of impacts. The gradual expected temperature rise may seem limited, with a likely range from 2°C to 4°C predicted for the coming century. However, a slightly higher temperature is only an indicator that is much more skewed. Along with the rising temperature, we can experience an increase in both frequency and intensity of extreme weather events such as prolonged droughts, floods, landslides, heat waves, and more intense storms; the spreading of insect-borne diseases such as malaria and dengue to new places where people are less immune to them; a decrease in crop yields in some areas due to extreme droughts or downpours and changes in timing and reliability of rainfall seasons; a global sea level rise of several centimeters per decade, which will affect coastal flooding, water supplies, tourism, and fisheries, and tens of millions of people will be forced to move inland; and melting glaciers, leading to water supply shortages. Climate change is here to stay and will accelerate. Although climate change is a global issue with impacts all over the world, those people with the least resources have the least capacity to adapt and therefore are the most vulnerable. Developing countries, more specifically its poorest inhabitants, do not have the means to cope with floods and other natural disasters; to make matters worse, their economies tend to be based on climate and/or weather-sensitive sectors such as agriculture and fisheries, which makes them all the more vulnerable.

The impact of underdevelopment, unplanned urbanization, and climate change is present in our everyday work; disasters are a development and humanitarian concern. A considerable incentive for rethinking disaster risk as an integral part of the development process comes from the aim of achieving the goals laid out in the Millennium Declaration. The Declaration sets forth a road map for human development supported by 191 nations. Eight Millennium Development Goals (MDGs) were agreed upon in 2000, which in turn have been broken down into 18 targets with 48 indicators for progress. Most goals are set for achievement by 2015. The MDGs have provided a focus for development efforts globally. While poverty has fallen and social indicators have improved, most countries will not meet the MDGs by 2015, and the existing gap between the rich and the poor will widen. Recently, the campaigns on poverty have resulted in key milestones on aid and debt relief. While positive, much more is needed if the MDGs are to become reality. These efforts to reduce poverty are vital for vulnerability reduction and strengthened resilience of communities to disasters.

Today 50% of the world's population live in urban centers, and by 2030 this is expected to increase to 60%. The majority of the largest cities, known as megacities, are in developing countries, where 90% of the population growth is urban in nature. Migration from rural to urban areas is often trigged by repeated natural disasters and the lack of livelihood opportunities. However, at the same time many megacities are built in areas where there is a heightened risk for earthquakes, floods, landslides, and other natural disasters. Many people living in large urban centers such as "slums" lack access to improved water, sanitation, security of tenure, durability of housing, and sufficient living area. This lack of access to basic services and livelihood leads to increasing risk of discrimination, social exclusion, and ultimately violence.

1.3   EMERGENCY PLANNING

The terrorist attacks of September 11, 2001, illustrated the need for all levels of government, the private sector, and nongovernmental agencies to prepare for, protect against, respond to, and recover from a wide spectrum of events that exceed the capabilities of any single entity. These events require a unified and coordinated approach to planning and emergency management.

Knowing how to plan for disasters is critical in emergency management. Planning can make the difference in mitigating against the effects of a disaster, including saving lives and protecting property and helping a community recover more quickly from a disaster. Developing an effective emergency operations plan (EOP) can have certain benefits for a municipality including the successful evacuation of its citizens and the ability to survive on its own without outside assistance for several days. The consequences of not having an emergency plan include the following: the need for immediate assistance and a higher number of casualties resulting from an attempted evacuation, with them being ineligible for the full amount of aid from upper-tier municipalities (regions or counties), provincial, state, or federal governments. In the United States, counties that do not have emergency plans cannot declare an emergency and are ineligible for any aid or for the full amount of aid. In Ontario, Canada, all municipalities must have an emergency plan in accordance with the Emergency Management and Civil Protection Act.

Emergency planning is not a one-time event. Rather, it is a continual cycle of planning, training, exercising, and revision that takes place throughout the five phases of the emergency management cycle: preparedness, prevention, response, recovery, and mitigation. The planning process does have one purpose: the development and maintenance of an up-to-date EOP. An EOP can be defined as a document describing how citizens and property will be protected in a disaster or emergency. Although the emergency planning process is cyclic, EOP development has a definite starting point. There are four steps in the emergency planning process: hazard analysis, EOP development, testing the plan, and plan maintenance and revision.

Step 1: The hazard analysis. It is the process by which hazards that threaten the community are identified, researched, and ranked according to the risks they pose and the areas and infrastructure that are vulnerable to damage from an event involving the hazards. The outcome of this step is a written hazard analysis that quantifies the overall risk to the community from each hazard. The hazard analysis component of the emergency planning process is covered in detail in Section 1.4.

Step 2: The development of an EOP. The EOP includes the basic plan, functional annexes, hazard-specific appendices, and implementing instructions. The outcome of this step is a completed plan, which is ready to be trained, exercised, and revised based on lessons learned from the exercises.

Step 3: Testing the plan through a series of training exercises. Training exercises of different types and varying complexity allow emergency managers to see what in the plan is unclear and what does not work. The outcomes of this step are lessons learned about weaknesses in the plan that can then be addressed in Step 4.

Step 4: Plan maintenance and revision. The outcome of this step is a revised EOP, based on current needs and resources, which may have changed since the development of the original EOP. After the EOP is developed, Steps 3 and 4 repeat in a continual cycle to keep the plan up to date. If you become aware that your community faces a new threat such as terrorism, the planning team will need to revisit Steps 1 and 2. Emergency planning is a team effort because disaster response requires coordination between many community agencies and organizations and different levels of government. Furthermore, different types of emergencies require different kinds of expertise and response capabilities. Thus, the first step in emergency planning is identification of all of the parties that should be involved.

Obviously, the specific individuals and organizations involved in response to an emergency will depend on the type of disaster. Law enforcement will probably have a role to play in most events, as will fire, emergency medical services (EMS), voluntary agencies (i.e., the Red Cross), and the media. On the other hand, hazardous materials (HazMat) personnel may or may not be involved in a given incident but should be involved in the planning process because they have specialized expertise that may be called on.

Getting all stakeholders to take an active interest in emergency planning can be a daunting task. To schedule meetings with so many participants may be even more difficult. It is critical, however, to have everyone's participation in the planning process and to have them feel ownership in the plan by involving them from the beginning. Also, their expertise and knowledge of their organizations' resources is crucial to developing an accurate plan that considers the entire community's needs and the resources that could be made available in an emergency.

It is definitely in the community's best interest to have the active participation of all stakeholders. The following are recommendations of what can be done to ensure that all stakeholders that should participate in the discussions do, so that a plan is formulated. Give the planning team plenty of notice of where and when the planning meeting will be held. If time permits, you might even survey the team members to find the time and place that will work for them. Provide information about team expectations ahead of time. Explain why participating on the planning team is important to the participants' agency and to the community itself. Show the participants how they will contribute to a more effective emergency response. Ask the Chief Administrative Officer (CAO) or their Chief of Staff to sign the meeting announcement. A directive from the executive office will carry the authority of the CAO and send a clear signal that the participants are expected to attend and that emergency planning is important to the community. Allow for flexibility in scheduling after the first meeting. Not all team members will need to attend all meetings. Some of the work can be completed by task forces or subcommittees. Where this is the case, gain concurrence on time frames and milestones but let the subcommittee members determine when it is most convenient to meet. In addition, emergency managers may wish to speak with their colleagues from adjacent municipalities to gain their ideas and inputs on how to gain and maintain interest in the planning process.

Working with personnel from other agencies and organizations requires collaboration. Collaboration is the process by which people work together as a team toward a common goal, in this case, the development of a community EOP. Successful collaboration requires a commitment to participate in shared decision making; a willingness to share information, resources, and tasks; and a professional sense of respect for individual team members. Collaboration can be made difficult by differences among agencies and organizations in terminology, experience, mission, and culture. It requires the flexibility from team members to reach an agreement on common terms and priorities and humility to learn from others' ways of doing things. Also, collaboration among the planning team members benefits the community by strengthening the overall response to the disaster. For example, collaboration can eliminate duplication of services, resulting in a more efficient response and expanding resource availability. It can further enhance problem solving through cross-pollination of ideas.

However, collaboration does not come automatically. Building a team that works well together takes time and effort and typically evolves through the following stages: forming, storming, norming, performing, and adjourning. Forming is when individual members come together as a team. During this stage the team members may be unfamiliar with each other and uncertain of their roles on the team. Storming is when team members become impatient and disillusioned and may disagree. Norming is when team members accept their roles and focus on the process. Performing is when team members work well together and make progress toward the goal. Adjourning is when the team's task is accomplished; team members may feel pride in their achievement and some sadness that the experience is ending. However, the planning team never really adjourns; team members meet less frequently as they plan and conduct exercises and revise the plan, but the core of the team remains intact.

The team leader plays a key role in the development of effective teams through all stages of the team's development. The team leader can initiate appropriate team-building activities that move the team through the stages and toward its goal. Other team roles besides that of the team leader include the task master who identifies the work to be done and motivates the team; the innovator who generates original ways to get the group's work done; the organizer who helps the group develop plans for getting the work done; the evaluator who analyzes ideas, suggestions, and plans made by the group; and the finisher who follows through on plans developed by the team.

An emergency operations planning team is on track when it displays the following characteristics: a common goal such as the development of an EOP, a leader who provides direction and guidance, open communication, constructive conflict resolution, mutual trust, and respect for each individuals contributions.

The EOP describes actions to be taken in response to natural, man-made, or technological hazards, detailing the tasks to be performed by specific organizational elements at projected times and places based on established objectives, assumptions, and assessment of capabilities. An EOP should be comprehensive in that it should cover all aspects of emergency prevention, preparedness, and response and addresses mitigation concerns as well. The EOP should take an all-hazard approach to emergency management and, therefore, be flexible enough to allow use in all emergencies, even unforeseen events. The EOP should also be risk-based and include hazard-specific information, based on the hazard analysis.

The purposes of the EOP are to give an overview of the community's emergency response organization and policies and to provide a general understanding of the community's approach to emergency response for all involved agencies and organizations. Although the plan provides the general approach to emergency response, it does not stand by itself. Rather, it forms the basis for the remainder of the plan, which also includes the functional annexes that address the performance of a particular broad task or function such as mass care or communications, and hazard-specific appendices that provide additional information specific to a particular hazard. In addition, each part of the EOP may have addenda in the form of standard operating procedures (SOPs), maps, charts, tables, forms, and checklists. These addenda may be included as attachments or incorporated by reference.

Although there is no mandatory format, the recommended format to ensure compatibility with other jurisdictions and levels of government for the municipal or local EOP includes the following components: introduction, purpose statement, signatory page, situation and assumptions, concept of operations, organization and assignment of responsibilities, administration and logistics, plan development and maintenance, and authorities and references.

The *introductory material* consists of the following elements: the promulgation document, the signature page, the dated title page and record of changes, the record of distribution, and the table of contents. The promulgation document is signed by the jurisdiction's chief elected official (CEO), affirming his or her support for the emergency management agency and planning process. It gives organizations the authority and responsibility to perform their tasks. It also mentions the tasked organizations' responsibility to prepare and maintain implementing instructions, gives notice of necessary EOP revisions, and commits to the training necessary to support the EOP. The signature page is signed by all partner organizations, demonstrating their commitment to EOP implementation. The dated title page and record of changes include date, description, and affected parts of the EOP. The record of distribution lists EOP recipients, facilitating and giving evidence of EOP distribution. Copies of the EOP should be numbered and recorded. The final component of the introductory material is the table of contents.

The *purpose statement* is a broad statement of what the EOP is meant to do and provides a synopsis of the EOP, annexes, and appendices. The purpose statement need not be complex but should include enough information to establish the direction for the remainder of the plan.

The *signature page* is signed by the CEO and the heads of each department, agency, or organization with responsibilities under the plan. The signature page demonstrates knowledge of and commitment to the EOP and its implementation. It provides accountability to the public for the response actions described in the plan.

The *situation statement* characterizes the planning environment, making it clear why emergency operation planning is necessary. It draws from the hazard analysis to narrow the scope of the EOP and includes the following types of information: hazards addressed by the plan, relative probability and impact, areas likely to be affected, vulnerable critical facilities, population distribution, special populations, interjurisdictional relationships, and maps.

The *assumptions statement* delineates what was assumed to be true when the EOP was developed. The assumptions statement shows the limits of the EOP and limiting liability. Within the assumptions statement it is helpful to list even "obvious" assumptions such as which identified hazards will occur, which individuals and organizations are familiar with the EOP, which individuals and organizations will execute their assigned responsibilities, what assistance may be needed and, if so, who and what will be available, and how executing the EOP will save lives and reduce damages.

The *concept of operations* section explains the community's overall approach to emergency response of what, when, and by whom. The concept of operations section describes the division of municipal, provincial/state, and federal responsibilities; when the EOP will be activated and when will it be inactivated; alert levels and basic actions that accompany each level; and the general sequence of actions before, during, and after an event. In addition, procedures and forms that are necessary to request outside assistance of various types are also included in this section.

The *organization and assignment of responsibilities* section lists the general areas of responsibility assigned by organization and position. It also identifies shared responsibilities and specifies which organization has primary responsibility and which has supportive roles. The organization and assignment of the responsibilities section specifies reporting relationships and lines of authority for an emergency response.

The *administration and logistics* section includes assumed resource needs for high-risk hazards; resource availability; mutual aid agreements; policies on augmenting response staff with public employees and volunteers; a statement that addresses liability issues; and resource management policies including acquisition, tracking, and financial recordkeeping.

Responsibility for the coordination of the development and revision of the EOP including annexes, appendices, and implementing instructions must be assigned to the appropriate persons. The *development and maintenance* section therefore describes the planning process; identifies the planning participants; assigns planning responsibilities; and describes the revision cycle, for example, training, exercising, review of lessons learned, and revisions.

The *authorities and reference* section cites the legal bases for emergency operations and activities, including laws, statutes, ordinances, executive orders, regulations, formal agreements, and predelegation of emergency authorities. This section also includes pertinent reference materials including related plans of other levels of government.

Annexes and appendices are different in content and address different topics. An annex explains how the community will carry out a broad function in any emergency such as warning or resource management. An appendix is a supplement to an annex that adds information about how to carry out the function in the face of a specific hazard. Thus every annex may have several appendices, with each addressing a particular hazard. Which hazard-specific appendices are included depends on the community's hazard analysis. For example, a community in California or British Columbia would probably include earthquake appendices in its EOP, a community in Florida or Southern Ontario would probably include flooding or flooding due to hurricane appendices, and a community in the Midwest or Prairie provinces would undoubtedly include appendices that address tornadoes and/or flooding. The decision about whether to develop an appendix rests with the planning team. In general, the organization of annexes and appendices parallels that of the basic plan. Specific sections can be developed to expand upon but not to repeat information that is contained in the basic plan.

It is important early in the planning process to choose the functions that will be included in the basic plan as annexes. Factors that should be considered when making these choices include the organizational structure of the state or province and local governments, capabilities of the jurisdiction's emergency services agencies, and the established policy with regard to the concept of operations. In addition, the communities' vulnerability to a hazard becomes important for developing an evocative EOP. Because communities vary so widely, no single listing of functional annexes can be prescribed for every single one. There are, however, eight core functions that are typically addressed in annexes in every EOP; they include direction and control, communications, warning, emergency public information (EPI), evacuation, mass care, health and medical, and resource management.

The *direction and control annex* allows a jurisdiction to analyze the situation and decide on the best response, direct the response teams, coordinate efforts with other jurisdictions, and make the best use of available resources. The *communications annex* provides detailed focus on the total communications system and how it will be used. The *warning annex* describes the warning systems in place and the responsibilities and procedures for using them. The *EPI* provides the procedure for giving the public accurate, timely, and useful information and instructions throughout the emergency period. Although the warning annex focuses on the procedures that the government uses to alert those at risk, an EPI annex deals with developing messages and accurate information, disseminating the information, and monitoring how the information is received. The warning system is one means for an EPI organization to get information out to the public; an EPI annex must address coordination with those responsible for the warning system. The *evacuation annex* describes the provisions that have been made to ensure the safe and orderly evacuation of people threatened by the hazards the jurisdiction faces. The *mass care annex* deals with the actions that are taken to protect evacuees and other disaster victims from the effects of the disaster, including providing temporary shelter, food, medical care, clothing, and other essential needs. The *health and medical annex* describes policies and procedures for mobilizing and managing health and medical services under emergency or disaster conditions. The *resource management annex* describes the means, organization, and process by which a jurisdiction will find, obtain, allocate, and distribute resources to satisfy needs that are generated by an emergency or disaster. In addition to the above-listed annexes, the EOP planning team should include annexes that make sense for the community. For example, if the community has a nuclear power plant, it may want to include an annex on radiological protection. Other functional annexes that may be included are damage assessment, search and rescue, emergency services, and aviation operations.

Hazard-specific appendices are attached to each functional annex to specify how a function should be carried out when faced with a particular hazard. Topics addressed in hazard-specific appendices include special planning requirements, priorities identified through hazard analysis, unique characteristics of the hazard requiring special attention, and special regulatory considerations.

Each annex or appendix as well as the basic plan may use implementing instructions. Implementing instructions are documents developed by individual agencies that provide detailed instructions for carrying out tasks assigned in the EOP. Implementing instructions may be included as attachments or referenced. Typically, implementing instructions are included in the EOP by reference only. Implementing instructions provide tools for carrying out the community's plan. They help ensure that those who are responsible for implementing the EOP are able to carry out their roles effectively.

Implementing instructions are used by all agency personnel who respond to disasters, whatever their function is. Implementing instructions are developed at the agency level for two reasons. Personnel from an agency with a specific function will have no idea how to tell another agency's personnel how to do their jobs. And agency personnel will be the persons who use the implementing instructions and therefore will know if they are helpful and effective. The implementing instructions used by agency personnel should support the agency's roles and responsibilities as described in the basic plan. Therefore, probably only some types of implementing instructions will be useful to the agency, depending on its function in disaster response.

There are several types of implementing instructions that an organization can develop, including SOPs, job aids, checklists, information cards, recordkeeping, and combination forms and maps. An example of the kind of forms that may be attached is a resource request form. Charts, tables, and checklists are appropriate for the organization and assignment of responsibilities annex, for example, a matrix of position responsibilities. The EOP planning team may use supporting documents as needed to clarify the contents of the plan, annex, or appendix. For example, the evacuation annex may be made clearer by attaching maps with evacuation routes marked. Since evacuation routes may change depending on the location of the hazard, maps may also be included in the hazard-specific appendices to the evacuation annex. Similarly, the locations of shelters and community centers may be marked on maps supporting the mass care annex.

SOPs provide step-by-step instructions for carrying out specific responsibilities; they describe who, what, where, when, and how. SOPs are appropriate for complex tasks requiring step-by-step instructions, tasks for which standards must be specified, and tasks for which documentation of performance protocols are required as protection against liability. To develop SOPs, an organization must develop a task list; determine who will perform each task, who will they report to, and who they will coordinate with; identify the step for each task; identify the standards for task completion; and test the procedures. SOPs are not static but dynamic and must be kept up-to-date through review and revision.

SOPs, or parts of SOPs, can be presented as job aids. A job aid is a written procedure that is intended to be used on the job while the task is being done. Job aids are appropriate for complex tasks, critical tasks that could result in serious consequences, tasks that are infrequently done, and procedures or personnel that change often. Job aids are also useful when conformity is needed among workers and/or across locations. Job aids should specify the task title, the purpose of the task, when to do the task, materials needed to perform the task, how to perform each step of the task, the desired result(s), standards to which the task must be performed, and how to check the work. A job aid may include examples, graphics, flowcharts, decision trees or tables, and a list of do's and don'ts.

Because job aids are designed to be used in the midst of completing a task, to be effective they must be clear. They should use action verbs and everyday language, highlight important information, and place warnings before the steps to which they apply. Formatting is also important when creating job aids. Numbering steps and using space, boxes, or lines to separate each step allows users to find their place easily within a document. It is important to note that job aids may not be useful for all tasks, especially simple tasks that are performed regularly, or must be accomplished quickly, from memory. If a task cannot be performed while referring to a job aid at the same time, a job aid is not appropriate for that particular task.

Checklists provide a list of tasks, steps, features, contents, or other items to be checked off. They often take the form of boxes to be checked off, for example, yes/no, done/not done, present/not present. They may also take the form of a rating scale. Checklists are useful for tasks that are made up of multiple steps or for when it might be necessary to document the completion of the steps. Checklists are less useful when observations must be recorded or when calculations or evaluations must be made.

Information cards provide information that is needed on the job in a convenient, often graphic form. Examples include reference lists, diagrams, labeled illustrations, charts, or tables. Information is sometimes summarized in matrix form such as a tax table. Information that might be usefully presented on information cards includes call-down rosters, contact lists, resource lists, organizational charts, task matrices, and equipment diagrams.

Common forms used as implementing instructions include recordkeeping forms on which calculations, observations, or other information such as damage assessments can be recorded and combination forms that serve multiple functions such as checklists with recordkeeping sections. Maps can also be used as implementing instructions to highlight geographic features and boundaries, jurisdictional boundaries, locations of key facilities, and transportation or evacuation routes. When a map is designed to show a particular feature, extraneous details are often eliminated. To be effective, implementing instructions must be appropriate for both the intended use and audience. They must also be complete in that they cover all of the components or steps. They should avoid jargon and ambiguity, be organized logically, for example, the way the task is actually done, and include instructions that identify the purpose and applicability of the particular implementing instruction. They should be sufficiently detailed in that they give all of the necessary information. They should also be up-to-date and the latest revision date should be included. They should be sufficient in scope in that, together, they cover each function fully. They should be identified in the EOP so that their existence is recorded. Implementing instructions can be incorporated or referenced.

1.4   ALL-HAZARDS APPROACH TO DISASTER MANAGEMENT

Hazards are conditions or situations that have the potential for causing harm to people or property. An all-hazards approach to disaster management is a systematic approach for identifying, analyzing, and estimating all natural, accidental, and malicious threats and hazards. The all-hazards approach to disaster management answers four key questions in a systematic way, as a basis for all emergency management planning: What can go wrong and how much warning time are we likely to have? How likely is it that the risk will happen? What are the consequences for specific stakeholders and society as a whole? How well prepared is the institution, collectively, to respond and recover from such a risk? By addressing these questions systematically, an all-hazards approach promotes the development of a management structure and processes and procedures throughout the four phases of emergency management that are applicable to every significant identified hazard. An all-hazards approach helps to balance and prioritize risk investments and actions, helps to identify interdependencies, promotes integration of lessons learned and the adoption of a forward-looking approach, and supports a consistent approach and enables cooperation.

The all-hazards approach informs all pillars of the emergency management planning process. Emergency management planning based on an all-hazards approach recognizes that the causes of emergencies can vary greatly, but many of the effects do not. Emergency management planning based on an all-hazards approach allows planners to address emergency functions common to all hazards in the basic plan instead of having unique plans for every type of hazard. All-hazards emergency management planning supports the identification of common tasks and who is responsible for accomplishing those tasks. A key component to the all-hazards approach is the hazard analysis. The steps in the hazard analysis process are to identify the hazards, profile each hazard, develop a community profile, determine vulnerability, and create and apply scenarios.

The first step is to develop a list of hazards that may occur in the community. This list is usually based on historical data about past events. Information about recent events is relatively easy to gather, while information about older events may be more difficult to find. There are many potential sources of hazard information. Some sources such as local newspapers and historical articles are readily available and kept in archives at the library. To get a more complete picture of the types of hazards that a community has faced historically, it may be necessary to check other sources such as other agencies, ministries and departments, local historical societies, and anecdotal information from long-time residents. There may be other sources of information, and emergency planners should take some time to check them so that their hazard analysis is as complete as possible. If the community has an existing hazard analysis, the best way to begin is by reviewing it and identifying any changes that may have occurred since it was developed or updated last. Some possible changes within or near the community that could cause hazard analysis information to change over time include new mitigation measures, for example, recently implemented regulations and new building codes, the opening or closing of facilities or structures that pose potential secondary hazards, for example, hazardous materials facilities and transport routes, and local development activities. There may be other long-term changes to investigate as well. These changes such as climatic changes in average temperature or rainfall/snowfall amounts are harder to track but could be very important to the hazard analysis.

The second step is to develop hazard profiles. Hazard profiles should address each hazard's magnitude, duration, seasonal pattern, and speed of onset. The availability of warnings is also a crucial part of the hazard profile.

The next step in the hazard analysis process is to combine hazard-specific information with a profile of the community to determine the community's vulnerability to the hazard or risk of damage from the hazard. Because different communities have different profiles, vulnerabilities to the same disaster vary. In developing a community's profile, the following factors must be taken into consideration: geography, property, infrastructure, demographics, and response organizations. Table 1.1 summarizes the key factors that are listed in a community profile. After gathering this information about the community, emergency managers can use it to develop the community's hazard analysis, as shown in Table 1.2.

After community and hazard profiles have been compiled, it is helpful to quantify the community's risk by merging the information so that the community can focus on the hazards that present the highest risk. Risk is the predicted impact that a hazard would have on the people, services, and specific facilities in the community. For example, during heavy rains, a specific road might be at risk of flooding, leading to restricted access to a critical facility.

Quantifying risk involves identifying the elements of the community including populations, facilities, and equipment that are potentially at risk from a specific hazard; developing response priorities; and assigning severity ratings. After compiling risk data into community risk profiles in surveying risk, it is helpful to develop response priorities. The following is a suggested hierarchy for setting priorities: priority one,

---

TABLE 1.1
Key Factors That Make Up a Community Profile

Key Community Factors

Geography	Property	Infrastructure	Demographics	Response Organizations
• Major geographic features • Typical weather patterns	• Numbers • Types • Ages • Building codes • Critical facilities • Potential secondary hazards	• Utilities construction, layout, access • Communication system layout, features, backups • Road systems • Air and water support	• Population size, distribution, concentrations • Numbers of people in vulnerable zones • Special populations • Animal populations	• Locations • Points of contact • Facilities • Services • Resources

TABLE 1.2

Use of Community Factors in Hazard Analysis

Type of Information	Used In
Geographic	• Predicting risk factors and the impact of potential hazards and secondary hazards
Property	• Projecting consequences of potential hazards to the local area
	• Identifying available resources (e.g., for sheltering)
Infrastructure	• Identifying points of vulnerability
	• Preparing evacuation routes, emergency communications, and project response and recovery requirements
Demographics	• Projecting consequences of disasters on the population
	• Disseminating warnings and public information
	• Planning evacuation and mass care
Response organizations	• Identifying response capabilities

life safety; priority two, essential facilities; and priority three, critical infrastructure. Life safety includes hazard areas, high-risk populations, and potential search and rescue situations. However, keep in mind that response personnel cannot respond if their own facilities are affected. Essential facilities include hospitals, emergency service buildings, shelters, and community centers. Critical infrastructure includes utilities, communication systems, and transportation systems. A severity rating or risk index should be assigned to each hazard that will predict, to the degree possible, the damage that can be expected in that community as a result of that hazard. This rating quantifies the expected impact of a specific hazard on people, essential facilities, property, and response assets. Table 1.3 gives examples of severity ratings that can be used in predicting the damage that can be expected from the impact of a specific hazard.

The next step is to develop a risk index for each hazard by assigning a value to each severity level. For example, the following values could be used to assign ratings to each severity level: 1, catastrophic; 2, critical; 3, limited; and 4, negligible. The ratings are assigned to each type of hazard data, which include magnitude, frequency of occurrence, speed of onset, community impact (severity rating), and special characteristics of the event. The severity level for each factor is then averaged to determine the overall risk level for that hazard.

The final step in the hazard analysis process is to develop scenarios for the top-ranked hazards or those that rank above a specified threshold that lay out the hazard's development into an emergency. Scenarios should be realistic and based on the community's hazard and risk data. To create a scenario, emergency managers need to be able to track the development of a specific type of emergency. A scenario should describe the initial warning of the event; the potential overall impact on the community; the potential impact of the event on specific community sectors; the potential consequences such as damage, casualties, and loss of services; and the actions and resources that would be needed to deal with the situation. Creating scenarios helps to identify situations that may exist in a disaster. These situations should be used to help ensure that the community is prepared should the hazard event occur.

TABLE 1.3
Examples of Severity Ratings for Predicting the Damage from the Impact of a Specific Hazard

Severity	Characteristics
Catastrophic	• >100 deaths
	• Complete shutdown of critical facilities for 30 days or more
	• >20% of property severely damaged
	• Total property damages exceeding $30 million
Critical	• <100 deaths
	• Injuries and/or illnesses result in permanent disability
	• Complete shutdown of critical facilities for at least 2 weeks
	• >10% of property severely damaged
	• Total property damage between $1 million and $30 million
Limited	• <10 deaths
	• Injuries and/or illnesses do not result in permanent disability
	• Complete shutdown of critical facilities for more than 1 week
	• <10% of property severely damaged
	• Total property damage less than $1 million
Negligible	• Injuries and/or illness treatable with first aid
	• Minor quality of life lost
	• Shutdown of critical facilities and services for 24 hours or less
	• <5% of property severely damaged

1.5 DECISION MAKING AND PROBLEM SOLVING

Decision making and problem solving are critically important skill areas for emergency managers, planners, first responders, voluntary agency (VOLAG) coordinators, and other professionals in emergency management. The ability to identify current and potential problems and to make sound, timely decisions before and during an emergency can literally affect the lives and well-being of the local citizenry. The decisions made can have an impact on the ability of response agencies to do their jobs and can make the difference in how quickly the community is able to recover from an event.

The ability to make sound, timely decisions during an emergency event is critical. Good problem solving and decision making can avert tragedy and help the community recover from the event more quickly. Conversely, poor decision making or the absence of decisions can potentially result in injury or death to victims or responders. But the repercussions do not stop there. Poor decisions in the early stages of an event can make the responders' job more difficult and more dangerous. In addition, they can give rise to much more critical or complex decisions later on, not to mention the effect this could have on community relations or their confidence in their first responders and public agencies. Good decision-making skills are one of the most critical assets an emergency management professional can have.

Let us now explore whether making ordinary day-to-day decisions or critical, time-sensitive decisions in an emergency using a standard problem-solving model will help ensure that your decisions are rational and logical. To begin, let us clarify what we mean by problem solving and decision making and how they relate to one another. Problem solving is a set of activities designed to analyze a situation systematically and generate, implement, and evaluate solutions. Decision making is a mechanism for making choices at each step of the problem-solving process. Decision making is part of problem solving, and occurs at every step of the problem-solving process.

Emergency decisions have their beginnings well before any emergency strikes. Often the number, type, and magnitude of decisions and problems that must be addressed during an emergency are a direct outgrowth of decisions that were or were not made at the outset of the emergency, or even before the emergency began.

Emergency management professionals have at their disposal a variety of resources designed to guide them in their decision making during emergencies. When a jurisdiction develops an EOP, SOPs, or other procedural documents, it provides the foundation for decision making that will occur during emergencies. Many decisions are made during the development of these documents, and there are real advantages to having made those decisions during the planning process rather than in the heat of an emergency.

The problem-solving model as described in this chapter contains five steps: identification of the problem, exploration of alternatives, selection of an alternative, implementation of the solution, and evaluation of the situation (Figure 1.1). The first step—the identification of the problem—includes delineating the problem parameters such as what is happening and not happening, who is involved, and what is at stake. The exploration of alternatives involves two parts: generating alternatives through working groups, steering committees, or surveys, and evaluating alternatives against a set of predefined criteria. Selection of the optimal solution is based on the evaluation of different alternatives. The implementation of the solution includes five parts, development of an action plan, determination of objectives, identification of needed resources, building of the plan, and implementation of the plan.

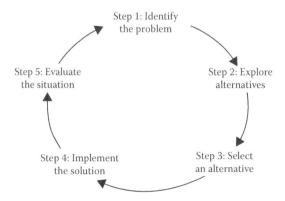

FIGURE 1.1   Decision-making model.

TABLE 1.4
Procedures and Criteria for Evaluating Different Alternatives

Procedures	Criteria
1. Identify constraints	• Technical (limited equipment or technology)
	• Political (legal restrictions or ordinances)
	• Economic (cost or capital restrictions)
	• Social (restrictions imposed by organized groups with special interests)
	• Human resources (limited ability of relevant people to understand or initiate certain actions)
	• Time (requirements that a solution be found within a prescribed time period, thereby eliminating consideration of long-range solutions)
2. Determine appropriateness	Does this solution fit the circumstances?
3. Verify adequacy	Will this option make enough of a difference to be worth doing?
4. Evaluate effectiveness	Will this option meet the objective?
5. Evaluate efficiency	What is the cost/benefit ratio of this option?
6. Determine side effects	What are the ramifications of this option?

The evaluation of the situation includes two parts: taking the implemented solution and monitoring its progress, and evaluating the results against the preestablished set of criteria. Criteria and procedures used for evaluating different alternatives and the optimal solution are listed in Table 1.4. Factors that can affect decision making include political factors, safety factors, financial factors, environmental factors, and ethical factors.

People have different styles of making decisions that depend on their personality or psychological type. Psychological type is a composite of an individual's preferences or preferred ways of taking in and organizing information.

People tend to favor one of four ways of approaching a problem: sensing (stability), intuition (innovation), thinking (effectiveness), and feeling (integrity). Although each approach has its strengths, it also has its blind spots. It is helpful to learn to ask the questions that all four approaches ask so as to arrive at more considered and, therefore, sounder decisions.

Decisions can also be made in one of four styles, depending on who is making the decision: individual, consultation group, or delegation. An important consideration in group decision making is avoiding "groupthink," in which group pressure produces a premature decision. Effective decision makers tend to have the following 10 characteristics: knowledge, initiative, advice seeking, selectivity, comprehensiveness, currency, flexibility, good judgment, calculated risk taking, and self-knowledge.

Ethics are a set of standards such as honesty, respect, and fairness that guide behavior. In the emergency management profession, ethical behavior is critical because disaster victims and coworkers must be able to depend on first responders and emergency personnel. Unethical practices include exceeding one's authority or making promises beyond the scope of one's authority, using their position for personal gain, and avoiding even the appearance of ethical violations. Ethical practices include placing the law and principle above private gain, acting impartially, protecting and conserving agency property, and putting forth an honest effort. Decisions that seem simple or routine in a day-to-day context may become difficult and have serious ethical implications during an emergency. Furthermore, a poor decision with ethical implications can escalate an emergency into an unmanageable situation as the emergency response progresses.

Ethical decision making requires being aware of one's own and the agency's ethical values and applying them whenever necessary. It involves being sensitive to the impact of decisions and being able to evaluate complex, ambiguous, and/or incomplete facts. Ethical decision making has three main components: commitment or motivation, consciousness or awareness, and competency or skill. Ethical commitment or motivation involves demonstrating a strong desire to act ethically and to do the right thing, especially when ethics imposes financial, social, or psychological costs. A crisis or emergency confronts us with many situations that test ethical commitment. Therefore, emergency managers need to be very clear about their own ethical values and have a strong understanding of ethical standards of conduct.

Ethical consciousness or awareness involves seeing and understanding the ethical implications of one's own behavior and applying their ethical values to their daily lives. It is important to understand that people's perceptions are their reality and so what the individual understands to be perfectly legal conduct may be perceived by the public as improper or inappropriate. Ethical competency or skill involves being competent in ethical decision-making skills, which include evaluation, creativity, and prediction. Evaluation is the ability to collect and evaluate relevant facts, and knowing when to stop collecting facts and to make prudent decisions based on incomplete and ambiguous facts. Creativity is the capacity to develop resourceful means of accomplishing goals in ways that avoid or minimize ethical problems. Prediction is the ability to foresee the potential consequences of conduct and assess the likelihood or risk that persons will be helped or harmed by an act. In turn, ethics can be applied to the decision-making model, specifically to step 3, that is, selection of an alternative. By applying ethics to the decision-making model, one can eliminate any alternatives that are unethical or give the appearance of being so.

1.6   LEADERSHIP AND INFLUENCE

A leader is one who sets direction and influences people to follow that direction. Leadership is critically important in emergency management, and lack of it can result in loss of public confidence, loss of property, and loss of life. The need for leadership applies across all of the phases of emergency management including prevention, preparedness, response, recovery, and mitigation. Part of being an effective leader is the ability to create an environment that encourages self-discovery and the testing of assumptions that may impede growth, change, and the development of a shared vision. Self-knowledge helps leaders develop their strengths as leaders.

There are three paradigms that relate to the development of leadership skills: technical paradigm or individual contributor, transactional paradigm or manager, and transformational paradigm or leader. These paradigms provide different ways of looking at the world. They are not mutually exclusive, and managers use all of them in one situation or another. But an effective leader has more of the attributes of the transformational paradigm. Transformational leaders view the organization as a moral system. They derive their credibility and power from behavioral integrity and core values. They are motivated to a higher purpose, approach challenges with a variety of perspectives, and give careful thought to the meaning of their actions. In turn, transformational leaders are focused and committed to a vision of the common good.

Leaders need to develop self-knowledge to grow. Part of this knowledge is being aware of the paradigms in which they operate and being willing to set aside comfortable behaviors to grow. It is also important for leaders to open up to others so as to build effective relationships. Three methods on how leaders can increase their self-knowledge are self-assessment, self-reflection, and soliciting authentic feedback. Another area in which leaders need to expand their self-knowledge is understanding how they think and what assumptions, biases, and beliefs they bring to their relationships.

The "ladder of inference" is a mental model that describes how we process our experiences and merge them with our own beliefs and assumptions. According to this model, as people move up the ladder their beliefs affect what they infer about what people observe and therefore become part of how people experience their own interaction with others. An obvious example of this would be only "hearing" that which supports their argument. But the process is usually much more subtle. A person's background influences the meanings that they ascribe to what they hear, which in turn can lead them to make assumptions. In fact, their beliefs affect which data they select in the first place. If they take the time to "walk" down the ladder of inference, they can learn a great deal about how their own beliefs, assumptions, background, culture, and other influences such as their own personal paradigm affect how they interpret what others say and how they interact with them. The ladder of inference is also a useful tool for reaching a better understanding of those they lead.

What happens when two individuals or a group of people sit down and discuss something such as an issue, a plan, a goal, or a problem? A healthy discussion will include an inquiry and advocacy. Inquiry involves talking with other people and learning from them. At this stage, members of the group are not judging, arguing, or trying to present their own viewpoint; they are just learning. During this phase, they should strive not only to hear the other person's words, but to learn about their mental models to understand where they are coming from and what they are really saying. A second aspect of communication, after the inquiry stage, is advocacy. Advocacy involves "selling" an idea or position or directing attention to certain facts members think are relevant. This is when a person begins to evaluate ideas, narrows the field, and works toward a consensus. This is also a time for observing their own thoughts, checking out their own ladder of inference. Inquiry requires that team or group members suspend assumptions. This does not mean laying them aside, but rather bringing them forward and making them explicit so that you and the others can explore their meaning and impact.

In a team context, inquiry and advocacy are commonly referred to as dialogue and discussion. During the dialogue phase, everyone should be in an inquiry mode, for example, sharing facts, ideas, and opinions without evaluating or defending them.

By the time they move to the discussion phase, everyone should have a common understanding of all of the facts and viewpoints. Then comes the discussion part, when everyone tries to determine what they believe in. The problem in many teams is that they tend to move too quickly to discussion, without adequate inquiry. This has the effect of stifling creative thinking and undermining trust. As a result, leaders need to be able to balance inquiry and advocacy. Leaders balance inquiry by using interaction with other people to learn, exchange ideas, and understand their perspectives. Leaders also need to balance advocacy by evaluating ideas, promoting one's own idea, and working toward a consensus.

Advocacy involves selling an idea or position or directing attention to certain facts you think are relevant. This is when you begin to evaluate ideas, narrow the field, and work toward consensus. An effective change process includes the following elements: a change in mind-set, purpose, process, predictable forces, structures, and sustainability. Throughout this process the leader must be able to advocate change and motivate others to take up the banner. They must be able to communicate what the organization wants to accomplish as a result of the change. The leader must be able to create a plan, implement and monitor the change, analyze its impact, and adjust. The leader must be able to identify potential resistance and deal with it. They must anticipate needed changes to the organization's systems, policies, action plans, and resources. The leader must be able to create a critical mass for change by identifying crucial supporters and meeting their needs. The leader must remain tuned to the human response to change and deal constructively with it.

To successfully facilitate change in an organization, the leader must be able to communicate the "four Ps" effectively: purpose (i.e., the why), picture (i.e., the vision), plan (i.e., the process), and part (i.e., each person's role in the vision and the process, and how the change will affect each one). Four rules for communicating the change are communicate first through actions, then communicate through words, recognize that perceptions will become distorted, and remember the "rule of six," that is, tell people six times in six ways. Finally, anticipate and allow for strong emotions among group or team members.

A leader cannot effectively facilitate change without mutual trust, but it is very difficult to build trust in a changing environment or to rebuild trust after it is lost. Therefore, one of the most critical aspects of the leader's job is to build an environment of trust. Trust is the very core of leadership, and trust must be earned. Trust in a relationship is based on mutual confidence that both parties will do what they say; communicate honestly; respect one another's knowledge, skills, and abilities; maintain confidentiality; and keep their interactions unguarded. To build trust, a leader must demonstrate conviction, courage, compassion, and community. A trusting relationship is a two-way street. An effective leader is both worthy of trust and able to trust others. A leader can build people's faith in his or her trustworthiness through predictable behavior, clear communication, keeping promises, and being forthright. Although the capacity for trusting others is influenced by history, the situation, the inherent risk, and other factors, a leader can expand the capacity for trust by giving people the benefit of the doubt and by looking for opportunities to stretch them beyond their comfort zone to demonstrate trust. Building trust is a slow process, and trust can be destroyed by a single event. When trust breaks down, a leader can use the following steps to begin to rebuild trust: by accepting responsibility, admitting the mistake, apologizing, acting to deal with the consequences, making amends, and attending to people's reactions and concerns.

In addition to building trust and facilitating change, an effective leader must be able to exert personal influence to achieve goals. An emergency management professional must be able to exercise influence in multiple directions. For example, upward with those of higher rank, laterally with peers in the same organization or response system, downward with subordinates, and outward with people outside the organization, including the media, the public, the business community, and others. There are three types of influence: position influence, domineering influence, and interpersonal influence. Although leaders may use all three types of influence in different situations, it is interpersonal influence that lays a foundation for trust, support, and collaboration. Effective interpersonal influence involves three elements: I, you, and we. Together, they add up to, "I am a trustworthy ally, you are a valuable resource, and we can accomplish this together."

An effective leader needs to be able to react appropriately to another person's point of view to foster a winning outcome. The ability to negotiate an agreement involves three skills: agreeing, constructive disagreement, and building on ideas. Agreeing involves stating what you like about the other person's idea and why you like it. Constructive disagreement is framing a disagreement with an idea in a way that preserves the person's self-esteem. This involves identifying the value of the idea, explaining your reservations, and discussing alternatives. Building on ideas acknowledges the connection and adds value. There are times when, in addition to personal influence, the leader needs political savvy, which is, literally, the ability to know people. Political savvy can be used in a positive way—for positive ends. It represents a balance between self-interest and organizational interest. Using influence well can benefit not only the manager but also the organization as a whole and the people the manager leads. Three building blocks for political savvy are an alliance mind-set, the ability to understand one's allies, and the ability to be an ally to others.

An alliance mind-set involves viewing others as potential allies. There are four rules for interacting with people as allies: assume that mutual respect exists, trust the other person and be someone whom he or she can trust, be open, and share information and look for mutual benefits. An effective leader is smart about who their allies are and what they care about. To develop this understanding, the leader needs to ask the question: Who are my allies? They may include those who may be affected, whose cooperation is needed, who could present obstacles, or who could help build support. The leader must also be aware of the concerns, interests, and motivations of their allies. "Walking a mile in the other person's shoes" will help them know how to influence them. The leader must also be aware of how their ideas relate to their allies' concerns. The leader needs to understand how their idea may help or hurt them, and how they could adjust their thinking to better accommodate their needs. Being an ally means invoking the principle of reciprocity, for example, "as we do things for others in the organization, they become more likely to help us in return." In turn, build goodwill every day and it will be there when you need it.

One of the most significant legacies that a leader can leave in an organization is to have created a new generation of leaders. Being able to create an environment that promotes the development of new leaders and at the same time nurtures one's own leadership skills is the hallmark of a truly effective leader. All of the strategies reviewed—developing self-knowledge, balancing inquiry and advocacy, and building trust—contribute to a leadership environment. In addition, a leader can encourage leadership development in others through such activities as building a shared vision, empowering others, creating a team environment, coaching, delegating, and mentoring.

1.7 EFFECTIVE COMMUNICATION SKILLS

An emergency manager must be a skilled communicator in order to achieve their objectives. They are required to convey information to a broad audience that includes public- and private-sector organizations, the media, disaster victims, and core-sponders. Even during nonemergency situations, they will need to rely on strong communication skills to coordinate with staff and to promote safety awareness.

The value of a model is that it simplifies a complex process. Communication is just such a complex process, and using a model will help send and receive communications and will help ensure that others respond as required in an emergency. Immediately before, during, and immediately after an emergency, emergency and response personnel must respond quickly. Time to communicate is limited, and often a specific message that must result in practical action must be relayed to a large group. A very simple model that sends the message efficiently and elicits the desired response is most useful. During the recovery phase, when sensitivity to the community's cultural values and attitudes is perhaps most important, a more complex, culturally based model may be more appropriate. Failure to discern attitudes, beliefs, values, and rules implicit in different groups could disenfranchise some citizens and harm the community's return to productivity and health. A cultural model is useful because it recognizes community members' shared interest in the community's future. Therefore, because communication is a complex process, it may be helpful to use one or more models to help ensure that communications are effective. Regardless of the model selected, a good model facilitates an efficient, two-way flow of communication and elicits the desired response.

Listening is a critical component of communication. When a person listens empathically, they do not just hear words, but they hear thoughts, beliefs, and feelings. Empathic listening is highly active and requires hard work. Emergency managers should work on their listening skills whenever they have the opportunity so that they feel comfortable with their ability to listen to and hear what people tell them. There are numerous variables involved in the communication process, including differences between the sender and the receiver, differences in communication styles, differences in previous experiences, and cultural differences.

In an emergency, people depend on information for physical and emotional comfort. To be effective, emergency communications must be timely, accurate, and clearly stated. There are considerable differences between day-to-day communications and emergency communications. For example, emergency information is important, timeliness is essential, warnings require response, barriers to communication are common in emergencies, and messages must be consistent to elicit the desired response. When communicating during emergency, emergency personnel should always present the information in sequence and present the reason for the message, the supporting information, and the conclusion. They should also avoid any jargon, codes, and acronyms. They should use common names for all personnel and facilities and omit unnecessary details. They should also speak in sync with other related authorities. Finally, they should keep the messages consistent across various media and word the message precisely, making every word count. Communication media range in complexity from handwritten notices to international satellite broadcasts. Selecting the appropriate media for both the message and the audience is essential to effective communication.

Whether people realize it or not and most often they do not, traditions shape the way people interact with each other. Cultural differences reflect internal beliefs and thought patterns that cause people to react differently to the same situation. Differences in age and sex, the presence of a disabling condition, and even the part of the country people live in can affect how people communicate. In turn, it is not realistic to become an expert on every culture that may be encountered. However, it is reasonable for one to learn about the populations that make up major parts of their community. For example, what groups are represented in the community? Where they are located? What are their needs? Making oneself aware of key cultural and other differences—both verbal and body language differences that one will need to address during an emergency—helps in understanding what to expect of the groups and whether their message is being communicated. Take into account cultural differences when addressing communications across cultures or those with special needs. Some things that one can do as a first step are do not assume sameness, do not assume that one understands what the other person means, and do not inadvertently cause the behavior.

Technology can assist an organization in meeting their communication goals. Selecting the best technology to support their message maximizes its impact. But choosing the wrong technology can interfere with their message, and there is no correlation between the complexity of the technology and the effectiveness of the communication. With every potential technology come limitations and cautions. Emergency personnel must always try to match the technology with the message's purpose and the audience. Emergency conditions can result in power outages or other conditions that may limit the choice of technology. Organizations must be prepared to choose lower technology methods to support their communication. They can use a mix of high-technology and low-technology tools to support their message. In choosing their methods, they must consider the message or what, the purpose or why, and the recipient or who.

One of the most important emergency management skills is oral communication. An emergency manager, planner, or responder may be asked to communicate critical information in a variety of ways. Each circumstance offers unique challenges and opportunities to match the verbal and nonverbal communication to the message and audience. Ensuring that the presentation matches the audience is critical to gaining the desired response. Matching messages and audiences will help to ensure communication success. The news media can be a strong ally in alerting and informing the public. It is important to establish credible and productive working relationships with representatives of the media. To minimize misunderstandings, build strong relationships with media representatives. More than half of face-to-face communication is exchanged through nonverbal cues. Because up to 65% of the meaning of a message is unspoken, it is imperative to learn to read nonverbal communication. Eyes, tone of voice, expression, volume, and gestures reflect attitudes, emotions, states of mind, and related messages. Nonverbal cues can have an impact equal to or stronger than the words that are spoken. When the speaker's body language is in sync with the verbal message, the message is reinforced. Listeners are likely to respond to this extra persuasion with increased respect, harmony, or trust. Speech anxiety is the single most common social anxiety in North America. Whatever the source of speech anxiety, taking these steps can help in reducing nervousness and in gaining control of the presentation through preparation, practice, and acceptance.

Three common elements to successful oral presentations are that the message matches the audience, the content and delivery match the purpose, and the delivery is clear and engaging. Matching the message to the audience begins with analyzing the needs of the audience. After the presenter determines the "who, what, when, where, and why" aspects, it is easy to determine the right message and the most effective delivery. It is likely that one will make one of two types of presentations: informational or motivational. To determine which type is appropriate, one should ask the question: "Am I relying on facts or shaping opinions?" Informational presentations transmit specific knowledge. They present information directly or through explanation and feature statistics or supporting research. Furthermore, the presenter's ideas are delivered in a logical sequence. Motivational presentations create awareness, change attitudes, or garner support. They tend to use concrete language to communicate abstract points, and more often than not use vivid and interesting language.

1.8   DEVELOPING AND MANAGING VOLUNTEERS

The use of volunteers has proven critical to emergency management. Both individual volunteers and established volunteer groups offer a wealth of skills and resources that can be used prior to, during, and after an emergency. Mobilizing the private sector can add significantly to emergency management programs. For an emergency management professional, the ability to work with volunteers before, during, and after an emergency can literally affect the lives and well-being of the local citizenry. Volunteers can impact, for better or worse, the ability of response agencies to do their jobs and can make a difference in how quickly the community is able to respond to and recover from a disaster.

A volunteer is an individual who, beyond the confines of paid employment and normal responsibilities, contributes time and service to assist in the accomplishment of a mission. There are four classifications of volunteers: professional, unskilled, spontaneous, and affiliated with a VOLAG. A VOLAG is an established organization whose mission is to provide services to the community through the use of trained volunteers. There are benefits to involving volunteers. For example, volunteers can provide services more cost-effectively and provide access to a broader range of expertise and experience. In addition, they increase paid staff members' effectiveness by enabling them to focus their efforts where they are most needed or by providing additional services. Volunteers can also provide resources for accomplishing maintenance tasks or upgrading which would otherwise be put on the back burner while immediate needs demand attention. Volunteers enable the agency to launch programs in areas in which paid staff lack expertise and can act as liaisons with the community to gain support for programs. Volunteers can also provide a direct line to private resources in the community, facilitate networking, and increase public awareness and program visibility. Challenges to working with volunteers often involve misperceptions or factors such as tension between volunteers and paid staff that can be alleviated with good planning and management. There are, however, real obstacles to involving volunteers such as a sparse population in rural areas.

Developing or maintaining an agency volunteer program requires someone to act as volunteer program director. Developing a solid volunteer program involves seven steps: analyzing agency and program needs, writing volunteer job descriptions, recruiting volunteers, placing volunteers, training volunteers, supervising and evaluating volunteers, and evaluating program.

Before deciding to develop an internal agency volunteer program, consider the needs of potential volunteers. Volunteer needs include the desire to make a difference, work with other people, and learn new skills. In addition, one must also consider the needs of the agency. Analyzing agency needs involves considering the agency's mission. The emergency manager would need to look at current staffing resources and areas of shortfall where volunteers could help or may be needed. In doing so, they will also need to describe the jobs that volunteers could do and the abilities and resources needed to do those jobs. Writing a well-defined job description is an important step because a good job description is the first step in the recruitment process. A good job description is also a tool for marketing the agency's need to potential volunteers and a focal point for the interview and screening process. The job description also serves as a basis for performance evaluation. It is also important to involve paid staff when planning for volunteer involvement. Soliciting their input on volunteer job descriptions will help alleviate tensions between paid and unpaid staff later on.

Once the agency needs analysis is complete, the next step is to write volunteer job descriptions. Points to consider when developing a job description include the purpose of the job, the job responsibilities and qualifications, and to whom the volunteer will report. In addition, the program director may also want to include the time commitment required for the job, the length of the appointment, who will provide support for the position, and the development opportunities, if any.

Recruitment can be broad based or targeted. Broad-based recruitment is used when there is a need for a large number of individuals for jobs that require commonly available skills. Targeted recruitment is used when looking for volunteers with specific skills to do specific jobs. Recruiters need to investigate the marketplace of potential volunteers in the community. The next step is to design a message based on what recruiters know about the volunteers they are trying to reach. They also need to choose a medium—for example, radio, television, Internet, direct mail—that will reach their targeted audience and determine where and when to deliver their message. An effective recruitment message contains the following elements: an interesting opening, a statement of the need or problem, a statement of how this job can meet the need or solve the problem, a statement to address the listener's question as to whether he or she can potentially do this job, how doing this job will benefit the volunteer, and a contact point for more information.

Placing of volunteers includes screening and interviewing. All volunteers should be screened. Those jobs that involve a higher risk such as working with money or children require more intense screening. Common screening tools include the application, reference checks, professional licenses, criminal background checks, and child abuse clearance checks. In addition, the interview is also a screening tool. If possible, the interviewer should be a volunteer, and the following tools should be used when conducting an interview: the potential volunteer's application, a form for recording the interview, a list of open-ended questions, and information about current volunteer opportunities. The goal of the interview should be to determine the applicant's skills and motivation for volunteering, matching the applicant with a position, and answering the applicant's questions. The interview can also be used as a screening tool for identifying undesirable candidates. During the interview, the interviewer should not ask any personal questions. Legally, the interviewer may not ask anything that is not directly related to the ability of the applicant to perform the specific volunteer job. The interview should result in a recommendation for further action, for placement, further screening, or rejection.

Good training is critical because it can save lives and protect the agency from lawsuits. There are two levels of training: agency orientation and job specific. Orientation to the agency should cover the agency's mission, the agency's values, agency procedures and issues, and the role of volunteers in the agency. These components can be addressed through a group presentation of a video and/or by various speakers. Training for the specific job should cover specific job responsibilities. It should also cover who the volunteer's immediate supervisor is and his or her expectations. The training should also cover authority and accountability, for example, what the volunteer can and cannot do without explicit direction, other team members' roles, and resources available to do the job. Training for the job is best accomplished by one-on-one mentoring. Unlike orientation, which is information based, training is skills based. Like orientation, training can be general or specific. General training includes training in such skills as communication, team building, problem solving and decision making, leadership and supervision, and stress management. Specific training teaches job-specific skills such as cardiopulmonary resuscitation (CPR). Training can and should be ongoing, for example, refresher skills training. One way to keep volunteers' training current is to certify those who have completed training and issue cards with expiration dates.

Good supervision is essential to volunteer success. A good supervisor delegates effectively, establishes performance expectations, acts as a coach and team builder, and communicates effectively. Good supervisors also know how to listen and are receptive to information from others. They assist staff in developing their skills and give constructive feedback and takes corrective action, when needed. They also recognize staff for their contributions. Recognition is a critical component of supervision because it is one of the keys to maintaining volunteer interest and, therefore, volunteer retention. Recognition can be formal or informal. Guidelines for evaluating volunteers include keeping evaluations confidential and making sure that comments are fair. Evaluation should focus on the work and not on the individual. In addition, supervisors should follow agency guidelines for disciplinary procedures. Disciplinary policy should be covered during orientation. Corrective action may include additional training or supervision, reassignment, suspension, or termination. Termination should be reserved for instances when other measures have failed or when there has been gross ethical misconduct. Volunteers should be made aware of grievance procedures to address complaints that cannot be resolved with their supervisors. Two-way feedback during the evaluation process allows emergency managers to evaluate their volunteer program as well as their volunteers. Evaluating the volunteer program regularly ensures that it is meeting the needs of the agency, community, and volunteers.

Some agencies may choose to use the volunteer services of local VOLAGs and community-based organizations (CBOs) instead of, or in addition to, maintaining their own volunteer program. The role of a VOLAG and/or CBO coordinator, as opposed to a volunteer program director, involves building relationships and acting as a liaison with local community VOLAGs and CBOs. In addition, they also collaborate with local community VOLAGs and CBOs to develop and exercise a plan for coordinating volunteer services in an emergency. In the United States, the National Voluntary Organizations Active in Disaster (NVOAD) is a consortium of recognized national voluntary organizations whose mission is to foster more effective service to people affected by disaster. NVOAD member agencies are coordinated by the state NVOAD.

Working with other agencies requires collaboration on a range of issues, including shared decision making; sharing information, resources, and tasks; flexibility in dealing with differences among agencies' terminology, experience, priorities, and culture; and respect and humility to learn from others' ways of doing things. Interagency collaboration benefits the community by eliminating duplication of services, expanding resource availability, and enhancing problem solving. In addition to VOLAGs, other sources of volunteers include CBOs; businesses, especially those with corporate volunteer programs; and professional associations such as doctors, engineers, and nurses. However, volunteers not associated with VOLAGs may need training in disaster response. Some of the special issues involved in working with volunteers include spontaneous volunteers; safety, risk management, and liability; and insurance and workers' compensation.

Spontaneous volunteers are people who show up unannounced to volunteer after a disaster and present both potential benefits such as supplementary assistance, and challenges such as their lack of screening and training. The preferred way to deal with spontaneous volunteers is according to a plan developed before a disaster occurs. Such a plan should address the important role that EPI plays in discouraging and/or channeling spontaneous volunteers. The media can be enlisted to publicize a phone number that those wishing to volunteer could call for instructions. Channeling spontaneous volunteers on site can be done by a volunteer coordinator with the use of a skills survey.

The law generally requires that a person act with the care that a reasonable person would exercise to prevent harm. In the United States, volunteers for government agencies may be subject to the Federal Tort Claims Act but are also covered by the Federal Volunteer Protection Act of 1997. In Canada, volunteers are subject to both federal and provincial legislation, but are also subject to municipal bylaws. If a volunteer assisting in the disaster response accidentally harms a person or their property, the law generally recognizes that the employer is responsible for the actions and inactions of his or her employee including volunteers. Most of the laws limiting the liability of volunteers are state or provincial laws. There are three keys to minimizing susceptibility to lawsuits, train volunteers well, supervise volunteers, and document their training and actions. Liability insurance is available to both organizations and individuals. In addition, some states and provinces provide Workers' Compensation or Workman's Insurance coverage to volunteers registered with an established agency.

Witnessing death, injury, and devastation can cause extreme stress reactions in some volunteers. In addition, working long hours and skipping meals can also contribute to volunteer workers' stress. Take these steps before, during, and after a disaster to manage stress. Before a disaster address stress during volunteer orientation and/or in stress management training seminars. During a disaster, volunteers should be appropriately matched to their job assignments, get regular meals and breaks, and rotate out at the end of a reasonable-length shift. After a disaster, invite a trained mental health professional to hold a critical incident stress debriefing.

2 Threat Ensemble, Vulnerability, and Risk Assessments

2.1 INTRODUCTION

Threats may be the result of natural events such as earthquakes and floods, accidents, or intentional acts to cause harm. Regardless of the nature of the threat, emergency managers and first responders have a responsibility to limit or manage risks from these threats to the extent possible. In Canada and the United States, the federal, state, and provincial governments have implemented emergency management plans in dealing with a variety of different threats and/or hazards.

One of the core challenges faced by emergency managers is how to prevent, mitigate, prepare, respond, and recover from different types of hazards. In addressing this challenge, emergency managers, particularly those working for municipalities, must adhere to strict criteria in developing their emergency plans. Part of the emergency planning process involves the following:

- Identifying what hazards exist in or near the community?
- How frequently do these hazards occur?
- How much damage can they cause?
- Which hazards pose the greatest threat?

Hazard identification or threat assessment, vulnerability, and risk assessment help guide the emergency manager to answer these questions. Threat, vulnerability, and risk assessments can help a community prepare for the worst and most likely hazards. In addition, they can save time by isolating any hazards that affect the community; allow for the creation of emergency plans, exercises, and training based on the most frequent and highest-risk scenarios; and help an organization's program become more proactive rather than reactive.

The terms threat, vulnerability, and risk are often used interchangeably when in fact they each have distinct and complementary definitions. For example, it is the union of a specific threat with a corresponding vulnerability that produces the potential harm for disaster. This concept was demonstrated during the ice storms of 1998 in Ontario and Quebec. The power grid damage that occurred was due to the vulnerability of the distribution towers to breaking from the weight of built-up ice. Had the towers been sturdier, or had the cables been buried, it would have been possible to tackle this vulnerability, and the threat (the ice storm itself) would not have caused power outages.

Similarly, a vulnerability to a threat does not necessarily mean that it will cause harm or damage. For example, many homes are built in low-lying areas (the vulnerability) that would flood if a river overflowed or spilled its banks (the threat). However, decades could go by without a flood ever occurring and any damage being done. Therefore, the convergence of a threat with vulnerability, combined with the knowledge or appreciation of a harmful consequence, which might emanate from them, produces the potential for harm; this is referred to as risk.

Threat assessments, vulnerability assessments, and risk assessments are tools that can be used to assess which hazards pose the greatest risk in terms of their probability of occurrence and their severity or potential impact. They are not intended to be used as prediction tools to determine which hazard will cause the next emergency. Prediction tools are sometimes referred to as *early warning systems* and are covered in detail in Chapter 5.

A risk management approach is adopted to undertake the threat, vulnerability, and risk assessments. There are four stages within the risk management approach: preparedness, mitigation, response, and recovery. Each of these stages works together to reduce the overall risk of harm. Preparedness includes those steps taken in advance of an event to prepare for its occurrence, including the development of effective policies, response plans, and procedures for how best to manage an emergency; the purchasing and stockpiling of emergency supplies; and the comprehensive training and exercising by emergency responders. The steps taken in the preparedness stage should be aimed at reducing the impacts and losses from inevitable failures or disruptions in the future.

Mitigation involves applying a series of sustained actions to reduce or eliminate the long-term impacts and risks associated with natural and human-induced disasters. Mitigation plans usually address vulnerability specifically, such as restricting homebuilding in flood plains, or a threat, for example, actively seeking out and capturing terrorists. When an event does take place, damage and harm can be minimized by responding quickly and in a coordinated fashion. Response activities are typically handled by first responders, including but not limited to firefighters, police, and ambulance services, but in serious emergencies other agencies and levels of government can also become involved, such as the military.

The recovery stage includes the efforts undertaken to repair and restore communities after an emergency. Once the immediate situation is stabilized, the affected community or organization needs to recover from the incident and rebuild. These activities are necessary to ensure the long-term viability of the community or organization and to prevent what in some cases can be serious long-term economic harm. As a result, comprehensive emergency management plans come with associated costs. These costs can include both financial and "opportunity costs," or the costs of choosing one potential option over another based on a series of factors. Understanding the full ramifications of these costs is essential in arriving at a complete risk management strategy.

2.2 THREAT ASSESSMENT

The first step in the risk management approach is identifying the threat or threat assessment. A threat assessment considers the full spectrum of threats, for example, natural, criminal, terrorist, and accidental, for a given community. The assessment should examine supporting information to evaluate the likelihood of occurrence for each threat. For natural threats, historical data concerning frequency of occurrence for given natural disasters such as tornadoes, hurricanes, floods, fire, or earthquakes can be used to determine the credibility of the given threat. For criminal threats, the crime rates in the surrounding area provide a good indicator of the type of criminal activity that may threaten a community. In addition, the type of assets and/or activities located within the community may also increase the attractiveness of a target in the eyes of the aggressor. The type of assets and/or activities located within a community will also relate directly to the likelihood of various types of accidents. For example, a community with a high concentration of industrial facilities will be at higher risk for industrial-type accidents, such as chemical spills or explosions, than one with a concentration of small retail businesses.

For terrorist threats, the attractiveness of the target is a primary consideration. In addition, the type of terrorist act may vary based on the potential adversary and the method of attack most likely to be successful for a given scenario. For example, a terrorist wishing to strike against the federal government may be more likely to attack a large federal building than to attack a multitenant office building containing a large number of commercial tenants and a few government tenants. However, if security at the large federal building makes mounting a successful attack too difficult, the terrorist may be diverted to a nearby facility that may not be as attractive from an occupancy perspective, but has a higher probability of success due to the absence of adequate security. In general, the likelihood of terrorist attacks cannot be quantified statistically since terrorism, by its very nature, is random. Hence, when considering terrorist threats, the concept of developing credible threat packages is important.

North America's size, varied topography, and relatively severe weather patterns make it particularly susceptible to the occurrence of natural disasters. Individuals, communities, and local, provincial, state, and federal governments face an increased risk of death, suffering, destruction, and cost from disasters, as a result of not only the potential increase in frequency of disaster events, but also the increased exposure to risk. Most noteworthy are floods, severe storms including severe winter storms, such as ice storms and hail storms, hurricanes and oft-accompanying tornadoes, geomagnetic storms, earthquakes and tsunamis, fires, and epidemics including the West Nile virus and foot-and-mouth disease.

Natural disasters can have a significant impact on infrastructure in a relatively short period of time. The primary point of impact tends to be the physical infrastructure, such as hydro lines and bridges. Damage is not always localized, often resulting in a ripple effect across a number of sectors. The ice storm in Central and Eastern Canada in 1998 is an excellent example of the degree of damage that can occur in a short time period.

According to the Canadian Disaster database, natural disasters have accounted for 70% of all disasters in Canadian history. Globally, flooding has been, by far, the greatest cause of disasters in the twentieth century, followed by severe storms. Although natural disasters are the result of naturally occurring extreme events, it must be reiterated that the impact of a disaster is the direct result of human influences and their impact on the environment. This influence can play a significant role in both the prevalence and scope of certain natural disasters. Climate change unless addressed by collaborative human action will inevitably, and significantly, alter our capacity to effectively manage the risks associated with natural disasters. Countries can also be susceptible to epidemic threats to their social and economic fabric: globally there are 14 million deaths each year due to infectious diseases. In addition, one-sixth of the world's population is affected by neglected, emerging, and reemerging diseases.

In recent years, the world has been experiencing dramatic shifts in climate. For example, in the past 20 years, Canada has experienced the greatest increases in average annual temperatures of any country and a commensurate rise in severe weather-related natural disasters. Associated with these shifts has been an increased occurrence of phenomena such as severe storms, floods, droughts, and forest fires. These occurrences have led to higher maintenance, cleanup, and insurance costs. Canadians currently spend in excess of US$12 billion per year in coping with and adapting to weather and weather-related disasters.

Climate change affects the frequency and severity of extreme weather events in several ways. It is generally accepted that additional warming will change the distribution of heat and thus the flow of energy through the climate system. This, in turn, will alter the circulation patterns of the atmosphere and the oceans, and it will also modify the hydrological cycle by which water is circulated between the Earth's surface and the atmosphere. As a result, the position of many of the world's major storm tracks could shift significantly.

In addition, it is expected that a warmer climate would affect the physical processes that generate different types of extreme weather events. For example, a rise in global temperatures could lead to an increase in the amount of water that is circulating through the hydrological cycle. Consequently, more moisture will be available in the atmosphere to precipitate as rain or snow. Current global climate change circulation models indicate that a warmer atmosphere will increase the amount of moisture transported into the middle and high latitudes of the Northern Hemisphere. These models also suggest that the additional precipitation will likely occur during heavier falls rather than in more days of precipitation. These models also suggest that the number and intensity of severe thunderstorms will increase in most areas in warmer climate.

An increase in spring and summer temperatures over the past two decades has led to an increase in the frequency of tornadoes in the Canadian Prairies and the Midwestern states. This hypothesis has been studied in a seminal work by Etkin, and his findings concluded that there was a positive correlation between the generally accepted outcomes of ongoing climate change and an increase in tornado activity.

Increases in wildfire have occurred in Yellowstone National Park from 1985 to 1990 because of climatic changes. In the boreal forests of central and northwestern

Canada, the area annually disturbed by fire and insects has doubled in the past two decades compared to the previous 50-year period. This is related to statistically significant winter, spring, and summer warming trends and probably more lightning. Other factors such as tree age and firefighting policy could also have contributed to this increase.

Different regions can be affected by rainstorm, ice-jam, and snowmelt floods. A shorter winter season under climate change may result in a reduced risk of snowmelt and ice-jam floods. The main concerns about increased flooding result from the fact that a warmer atmosphere can hold more moisture, and precipitation is expected to increase as a result. As well, precipitation is expected to become more intense over smaller areas, which suggests greater flooding problems especially in smaller catchment areas.

There is also a concern that with an increase in heavier rainfall events, the number of dry days between events may increase and drought will become more severe. This effect could be worsened by higher air temperatures and increased evaporation. Some general circulation model studies show reduced soil moisture values over the mid-North American continent, suggesting more frequent droughts in the future. In Canadian history, the Saguenay floods of 1996 are considered to have been the worst to date. These floods caused 10 deaths, destroyed 1,718 houses and 900 cottages, displaced 16,000 people, and resulted in US$800 million in damages.

For floods and droughts, other forms of change such as land use are also important factors in the changing severity of extreme weather events. For example, the reduction of woody vegetation, the urbanization of watersheds, increased areas of impermeable surfaces for highways, and some other land use changes increase the amount of precipitation that quickly becomes surface runoff. Small rivers and streams in affected regions become increasingly "flashy" with higher peak flows and less flow during dry periods.

Increasing damage from floods and other extreme weather-related hazards is expected in the future. For example, there will be more people living in larger but fewer urban centers. In the event of flooding, physical damage to property, especially in urban areas, is the major cause of physical losses. Damage to crops, livestock, and agricultural infrastructure can also be high in rural areas.

Changing climatic patterns with an associated increase in extreme flooding events are expected. Different regions can be affected by rainstorm, ice-jam, and snowmelt floods. A shorter winter season because of climate change may result in a reduced snowpack in many areas and thereby there would be a reduced risk of snowmelt and ice-jam floods. The main concerns about increased flooding result from the fact that a warmer atmosphere can hold more moisture, and, as a result, precipitation is expected to increase. However, precipitation is expected to become more intense over smaller areas, which suggests greater flooding problems especially in smaller catchment areas. For instance, precipitation over the Great Lakes was relatively higher and more variable for the period between 1940 and 1990 compared to the period between 1900 and 1940.

Aging infrastructure that is prone to damage will increase loss levels. In addition, as populations continue to grow in North America, more demand is placed on the aging infrastructure to provide essential services including drinking water, heat, and electricity. During the 1960s, governments in Canada spent 4.5%–5% of GDP on infrastructure projects. That amount has now declined to 2%. These projects include many of the facilities that are integral parts of our urban systems including roads, bridges, water distribution systems, sewer networks, public buildings, dams, and dykes. Many of these water structures are now 40–55 years old and may require major maintenance in the near and medium term. It is unclear how this activity will be financed. In the absence of required maintenance, all structures become more prone to damages. The flood infrastructure (both concrete and earthen) that now supports human settlements was often developed at the expense of naturally occurring flood-mitigating infrastructure, such as wetlands, floodplains, and other natural areas. Given the current trend in damages, it is unclear if the economic benefits associated with traditional settlements will continue to outweigh the costs over the long term.

Severe storms can significantly affect the reliability and overall safety and security of infrastructure and services. Although forecasting technology has allowed experts to accurately track and predict the severity of storms, overall understanding of their impact upon critical infrastructure remains limited. Moreover, current predictive methodologies regarding the cascading effects of a critical infrastructure incapacity or failure are limited.

The infrastructure sectors most prone to some level of incapacity due to certain severe storms have been telecommunications, energy service delivery, transportation, and emergency services. Transportation networks, especially in major urban centers, can become choked, whereas in rural communities, roadways and access points can become entirely blocked, given the lack of transportation infrastructure redundancy. Emergency services are also particularly susceptible to the effects of severe storms as wide-impact storms can place tremendous strains on communities with limited emergency response services.

Geomagnetic storms, although infrequent, have the potential to severely impair infrastructure and services. They occur when sunspots on the surface of the sun erupt and impact the Earth's magnetosphere, thereby disturbing solar wind and reducing the global magnetic field. In North America, it has been demonstrated that power systems, pipelines, and communications are at risk from the damaging effects of geomagnetic storms. Consequences of geomagnetic storm activity can include widespread power failures, pipeline corrosion, the shutdown of cable systems, an increased drag on satellites, inaccurate navigational sensors, and the potential loss of millions of dollars in revenue. Geomagnetic storms also disrupt telephone and cable lines, submarine cables, and the geosynchronous satellites used for telecommunications and global positioning systems.

Ongoing research by scientists, engineers, and emergency preparedness officials has produced quantifiable insights into what effects earthquakes can have on various structures. This research has resulted in improvements to the National Building Code of Canada, with modern buildings in earthquake-prone areas having built-in earthquake resistance to help limit damage and injuries. Increasing the focus on the vulnerability of essential lifeline services (including power lines, water supply, sewage disposal, communications, oil and gas pipelines, and transportation facilities) both within a building and throughout the urban centers is required. Interruption of such services due to an earthquake can pose severe hardship to a community or a threat to the health and safety of the population.

Should a long-predicted earthquake of extreme magnitude occur on the West Coast, its effect could be catastrophic to the critical infrastructure network of the region. Moreover, given the relative unlikelihood of major seismic activity on the East Coast, mitigative activity on the threat of earthquakes on critical infrastructure has been limited, and therefore the potential for damage could be considered high. Besides providing adequate strength and ductility to resist ground motion, it may be advisable to introduce multiple paths or loops in a distribution network in order to provide alternate paths of supply. These redundancy measures will limit the regions that can be affected by a failure or interruption at one point of a network.

Tsunamis are relatively rare compared to other natural hazards; however, they do present a threat to maritime infrastructures, notably ports and shipping yards. They also present a general threat to the transport and telecommunication sectors of those regions vulnerable to tsunamis.

Forest fires can lead to shutdowns in power for extended periods when the fires get too close to transmission lines. Forest fires present a significant threat to critical infrastructure, particularly to oil pipelines, and to energy transmission and telecommunications lines. They also place a strain on the services and safety sectors in affected areas.

Over the course of the twenty-first century, climate change will continue and even accelerate. Changes will likely bring shifting weather and climate patterns, resulting in more extreme weather events. All of these symptoms may present additional stresses on critical infrastructure. Expansion of urban development and economic activity into more remote parts of the country may also increase the risk of disaster. As the urbanization of underdeveloped lands continues, pressure may be placed upon infrastructure that lacks the redundancy and robustness to safely support these bourgeoning communities.

Furthermore, the increasing shift to wireless technology may make the telecommunications and information infrastructure less vulnerable to storms and other natural events, because the system will be less reliant on transmission lines. Wireless technology is still reliant on transmission towers, but the time and costs related to repairing the system will be reduced.

Accidents can interrupt service and supply of telecommunications and energy, disrupt transportation networks, such as creating bottlenecks in rail networks, dislocating hundreds or thousands of people, and cause injury or death. For telecommunications and power grids, a failure in one part of the network can cause cascading failures across a much wider area.

The increasing interconnectedness of critical infrastructure means the impact of an accident is greater and more widespread than before. Where people are forced to evacuate their places of business, economic and financial losses can be considerable. For example, the accident and fire at the Bell station in Toronto in 1999 affected more than 113,000 phone lines. The Hospital for Sick Children was forced to use two-way radios, 9-1-1 service was reduced, traffic lights at 550 intersections were affected, and businesses were unable to take orders or contact staff. Trading value on the Toronto Stock Exchange (TSE) fell by nearly US$1 billion, and the exchange was forced to close its derivatives trading floor. The online banking system was also affected by the blackout, leaving consumers in the downtown core without access to bank machines or direct payment services.

The Great Northeast Blackout of 1965 left 30 million people across 200,000 km² without power. Urban centers, which are heavily dependent on electricity for lighting, transportation, elevators, and traffic lights, were paralyzed. Hundreds of thousands of commuters were stuck in Manhattan and forced to camp out in hotel lobbies or their offices. Ten thousand people were trapped between stations in the subway system. Had the blackout lasted into a second day, serious logistical, health, and security problems could have potentially developed.

Accidents that lead to large numbers of people injured, killed, or displaced have the potential to overwhelm the health-care system. Although no one was killed in the Mississauga train derailment of 1979, 139 ambulances and several buses were held on standby for an expected flood of casualties. Had there been large numbers of injured, they would have had to travel to outlying hospitals for treatment, because three local hospitals had been evacuated.

Infrastructure has long been a target for malicious attack, whether for criminal, military, or political purposes. Infrastructure supports society, delivering a range of services upon which individuals and other sectors depend. Any damage or interruption causes ripples across the system. Attacking infrastructure, therefore, has a "force multiplier" effect, allowing a small attack to achieve a much greater impact. For this reason, critical infrastructure and networks have historically proven to be appealing targets for vandals, criminals, and terrorists.

Physical attacks on critical infrastructure may impede the delivery of services, such as power, transportation, and health care. Because such a disruption has a major impact on a wide cross section of society, an attack on the physical infrastructure magnifies the impact of the initial attack. For example, the explosion at the Utah Power electrical plant during the 2001 Salt Lake City Olympic Games caused a power outage that left 33,000 homes in Salt Lake, Davis, and Tooele counties without power for nearly an hour and later sparked a fire at the Tesoro oil refinery in North Salt Lake. Smoke from the fire was so thick that Interstate 15 had to be closed. During the power outage, the airport ran on emergency generators, although flight operations were not affected.

In the immediate aftermath of the September 11, 2001, attacks in New York City, the local emergency services sector was dealt a serious blow when hundreds of responders were killed under the collapsing towers of the World Trade Center (WTC). Local communication was disrupted, leading several wireless carriers to donate mobile phones and pagers to emergency personnel. The attacks hindered local emergency transportation and taxed the resources of the local health infrastructure.

The attacks of September 11, 2001, impacted critical infrastructure in two ways: First, critical infrastructure facilities and operations were directly disrupted by the physical impact of the attacks. Second, the decisions that regulators, owners, and users made in response to the attacks also impacted critical infrastructure. For example, the U.S. government was responsible, through the Federal Aviation Authority (FAA), for issuing the first-ever national grounding of commercial aircraft immediately following the attacks. Owners of infrastructure such as financial markets and market participants altered the financial and banking sector by deciding to temporarily close key markets as a safety precaution. Increased demand for telephone and Internet connections forced carriers to truncate their services to avoid crashing their networks.

There is much speculation as to the degree to which terrorist organizations and nation states are actively seeking out chemical and biological weapons. Following the initial September 11 attacks, the United States was once again put on heightened alert by the anthrax outbreaks. Smallpox has also become a sensitive issue, with several countries reportedly holding stores of the deadly disease.

Anthrax is an acute infectious disease caused by the spore-forming bacterium *Bacillus anthracis*. Anthrax most commonly occurs in wild and domestic lower vertebrates (cattle, sheep, goats, camels, antelopes, and other herbivores), but it can also occur in humans when they are exposed to infected animals or tissue from infected animals. Although the inhaled form of anthrax is often cited as a potential biological weapon, there are limitations to its use. The bacteria are found in nature, but it is difficult to find strains that will cause serious disease. And once such a strain is found, it is dangerous to handle.

Subsequent to the September 11 attacks in New York and Washington, U.S. law enforcement and health officials discovered that *B. anthracis* spores had been intentionally distributed through the postal system, causing 22 people to be infected with anthrax, leading to the deaths of 5 of them. Despite the panic that ensued after several envelopes containing anthrax spores were received through the mail in the United States, the bacteria in letters does not make an effective biological weapon. Spores of anthrax tend to clump together and fall to the ground, so they are not easily inhaled. When sent through the mail, the spores are often mixed with light powder that can be inhaled, but sometimes the spores are simply spread on the paper. To be released as a weapon, the clumps of anthrax spores would have to be ground down to a size that can be inhaled easily, freeze-dried, and delivered in an aerosol spray from an aircraft. A standard crop-dusting plane, however, would not be suitable. The spraying equipment used to deliver insecticide on to fields produces droplets that are too large to be absorbed through the lungs. Even after it is delivered, the bacteria's behavior is somewhat unpredictable. It normally has an incubation period of up to 7 days, but could take up to 60 days to develop. And unlike other potential biological weapons, such as smallpox or plague, anthrax cannot be transmitted from person to person, so it does not spread through a population after it has been released.

If a virulent strain of the anthrax bacteria got into the Canadian agricultural system, its effects could be far reaching and devastating. Anthrax is a named disease under the Health of Animals Act, and the general protocol for the release of the disease into a herd of cattle calls for the infected animals to be destroyed. However, those animals found to be uninfected—though in contact with the infected carrier—would merely be required to be quarantined. Though this technique is medically sound, it might not alleviate public health concerns.

Smallpox was officially declared eradicated in 1980 by the World Health Organization (WHO). However, that diagnosis was apparently premature as the virus is now considered one of the world's top six viruses that pose a threat to public health. Since naturally occurring smallpox virus has been eradicated, the type of smallpox that would presumably be released as a biological agent would be a genetically altered version with properties significantly different than those found in the natural species.

A biological attack using the smallpox virus has the potential to be devastating and would, according to some estimates, kill ~30% of those whom it infects. In a worst-case scenario, an outbreak of smallpox could negatively affect health-care services, transportation, government, and the economy. Some reports have stated that it could take up to 1 year to recover from a smallpox attack.

At present, the threat level of a smallpox outbreak is considered low. The stocks of the virus are located in two places in the world: the Centers for Disease Control and Prevention (CDC) in Atlanta, Georgia, and the State Research Centre of Virology and Biotechnology in Novosibirsk, Russia. However, it is not known if terrorists or other states have obtained the virus. It is only known that smallpox stocks do exist in the world and that terrorist organizations may view the virus to be an effective weapon of mass destruction.

Cyberattacks on critical infrastructure cost billions of dollars in lost intellectual property, maintenance and repair, lost revenue, and increased security. Cyberattacks can corrupt information through viruses, worms, and data tampering; disclose confidential information and trade secrets; steal telephone and Internet services; and interrupt a company's ability to deliver services to its clients.

Beyond the direct impact, cyberattacks reduce the public's confidence in the security of Internet transactions and e-commerce, damaging corporate reputations and reducing the efficiency of the economy. It is estimated that most attacks are never reported because the victims fear a loss of consumer confidence and damage to their reputations or are afraid of encouraging more attacks.

Due to the cross-cutting and rapidly changing nature of the threat and limitations in the ability to detect, monitor, and report, it is difficult to establish a complete picture of the computer-based threat to critical infrastructure. As a result, there are a number of uncertainties in the data contained in threat assessments and qualitative information remains anecdotal, making it difficult for the intelligence and enforcement community to generate effective analysis of the changing nature of the threat and degree of risk.

Traditional methodologies for addressing threats are often ineffective in detecting and analyzing computer-based threats. Technology impacts strategic analysis in terms of shorter time frames to indicate changing trends relating to threat activity. Experts have predicted that new breeds of computer attacks, such as worms and viruses, which are capable of knocking out millions of computers around the Internet in a matter of minutes, are likely to emerge. New worm variants have provided security experts with new challenges. The Linux-based Slapper worm included an innovation that is likely to reappear in a more dangerous form in the future: it establishes a peer-to-peer network among affected servers, enabling a hacker to take over the servers and use them to attack another web location, known as a distributed denial of service (DDoS) attack.

Four factors contribute to a community's vulnerability to the vast spectrum of threats. First, a community's population, built environment, and wealth are increasingly concentrated in a small number of highly vulnerable areas and many such communities are at risk from multiple hazards. Second, climate change may increase the frequency and severity of extreme weather events. Third, a community-built environment is aging and is more susceptible to damage. Fourth, communities are increasingly more reliant on advanced technologies that are frequently disrupted during disasters.

Global climate change is providing new and distinct challenges for emergency managers globally. Aging infrastructure should be tested to cope with increasing devastation caused by natural phenomena. Therefore, a multipronged approach to infrastructure protection is needed.

The increased urbanization of communities and the concomitant need for additional vital, uninterrupted services for bourgeoning cities raises the possibility that our critical infrastructure is more vulnerable than ever. Moreover, we remain equally susceptible to human error, mechanical failures, and computer programming errors.

The threat environment has become more complicated with the advent of new technologies that have empowered all people with the ability to more efficiently access and manipulate computer-based information stores. This presents new challenges in the form of cyber-based threat actors who, be it accidentally or maliciously, can impose significant costs on private and public computer networks.

Finally, significant events, such as the terrorist attacks of September 11, have confirmed what the security and intelligence community had long feared, which is that malicious actors, motivated by a multitude of reasons, would be willing to use deadly force to strike at the fabric of postindustrial society. Those events have altered the way in which emergency management professionals and policy makers conduct their affairs because the possibility, regardless of how remote, that an event on an equally grand scale might occur again precipitates the need for robust and flexible mitigation, preparedness, and response and recovery plans. The future safety, security, and economic well-being of a community will depend on its ability to predict and respond to threats to their most critical assets. Actions to secure a community should be ongoing and will have to evolve in relation to development and implementation of new technologies.

2.3   VULNERABILITY ASSESSMENT

Once likely threats are identified, a vulnerability assessment must be performed. Communities must identify exposure to hazard impacts to proactively address emergency response, disaster recovery, and hazard mitigation and incorporate sustainable development practices into comprehensive planning. Hazard mitigation, an important part of sustainable development, eliminates or minimizes disaster-related damages and empowers communities to respond to and recover more quickly from disasters. The process of assessing the vulnerability of an area entails determining the following: Who is vulnerable? What is vulnerable? Where is it located? In determining the vulnerability of a population, indicators need to be established to identify areas with multiple criteria for assessing vulnerability, which requires specific considerations for the population at risk.

There are many different aspects that can be incorporated into a vulnerability index (VI), such as monetary vulnerability, social vulnerability, criticality, economic vulnerability, environmental vulnerability, and critical infrastructure vulnerability. Criticality refers to a community's accessibility to emergency services such as fire, police, ambulance, and medical care. Economic vulnerability is the cost incurred or money lost through loss of production, distribution, and consumption. Environmental vulnerability pertains to environmentally sensitive areas (ESAs), areas that have been identified as being susceptible to potential disasters. From an environmental perspective, naturally occurring events, such as flooding, due to spring "freshet" are not always viewed as a hazard, since they provide "washout" and/or replenishment of sediments for aquatic habitat. Critical infrastructure vulnerability is a measure of the effect of damage to critical infrastructure, including telecommunications, electricity, drinking water, and sewer services.

A disaster will only occur if an extreme event strikes a vulnerable population. However, what is meant by vulnerability needs to be specified. In the traditional meaning, a disaster is a state in which a population, population group, or individual is not able to cope or is not able to overcome the adverse effects of an extreme event without outside help. When examining extreme events, there is the primary cause of the event and the impact or magnitude of the event, which is determined by human influences. For example, the definition of a disaster as a state where people at risk can no longer help themselves conforms to the modern view of a disaster as a social event, where people at risk are vulnerable to an extreme event because of their social conditions. According to this view, disaster management is not only a technical task, but also a social task.

Vulnerability is a function of exposure, sensitivity, and adaptive capacity. Differential vulnerabilities exist across different systems based on inherent response abilities. Both humans and natural systems are affected. The challenge is to define factors (indices) that quantify the impact on human vulnerability of disasters and obtain a quantitative basis for decision making. The United Nations defines this factor as the "index of human security." An indicator, comprising a single data point (or a single variable) or an output value from a set of data (aggregation of variables), is a quantity that describes a system or process that has significance beyond the face value of its components. It aims to communicate information on the system or process. The dominant criterion behind an indicator's specification is scientific and/or engineering knowledge and judgment.

An index is a mathematical aggregation of variables or indicators, often across different measurement units, so that the result is dimensionless. The purpose of an index is to provide compact and targeted information for management and policy development and decisions. The problem of combining the individual components is overcome by scaling and weighting processes, which are not absolute, but reflect social preferences.

Vulnerability is a relationship between a purposive system and its environment, where that environment varies over time. Which environmental perturbations are significant depends on the objectives of the system as only those perturbations that can inhibit the achievement of these objectives are significant. In the past, physical aspects of vulnerability—the spatial distribution of populations and infrastructure in relation to a hazard—tended to receive more attention in hazards research. But the social aspects of vulnerability are increasingly being recognized. For individuals, the susceptibility to hazards depends largely on behavior, well-being, and the resources people have to enable them to avoid and recover from harm. These, in turn, are largely determined by wider social, economic, and political patterns and processes that differentiate how hazards impact on people and human systems. Analyses of vulnerability increasingly highlight its socially constructed nature, underlining the importance of understanding how sociopolitical processes can create vulnerability and thereby create "disaster."

Vulnerability of a community susceptible to hazards can be defined by four different types of vulnerability: critical vulnerability, monetary vulnerability, social vulnerability, and critical infrastructure vulnerability. Critical vulnerability focuses on determining the vulnerabilities of key individual facilities, lifelines, or resources within the community. Because these facilities play a central role in disaster response and recovery, it is important to protect critical facilities to ensure that service interruption is reduced or eliminated. Critical facilities include police stations, fire and rescue facilities, hospitals, shelters, schools, nursing homes, and any other structures deemed essential by the community.

Monetary vulnerability is defined as the cost of damages or the cost of reestablishing previous conditions after a disaster has passed. Monetary vulnerability has well-established methodologies associated with it and has been studied fairly extensively. Damages can be assessed based on costs incurred from previous events or historical disasters. In addition, damages can be determined from extensive interviews with homeowners to identify the type and condition of housing and what would be damaged inside the house at different magnitudes. The former approach is likely to underestimate damages as a result of people not accounting for personal time spent cleaning up after an event, but is likely to overestimate damages based on insurance claims from individuals who overstate the damages incurred. The latter approach is labor and time intensive for data collection and requires people to hypothesize about their behavior during an event, which may be very different under the stress of the actual situation. Another approach is to use lifestyle classes rather than residence classes to calculate damage costs. However, the difficulty here is to relate the different lifestyle classes to each residence damaged, a relatively labor- and time-intensive task.

Damages are usually divided into direct and indirect damages. The direct damages comprise the tangible damages to property that result from direct contact with the event at hand, while the indirect damages are because of interruption to physical and economic linkages, such as the interruption of the electrical grid system and loss of income or business profits. Another intangible damage is anxiety. It has been argued that anxiety affects productivity, which in turn affects the gross national product (GNP).

Social vulnerability is defined as the condition of persons at risk, their integration into the community, and their access to vital services. Social vulnerability is partially the product of social inequalities to those social factors that influence or shape the susceptibility of various groups to harm and that also govern their ability to respond. It is, however, important to note that social vulnerability is not registered by exposure to hazards alone, but also resides in the sensitivity and resilience of the system to prepare, cope, and recover from such hazards. However, it is also important to note that a focus limited to the stresses associated with a particular vulnerability analysis is also insufficient for understanding the impact on and responses of the affected system or its components.

Two of the principal archetypal reduced-form models of social vulnerability are the risk–hazard (RH) model and the pressure and release (PAR) model. Initial RH models sought to understand the impact of a hazard as a function of exposure to the hazardous event and the sensitivity of the entity exposed. Applications of this model in environmental and climate impact assessments generally emphasized exposure and sensitivity to perturbations and stressors and worked from the hazard to the impacts. The model does not address the distinction among exposed subsystems and components that lead to significant variations in the consequences of the hazards or the role of the political economy in shaping differential exposure and consequences. This led to the development of the PAR model.

The PAR model understands a disaster as the intersection between socioeconomic pressure and physical exposure. Risk is explicitly defined as a function of the perturbation, stressor, or stress and the vulnerability of the exposed unit. In this way, it directs attention to the conditions that make exposure unsafe, leading to vulnerability and to the causes creating these conditions. Used primarily to address social groups facing disaster events, the model emphasizes distinctions in vulnerability by different exposure units such as social class and ethnicity. The model distinguishes between three components on the social side, that is, root causes, dynamic pressures, and unsafe conditions, and one component on the natural side, that is, the natural hazards itself. The principal root causes include "economic, demographic and political processes," which affect the allocation and distribution of resources between different groups of people. Dynamic pressures translate economic and political processes in local circumstances (e.g., migration patterns). Unsafe conditions are the specific forms in which vulnerability is expressed in time and space, such as those induced by the physical environment, local economy, or social relations. Although explicitly highlighting vulnerability, the PAR model appears insufficiently comprehensive for the broader concerns of sustainability science. The model also tends to underplay feedback beyond the system of analysis that the integrative RH models included.

Dwyer et al. undertook detailed questionnaires to identify and quantify factors that contribute to vulnerability to a natural hazard. Fifteen rules were identified for characterizing vulnerability. These rules were applied using synthetic population estimation to generate vulnerability maps for different scenarios. The scenarios were based on the severity of injury and direct damages to homes given a damage event. Age, income, gender, employment, residence type, household type, tenure type, health insurance, house insurance, car ownership, disability, English-language skills, debt, and/or savings were identified as playing a role in household vulnerability.

Complete destruction of a residence is the great equalizer, making everyone vulnerable. Apart from this, the rules demonstrated that it is a combination of attributes that make a person/household vulnerable. The relative importance of individual indicators was such that injuries, followed by damage to the house, were at the top. House insurance, income, and tenure type were next, followed by age, whereas household type, health insurance, residence type, gender, and disability were the lowest-scoring indicators.

Vulnerability indices may assume a variety of formats. Relevant indices can be aggregated to form a single index, or separate indices can be developed to represent different vulnerabilities. Additionally, an index can be constructed through geographical groupings or by profiling individual geographic entities. The Social Flood Vulnerability Index (SFVI) is a composite additive index that incorporates both social characteristics and financial deprivation indicators. The Tapsell et al. study, based on the analysis of other studies undertaken in the United Kingdom, identifies age and financial status as common key indicators. The Townsend Index was used to identify financially deprived households. Another financial deprivation index is the Carstairs Index, which incorporates car ownership, household crowding, unemployment, and a social class variable. The SFVI consists of the Townsend Index, weighted at 0.25, a "long-term sick" variable (residents with limiting, long-term illness as a percentage of the population), single parents, and the elderly (age 75 and over). The final index is calculated by summing the standardized Z-scores.

Rygel et al. used poverty, gender, race and ethnicity, age, and disabilities to develop a VI. Composite indices can have inherent problems. When averaging factors, the importance of an individual factor is reduced and an overall index may classify an area as "not vulnerable," while an individual factor may indicate "extreme vulnerability." When weighting factors, it is difficult to identify the weights to be applied. The assignments can be subjective, relying on expert judgment, rather than quantitative. Additionally, weights cannot account for variations in the relative importance of factors over space and time. In order to avoid the problems associated with averaging or weighting, a method that undertook the principal component analyses of 57 variables was developed. The top three components that were used in the development of the index are poverty, immigrants, and old age/disabilities. To avoid averaging the three parameters for each geographic area, Pareto rankings were used to classify the regions according to the significance of all three components.

St. Bernard included life expectancy at birth, security, social order, and governance indicators, for example, indictable crimes per 100,000 population; resource allocation indicators, for example, the proportion of children belonging to the poorest quintiles; and communications architecture, for example, computer literacy rate.

Health is an important aspect of vulnerability. After an event, an individual's health can be affected by the stress and trauma associated with the loss of possessions. Enteric illnesses can be prevalent and cramped conditions in evacuation shelters can increase transmission of these illnesses. During a disaster and particularly during the recovery phase, preexisting medical conditions can be exacerbated; this may include but is not limited to heart and respiratory problems. It can be difficult to obtain appointments for, or travel to, physician's offices, and this can aggravate underlying conditions.

A common theme among the different studies for weighing the indicators of social vulnerability is that the approach itself is relatively subjective since it is based on professional opinion or judgment and is therefore inherently biased. Some authors criticize the conceptualization of social vulnerability for overemphasizing the social, political, and economic processes and structures that lead to vulnerable conditions. Inherent in such a view is the tendency to understand people as passive victims and to neglect the subjective and intersubjective interpretation and perception of disastrous events. Bankoff criticizes the very basis of the concept, since in his view it is shaped by a knowledge system that was developed and formed within the academic environment of Western countries and therefore inevitably represents values and principles of that culture. According to Bankoff, the ultimate aim underlying this concept is to depict large parts of the world as dangerous and hostile to provide further justification for interference and intervention.

As previously defined, critical infrastructure vulnerability is a measure of the effect of damage to critical infrastructure, including telecommunication and information networks, the national electrical grid, oil and natural gas systems, water systems, transportation networks, and banking and financial systems. Because of the linkages and interdependencies between these various systems and networks, many of the dimensions and indicators identified for social vulnerability can also be incorporated into critical infrastructure vulnerability.

The VI of a community to a particular event can be determined using Monte Carlo simulation. Weightings (γ) can be assigned to each data category in determining the critical, monetary, social, and critical infrastructure vulnerabilities (Equation 2.1). The total number of separate variables used for estimating the individual vulnerability indices depends on the number of dimensions and/or indicators used to define the community. This number is often constrained by the availability of physical and census data for the community. The weighting assigned to each variable can be expressed as a fraction from 0 to 1. A weighting value of 1 assigned to any variable means that the total weighting for determining the VI is carried by that single variable alone. A value of 1 over n means that the weightings are uniformly distributed across all data categories, where n is the number of data variables used to define the VI:

$$VI_S = VI_1\gamma_1 + VI_2\gamma_2 + VI_3\gamma_3 + \cdots + VI_n\gamma_n \tag{2.1}$$

Monte Carlo simulation can then be used to determine the actual weightings. The Monte Carlo method is based on the generation of multiple trials to determine the expected value of a random variable. The basis for the method is provided by the following equation:

$$Pr\left\{\left|\frac{1}{N}\sum_N \varepsilon - \mu\right| < \frac{3\sigma}{\sqrt{N}}\right\} \approx 99.8\% \tag{2.2}$$

where:

Pr is the probability

The Monte Carlo method provides an estimate of the expected value of a random variable (weightings, γ), and also predicts the estimation error (ε), which is proportional to the number of iterations (N). The total error is given by the following equation:

$$\varepsilon = \frac{3\sigma}{\sqrt{N}} \tag{2.3}$$

where:

sigma (σ) is the standard deviation of the random variable

The upper bound of sigma (σ) can be estimated by calculating the standard deviation of the maximum, minimum, and average values of the random variable. A gross estimation of the random variable is the average of the maximum and minimum values. An absolute error of 2% is the average divided by 50. The average, median, standard deviation, and maximum and minimum values are determined for the entire population of the random variable (γ). A total of n separate distributions of the random variable (γ) are determined. The average value of the entire population of the random variable (γ) is selected as the maximum value of the weighting (γ) within the distribution series. Subsequent values in the distribution series are determined using a logarithmic decay on either side of the maximum weighting (γ) value:

$$\gamma_n = \frac{1 - \mu}{LN(9) - LN(1)} SQN + \mu \qquad (2.4)$$

where:
  γ_n is the value of the weighting (γ) at the nth position (1 through n) within the distribution series
  μ is the maximum value of the weighting (γ) within the distribution series
  SQN is the sequence position number (1 through n) within the distribution series
  LN is the log-normal

The overall and individual vulnerability indices are calculated for each of the distribution series, 1 through n. The social VI is independent of the magnitude of the event since all households and/or businesses within a vulnerable or damage-susceptible area will be impacted to some level or degree during an extreme event regardless of magnitude. Therefore, the social VI is calculated for the entire area and not for any one specific event.

The total VI for each vulnerable area is estimated by applying weighting factors (γ) to the monetary vulnerability, social vulnerability, critical vulnerability, and critical infrastructure vulnerability. Monte Carlo simulation is used again to estimate the weighting factors. The weighting factors (γ) are estimated using the same methodology used to estimate the weighting factors for the individual vulnerability indices, the difference being that the distribution sequences are limited to the number of vulnerability indices used to determine the weighting factors for the total VI as opposed to the number of data categories and/or variables that were used to determine the individual vulnerability indices.

2.4   RISK ASSESSMENT

Once the vulnerability assessment has been completed the next step is to undertake a risk assessment of the vulnerable area and/or community. The risk index for each vulnerable area is computed by taking the product of the probability of an event of magnitude x occurring in any given year over a time period of n years and the vulnerability of the damage area that would result from that event:

$$R = 1 - [1 - P(X \geq x_T)]^n \qquad (2.5)$$

$$RI = R \times (VI) \tag{2.6}$$

where:

$P(X \geq x_T)$ is the probability that event X of magnitude x_T will be exceed in any given year

T is the return period

n is the expected life of the vulnerable area when the return period T is exceeded

R is the probability that an event $x \geq x_T$ will occur at least once in n years

RI is the risk index

Empirical descriptors of "very high," "high," "moderate," and "low" may be used to interpret and define the risk index values for each vulnerable area. For areas that are defined as being "very high" or "high," measures recommended to mitigate risks should be implemented immediately; for areas described as being "moderate," measures recommended to mitigate risks should be implemented in the near future; and for areas described as being "low," recommended measures should enhance the overall resilience of the area but not be a priority in relation to other higher risk areas.

Based on the findings from the risk assessment, the next step in the process is to identify mitigation options that will lower the various levels of risk. If minimum standard measures for a given vulnerable area are not currently present, these measures should automatically be included in the upgrade recommendations. Additional measure upgrades above the minimum standards should be recommended as necessary to address the specific threats identified for the area and/or community. The estimated capital cost of implementing the recommended mitigation options is usually provided in the threat and/or vulnerability assessment report. The estimated installation and operating costs for the recommended mitigation options are also usually provided in the threat and vulnerability assessment report. All operating costs are customarily estimated on a per annual basis.

The implementation of the recommended mitigation options should have a positive effect on the overall vulnerability and risk indices. The final step in the process is to reevaluate these two ratings for each threat in light of the recommended upgrades. For example, the implementation of technical guidelines and policies for managing floodplains will not prevent floods from occurring, but it will reduce the impact of floods by limiting the number of buildings and residents within flood-susceptible lands. Therefore, the impact of losses as a result of flooding would improve, but the VI would stay the same.

3 Critical Infrastructure Protection

3.1 INTRODUCTION

Critical infrastructure protection (CIP) is a concept that relates to the preparedness and response to serious incidents that involve the critical infrastructure of a region or nation. The American Presidential Directive PDD-63 of May 1998 set up a national program for CIP in the United States. In Europe, the equivalent "European Program for Critical Infrastructure Protection" (EPCIP) refers to the doctrine or specific programs created as a result of the European Commission's directive EU COM(2006) 786, which designates European critical infrastructure that, in case of fault, incident, or attack, could impact both the country where it is hosted and at least one other European member state. Member states are obliged to adopt the 2006 directive into their national statutes.

On March 9, 1999, Deputy Defense Secretary John Hamre warned the U.S. Congress of a cyberterrorist attack "electronic Pearl Harbor," saying, "It is not going to be against Navy ships sitting in a Navy shipyard. It is going to be against commercial infrastructure." Later this fear was qualified by President Clinton (Federation of American Scientists 2000) after reports of actual cyberterrorist attacks in 2000: "I think it was an alarm. I don't think it was Pearl Harbor. We lost our Pacific fleet at Pearl Harbor. I don't think the analogous loss was that great." There are many examples of computer systems that have been hacked or victims of extortion. In the past, the systems and networks of the infrastructure elements were physically and logically independent and separate. They had little interaction or connection with each other or other sectors of the infrastructure. With advances in technology, the systems within each sector became automated, and interlinked through computers and communications facilities. As a result, the flow of electricity, oil, gas, and telecommunications throughout the country is linked—albeit sometimes indirectly—but the resulting linkages blur traditional security borders.

While this increased reliance on interlinked capabilities helps make the economy and the nation more efficient and perhaps stronger, it also makes the country more vulnerable to disruption and attack. This interdependent and interrelated infrastructure is more vulnerable to physical and cyberdisruptions because it has become a complex system with single points of failure. In the past an incident that would have been an isolated failure can now cause widespread disruption because of cascading effects. For example, capabilities within the information and communication sector have enabled the United States to reshape its government and business processes, while becoming increasingly software driven. One catastrophic failure in this sector now has the potential to bring down multiple systems including air traffic control, emergency services, banking, trains, electrical power, and dam control.

Elements of the infrastructure are also considered possible targets of terrorism. Traditionally, critical infrastructure elements have been lucrative targets for anyone wanting to attack another country. Now, because infrastructure has become a national lifeline, terrorists can achieve high economic and political value by attacking elements of it. Disrupting or even disabling infrastructure may reduce the ability to defend the nation, erode public confidence in critical services, and reduce economic strength. Additionally, well-chosen terrorist attacks can become easier and less costly than traditional warfare because of the interdependence of infrastructure elements. These infrastructure elements can become easier targets where there is a low probability of detection.

The elements of infrastructure are also increasingly vulnerable to a dangerous mix of traditional and nontraditional types of threats. Traditional and nontraditional threats include equipment failures, human error, weather and natural causes, physical attacks, and cyberattacks. For each of these threats the cascading effect caused by single points of failure has the potential to pose dire and far-reaching consequences.

There are fears that the frequency and severity of critical infrastructure incidents will increase in the future. Although efforts are under way, there is no unified national capability to protect the interrelated aspects of a country's infrastructure. One reason for this is that a good understanding of the interrelationships does not exist. There is also no consensus on how the elements of the infrastructure mesh together or on how each element functions and affects the others. Securing national infrastructure depends on understanding the relationships among its elements.

CIP requires the development of a national capability to identify and monitor the critical elements and to determine when and if the elements are under attack or are the victims of destructive natural occurrences. CIP is important because it is the link between risk management and infrastructure assurance. It provides the capability needed to eliminate potential vulnerabilities in the critical infrastructure. CIP practitioners determine vulnerabilities and analyze alternatives in order to prepare for incidents. They focus on improving the capability to detect and warn of impending attacks on, and system failures within, the critical elements of the national infrastructure. This chapter provides an overview and discussion of the different critical infrastructure sectors and organizational responsibilities including buildings and/or facilities, water supply systems, energy facilities, agricultural and food systems, aviation systems, maritime assets, land transportation systems, and cyber systems.

3.2   BUILDING SECURITY

The design and construction of secure and safe buildings continues to be the primary goal for owners, architects, engineers, project managers, and other stakeholders, which may include but is not limited to construction managers, developers, facilities managers, code officials, fire marshals, building inspectors, public officials, emergency managers, law enforcement agencies, lenders, insurers, and product manufacturers. The development of safe and secure buildings can often be a challenge due to funding limitations, resistance from the occupants due to the impact on operations, productivity, and accessibility, and the impact on the surrounding environment and building architecture due to perimeter security, hardening, and standoff requirements. Understanding how on-site security can impact a building's safety is important as well. A balance between security, safety, and other design objectives along with the needs of the facility can be attained through the establishment of an integrated design process. An integrated design process requires that all design team members have a clear understanding of each other's goals and objectives. Through this collaboration members can overcome the above challenges. This in turn will lead to the development of a solution that addresses all of the requirements.

Designing buildings for security and safety requires a proactive approach that anticipates and then protects the building occupants, resources, structure, and continuity of operations from multiple hazards. The first step in this process is to understand the various risks they pose. There are a number of defined assessment types to consider that will lead the project team in making security and safety design decisions (see Chapter 2). This section identifies the resources and/or assets to be protected with respect to building security, highlights the possible "threats," and establishes a likely consequence of occurrence or "risk." This assessment is weighed against the vulnerabilities that are specific to the facility and through these assessments building owners and other invested parties select the most appropriate safety and security measures to implement. The selection of safety and security measures is based upon the security requirements for the buildings, the acceptable levels of risk for the building, the cost-effectiveness of the measures proposed, the overall life cycle cost, and the impact the proposed measures will have on the design, construction, and operational use of the building.

The objective of hazard mitigation is to minimize loss of life, property, and function due to disasters. Designing buildings to resist hazard(s) begins with a comprehensive risk assessment. The process includes identification of the hazards present and an assessment of their potential impacts and effects on the building and surrounding environment. When hazard mitigation is implemented in a risk-based approach, every dollar spent on mitigation actions results in disaster losses of $4 being avoided on average.

Different organizations use varying terms to refer to the components of a risk assessment. For example, law enforcement and intelligence communities refer to intentional adversarial acts such as sabotage and terrorism as threats, whereas unintentional naturally occurring events such as hurricanes and floods are referred to as "hazards" by emergency managers; however, both are forces that have the potential to cause damage, casualties, and loss of function of the building. Regardless of which organization is conducting the risk assessment, the process of identifying what can happen at a building's location, how it can affect the physical structure, and what the consequences are remains the same from one application to the next.

There are times when design requirements addressing various security threats and hazards are in direct conflict with other design objectives. For example, blast-resistive glazing on windows to ensure structural integrity may impede emergency egress during a fire; access control measures that prevent intrusion may restrict emergency egress; and reducing light pollution to obtain Leadership in Energy and Environmental Design (LEED) certification might conflict with the aim of providing ample lighting to meet security objectives. Similarly, site design and security can complement each other as in the case of a storm water management requirement that doubles as a vehicle barrier. Effective communication between design team members throughout the design process is necessary in order to achieve the common goal of a safe and secure building.

When implementing security and safety measures, a balance between operational, technical, and physical safety methods is necessary. For example, to protect a given facility from unwanted intruders, an operational approach to security would be the deployment of armed/unarmed security guards (24/7) depending on the risk; a technical approach may include ample lighting, intrusion detection systems, surveillance cameras, and warning sirens; and a physical approach may include locked doorways and vehicle barriers. In practice, a "multibarrier" approach to security is adopted and a combination of methods is employed, whereby a deficiency in one area may be compensated by a greater emphasis in another.

Risk reduction strategies can also be characterized as either structural or nonstructural. Structural mitigation measures focus on those building components that carry gravity, wind, seismic, and other loads, such as columns, beams, foundations, and braces. Examples of structural mitigation measures include building materials, construction methods, building code compliance, and site planning. Nonstructural measures focus on risks arising from damage to non-load-bearing building components, including architectural elements such as partitions, decorative ornamentation, and cladding; mechanical, electrical, and plumbing components such as heating, ventilation, and air conditioning (HVAC), life safety, and utility systems; and furniture, fixtures, and equipment such as desks, shelves, and other material contents. Nonstructural mitigation measures are prescriptive and engineered or nonengineered and may involve securing the above elements to the structure or keep them in a position that minimizes damage and disruption to building function. Nonstructural components, including building contents, account for over three-quarters of the cost of a building; this figure can be higher for specialized occupancies such as medical facilities. Structural and nonstructural components can potentially interact during an incident, requiring an approach to implementing a comprehensive plan of structural and nonstructural mitigation actions.

According to the World Building Design Guide (WBDG), the four fundamental principles of all-hazard building design are plan for fire protection, ensure occupant safety and health, prevent natural hazards and provide security, and provide security for building occupants and assets. Planning for fire protection for a building involves a systems approach that enables the designer to analyze all of the building's components as a total building fire safety system package.

Some injuries and illnesses are related to unsafe or unhealthy building design and operations. These can be prevented by measures that account for issues such as indoor air quality, electrical safety, fall protection, ergonomics, and accident prevention. Annual damage costs for recovery efforts, including the repair of damaged buildings, from the impact of hurricanes, floods, earthquakes, tornados, blizzards, and other natural disasters, are in the billions of dollars for both the United States and Canada. A significant percentage of this amount could be alleviated if designers properly anticipated the risks and frequency associated with major natural hazards.

Effective secure building design involves implementing countermeasures to deter, detect, delay, and respond to attacks from human aggressors. It provides for mitigating measures to limit hazards to prevent catastrophic damage and provide resiliency should an emergency occur. The United States has the highest fire losses in terms of both frequency and total losses of any modern technological society. In Canada, although the frequency of fires has remained relatively low than in the United States and has been decreasing steadily, the number of fatalities due to fire still remains high. This is an indication of the lethality of fires despite the decrease in their occurrence. As a result, new facilities and renovation projects need to be designed to incorporate efficient, cost-effective passive, and automatic fire protection systems. These systems are effective in detecting, containing, and controlling, and/or extinguishing a fire event in the early stages.

Engineers who specialize in fire protection must be involved in all aspects of the design in order to ensure a reasonable degree of protection of human life from fire and the products of combustion as well as to reduce the potential loss from fire. Planning for fire protection in and around a building involves knowing the four sources of fire: natural, man-made, wildfire, and incidental. It also involves taking an integrated systems approach that enables the designer to analyze all of the building's components as a total building fire safety system package.

This type of analysis requires more than code compliance or meeting the minimum legal responsibilities for protecting a building. Building and fire codes are intended to protect against loss of life and limit fire impact on the community. They do not necessarily protect the mission or assets, or solve problems brought upon by new projects with unique circumstances. Therefore, it is necessary for the design team to integrate code requirements with other fire safety measures as well as other design strategies to achieve a balanced design that will provide the desired levels of safety including evacuation, recovery, and egress from smoke.

The success of any complex project hinges on getting all the stakeholders, owners, designers, and consultants working together in a collaborative manner to achieve performance-based design solutions. Several guidelines have been published by the Society of Fire Protection Engineers in collaboration with the National Fire Protection Association on performance-based building design. Issues that need to be addressed in developing a successful fire protection design usually include the design team, design standards and criteria, and site planning requirements.

An engineer with significant experience in fire protection analysis and design should be part of the design team. The fire protection engineer should be involved in all phases of design, from planning to occupancy. Design standards and criteria such as the building codes are to be utilized by the design team, including statutory requirements, voluntary requirements addressing the owner's performance needs, and requirements that are sometimes imposed by insurance carriers on commercial projects. A quality site plan will integrate performance requirements associated with fire department access, suppression, and separation distances and site or building security.

Buildings should be designed with simple layouts that enable firefighters to locate an area quickly. The building site plan should provide rapid access to various features such as fire department connections (FDCs), hose valves, elevators and stairs, enunciators, and key boxes. The property should accommodate the access of the fire apparatus into and around the building site; comply with local authorities having jurisdiction to accommodate the access of fire apparatus into and around the building site; and coordinate the access control point layout. Fire hydrants should also be located on the property. Furthermore, the fire protection engineer should coordinate with security engineers to ensure that fire protection measures and site plan design are coordinated and integrated with security protection measures. Depending on the size of the project and budget, the fire protection engineer and security design engineer may be one and the same person.

When considering the design and implementation of fire protection measures, there are a number of requirements that need to be taken into consideration, including building construction requirements, egress requirements, fire detection and notification system requirements, emergency system requirements, and special fire protection requirements.

Building construction requirements will address the following elements (Figure 3.1):

- Construction type, allowable height, and area
- Exposures or separation requirements
- Fire ratings, materials, and systems
- Occupancy types
- Interior finishes
- Exit stairway enclosures

FIGURE 3.1   Interior exit.

Egress requirements will address the following elements:

- Exit stairway remoteness
- Exit discharge
- Areas of refuge
- Accessible exits
- Door locking arrangements (security interface)

When planning for access to the facility, the following design elements must be taken into consideration:

- Limit travel distances for people with disabilities from site arrival points, such as public sidewalks, public transportation stops, and parking, to accessible building and facility entrances.
- Provide accessible routes that require minimum effort. Wherever possible, provide walks with no more than 1:20 running slope over ramps.
- Design ramps with lower running slopes and not at the maximum 1:12 permitted when they are required.
- Provide equivalent access and travel options to people with disabilities and also provide equivalent, safe, easy, and compliant access to the facility while maximizing security.

When planning for access to spaces within the facility, plan the layout to best accommodate all persons, including people with disabilities:

- Layout spaces wherever possible to limit travel distance between elements within the space.
- Group and centralize spaces to limit the extent of travel required between the spaces. For instance, in a residential garden-style apartment complex, centrally locating the swimming pool and associated amenities is preferred to locating the swimming pool and amenities at one end of the site.
- Limit the need for travel between levels and the reliance on elevators and lifts, if possible.
- Limit the need for vertical access on sites and within one- and two-story facilities and between intermediate levels of high-rise facilities, thereby reducing the likelihood of interruptions in the accessible routes due to elevator or lift malfunctions and thus limiting the need to install ramps.
- Where vertical access is required between two levels, consider ramps designed at low grade to permit guaranteed access. Ramps can often be creatively integrated into designs without negatively impacting the aesthetics. For instance, integrating ramped access into a swimming pool can avoid the need to install an industrial-looking pool lift, which can clearly stand out from the overall design.
- Choose lighting options that accommodate people with low vision.

Fire detection and notification system requirements will address the following elements:

- Detection
- Notification
- Survivability of systems

Fire suppression requirements will address the following elements:

- Water supply
- Type of automatic fire extinguishing system
- Water-based fire extinguishing system
- Non-water-based fire extinguishing system
- Standpipes and fire department hose outlets

Emergency power, lighting, and exit signage will address the following elements:

- Survivability of systems
- Electrical safety
- Distributed energy resources

Special fire protection requirements will address the following elements (Figure 3.2):

- Engineered smoke control systems
- Fireproofing and firestopping
- Atrium spaces
- Mission-critical facility needs

FIGURE 3.2   Fire department hose outlets.

Modern buildings are considered safe and healthy working environments; however, the potential for occupational illnesses and injuries requires architects, engineers, and facility managers to design and maintain buildings and processes that ensure occupant safety and health. Building designs must focus on eliminating or preventing hazards to personnel rather than rely on personal protective equipment and administrative or process procedures for prevention. Protecting the health and safety of building occupants has expanded beyond disease prevention and nuisance control to include mental as well as physical health and safeguarding the ecological health of the work environment. An integrated approach includes work flow processes analysis and hazard recognition to develop solutions that provide healthy built environments besides meeting other project requirements. In addition, consideration of health and safety issues should be an integral part of all phases of a building's life cycle, including planning, design, construction, operations and maintenance, renovation, and final disposal.

To provide a healthy and safe working environment, the following issues must be addressed by the design team:

- Provide designs that eliminate or reduce hazards in the workplace to prevent mishaps and reduce reliance on personal protective equipment.
- Prevent occupational injuries and illnesses.
- Prevent falls from heights.
- Prevent slips, trips, and falls.
- Ensure electrical safety from turnover through operations and maintenance.
- Ensure that modifications are in conformance with life safety codes and standards and are documented.
- Eliminate exposure to hazardous materials [e.g., volatile organic compounds (VOCs) and formaldehyde and lead and asbestos in older buildings].
- Provide good indoor air quality and adequate ventilation.
- Analyze work requirements and provide ergonomic workplaces to prevent work-related musculoskeletal disorders.
- Perform proper building operations and maintenance.

When addressing the above issues, the following recommendations are to be undertaken:

- Provide designs that eliminate or reduce hazards in the workplace to prevent mishaps.
  - Provide designs in accordance with good practice as well as applicable building, fire, safety, and health codes and regulations. See Standards and Code Organizations found at the end of this section.
  - Conduct preliminary hazard analyses and design reviews to eliminate or mitigate hazards at the workplace.
  - Use registered design professionals and accredited safety professionals to ensure compliance with safety standards and codes.
  - Provide engineering controls in place rather than rely on personal protective equipment or administrative work procedures to prevent mishaps.

- Integrate safety mechanisms, such as built-in anchors or tie-off points, into the building design, especially for large mechanical systems.
- Design a means for safely cleaning and maintaining interior spaces and building exteriors.
- Provide for receiving, storing, and handling of materials, such as combustibles, cleaning products, office supplies, and perishables.
- Prevent occupational injuries and illnesses
  - Consider work practices and the employee's physical requirements, and eliminate confined spaces when designing buildings and processes.
  - Design for safe replacement and modifications of equipment to reduce the risk of injury to operations and maintenance staff.
  - Comply with applicable regulatory requirements such as the Occupational Safety and Health Administration (OSHA) standards.
  - Provide proper ventilation under all circumstances, and allow for natural lighting where possible.
  - Mitigate noise hazards from equipment and processes.
  - Designate safe locations for installation of radio-frequency equipment such as antennas on rooftop penthouses.
- Prevent falls from heights.
  - Provide guardrails and barriers that will prevent falls from heights in both interior and exterior spaces.
  - Provide fall protection for all maintenance personnel especially for roof-mounted equipment such as HVAC equipment and cooling towers.
  - Provide certified tie-off points for fall arrest systems.
- Prevent slips, trips, and falls.
  - Provide interior and exterior floor surfaces that do not pose slip or trip hazards.
  - Select exterior walking surface materials that are not susceptible to changes in elevation as a result of freeze/thaw cycles.
  - Provide adequate illumination, both natural and artificial, for all interior and exterior areas.
  - Comply with all regulatory and statutory requirements.
- Ensure electrical safety
  - Ensure compliance with the national/state/provincial electrical codes.
  - Provide adequate space for maintenance, repair, and expansion in electrical rooms and closets.
  - Provide adequate drainage and/or containment from areas with energized electrical equipment.
  - Evaluate all areas where ground fault circuit interruption (GFI) and arc fault interruption (AFI) devices may be needed.
  - Consider the response of emergency personnel in cases of fires and natural disasters.
  - Label all electrical control panels and circuits.
  - Install nonconductive flooring at service locations for high-voltage equipment.
  - Specify high-visibility colors for high-voltage ducts and conduits.

- Eliminate exposure to hazardous materials.
  - Identify, isolate, remove, or manage in place hazardous materials such as lead and asbestos.
  - Consider use of sampling techniques for hazardous substances in all phases of the project to include planning, design, construction, and maintenance.
  - Consider occupant operations and materials in designing ventilation and drainage systems.
  - Incorporate integrated pest management (IPM) concepts and requirements into facility design and construction (e.g., use of proper door sweeps, lighting, trash compactors) and require that the use of IPM be performed by qualified personnel during all phases of construction and after the facility is completed. This should include not only interior pest management, but landscape and turf pest management as well.
  - Provide adequate space for hazardous materials storage compartments and segregate hazardous materials to avoid incompatibility.
  - Substitute high-hazardous products with those of lower toxicity/physical properties.
- Provide good indoor air quality and adequate ventilation.
  - Consider ventilation systems that will exceed minimum standards.
  - Recognize and provide specially designed industrial ventilation for all industrial processes to remove potential contaminants from the breathing zone.
  - Design separate ventilation systems for industrial and hazardous areas within a building.
  - Consider the use of carbon monoxide (CO) monitoring equipment if there are CO sources, such as fuel-burning equipment or garages, in the building.
  - Specify materials and furnishings that are low emitters of indoor air contaminants such as VOCs.
  - Consider the indoor relative humidity in the design of the ventilation system.
  - Avoid interior insulation of ductwork.
  - Locate outside air intakes to minimize entrainment of exhaust fumes and other odors (e.g., vehicle exhaust, grass cutting and ground maintenance activities, industrial pollutant sources, cooling tower blow-offs, and sewage ejector pits).
  - Ensure the integrity of the building envelope, including caulks and seals, to preclude water intrusion that may contribute to mold growth.
  - Prevent return air plenums/systems from entraining air from unintended spaces.
  - Provide air barriers at interior walls between thermally different spaces to prevent mold and mildew.
- Provide ergonomic workplaces and furniture to prevent work-related musculoskeletal disorders
  - Design workplaces that make the job fit the person.
  - Select furnishings, chairs, and equipment that are ergonomically designed and approved for that use.

- Design equipment and furnishings reflective of work practices in an effort to eliminate repetitive motions and vibrations as well as prevent strains and sprains.
- Consider using worker comfort surveys in the design phase to help eliminate work-related musculoskeletal disorders.
- Accept the principle that one size does not fit all employees.
- Consider providing break areas to allow the employees to temporarily leave the workplace.
- Minimize lighting glare on computer monitor screens.
- Provide task lighting at workstations to minimize eye fatigue.
- Perform proper building operations and maintenance
  - Proper preventative maintenance (PM) not only improves the useful life of the systems and building structures, but it can lend to good indoor air quality and prevent "sick building" syndromes.
  - Ensure that all maintenance and operation documentation, especially an equipment inventory, is submitted to the building owner or operator prior to building occupancy.
  - Follow manufacturer recommendations for proper building operations and maintenance.
  - Include safety training of operator personnel as part of the construction contractor's deliverables.
  - Require the use of IPM for all pest management services, interior and exterior of the building.
  - Require building maintenance personnel to maintain the HVAC air infiltration devices and condensate water biocides appropriately.
  - Monitor chemical inventories to identify opportunities to substitute green products.
  - Consider incorporating continuous commissioning into your building maintenance program.

Potential exposure of building occupants to molds from contaminated HVAC systems, especially during maintenance and renovation projects, remains a serious concern. Reaction to exposure can range from negligible to severe among building occupants and can frequently be very difficult to definitively identify as a causal factor for occupants' symptoms. Special care must be exercised in the HVAC design, especially to prevent excessive humidity in system components.

Fiberglass is used extensively in building construction, especially for insulation and sound attenuation in HVAC systems. Considerable concern exists regarding the potential adverse health effects of inhaling fiberglass fibers. A number of studies are currently investigating the long-term effects of inhalation exposure to fiberglass. At a minimum, fiberglass exposed to the air stream in an HVAC system will shed particles and serve as a matrix for collecting dust and dirt that act as a substrate for microbial growth.

Contamination of domestic hot water systems, cooling towers, and condensate pans continues to result in infections of building occupants on a regular basis. The results of such infections can range from mild to fatal and affect one or many

FIGURE 3.3   Air France crash in Toronto, Ontario (August 2, 2005, storm).

employees. They invariably result in employee apprehension and media attention. Mechanical engineers must be vigilant to avoid system designs that may promote the growth of bacteria (e.g., *Legionella* spp.).

Buildings in any geographic location are subject to a wide variety of natural phenomena such as windstorms, floods, earthquakes, and other hazards. While the occurrence of these incidents cannot be precisely predicted, their impact is well understood and can be managed through a program of hazard-risk mitigation planning (Figure 3.3).

Once the overall risk to the facility has been determined, mitigation measures should be identified, prioritized, and implemented. The impact of natural hazards and the costs of the disasters they cause are reduced through the implementation of mitigation measures. Integrating mitigation measures into new construction is more cost-effective than retrofitting existing structures. All-hazard mitigation is the most cost-effective approach to identifying risks, maximizing the protective effect of mitigation measures, and optimizing all-hazard design techniques with other building technologies.

To mitigate losses of life, property, and function is to design buildings that are disaster resistant. A variety of techniques are available to mitigate the effects of natural hazards on buildings. Depending on the hazards identified, the location and construction type of a proposed building or facility, and the specific performance requirements for the building, the structure can be designed to resist hazard effects such as induced loads. During the building's life cycle, opportunities to further reduce the risk from natural hazards may exist when renovation projects and repairs of the existing structure are undertaken. When incorporating disaster reduction measures into the building design, some or all of the issues outlined below should be considered in order to protect lives, properties, and operations from damages caused by natural hazards (Figure 3.4).

The design of an earthquake-resistant building is influenced by the level of seismic resistance desired. This can range from prevention of nonstructural damage in frequent minor ground shaking to prevention of structural damage and minimization of nonstructural damage in occasional moderate ground shaking, and even avoidance of collapse or serious damage in rare major ground shaking. These performance objectives can be accomplished through a variety of mitigation measures such as structural components like shear walls, braced frames, moment-resisting frames, and diaphragms; base isolation; energy-dissipating devices such as viscous-elastic dampers, elastomeric dampers, and hysteretic-loop dampers; and bracing of

FIGURE 3.4   Earthquake damage in British Columbia.

FIGURE 3.5   Tornado damage in Petawawa, Ontario.

nonstructural components. However, the primary focus of earthquake design is life safety, getting people out of the building safely, not the ability of a building to withstand the effects of an earthquake (Figure 3.5).

The key strategy to protecting a building from high winds caused by tornados, hurricanes, and gust fronts is to maintain the integrity of the building envelope, including roofs and windows, and to design the structure to withstand the expected lateral and uplift forces. For example, roof trusses and gables must be braced; hurricane straps must be used to strengthen the connection between the roof and walls and walls and foundation; and doors and windows must be protected by covering and/or bracing. The load path and connectors are just as important as bracing. When

FIGURE 3.6 Flooding in Toronto, Ontario (August 19, 2005, event).

planning renovation projects, designers should consider opportunities to upgrade the roof structure and covering, and enhance the protection of openings by considering the addition of impact-resistant windows, doors, and louvers (Figure 3.6).

Flood mitigation is best achieved by hazard avoidance, that is, by selecting a site away from coastal, estuarine, and riverine floodplains. Still flooding and velocity flooding also present hazards. Riverine hazards are associated with flooding from stream networks. Coastal hazards are associated with flooding from oceans or lakes. Still-water events are characterized by rising water with no horizontal movement. Velocity events are characterized by fast-moving waters at any depth. If buildings are sited in flood-prone locations, they should be elevated above expected flood levels to reduce the chances of flooding and to limit the potential damage to the building and its contents when it is flooded. Flood mitigation techniques include elevating the building so that the lowest floor elevation is above the flood level; dry floodproofing, or making the building watertight to prevent water entry; wet floodproofing, or making uninhabited or noncritical parts of the building resistant to water damage; relocation of the building; and the incorporation of floodwalls into site design to keep water away from the building. In addition, levees require a significant amount of care and are discouraged as a mitigation measure. Furthermore, berms for frequent events are usually a better choice. However, the choice of mitigation option will be mandated by regional/state/provincial regulations.

One of the primary performance requirements for any building is that it should keep the interior space dry. All roofs and walls must therefore shed rainwater, and design requirements are the same everywhere in this respect. For example, roof drainage design must minimize the possibility of ponding water, and existing buildings with flat roofs must be inspected to determine compliance with this requirement. Recommendations for addressing rainfall and wind-driven rain can be found in the International Building Code Series.

FIGURE 3.7 Subsidence due to intense storm in Orangeville, Ontario.

Ground subsidence can result from mining, sinkholes, underground fluid withdrawal, hydrocompaction, and organic soil drainage and oxidation. Subsidence mitigation can best be achieved through careful site selection, including a geotechnical study of the site. In subsidence-prone areas, foundations must be appropriately constructed, basements and other below-ground projections must be minimized, and utility lines and connections must be stress resistant. When retrofitting structures to be more subsidence-resistant earth reinforcement techniques and materials such as shear walls, geofabrics, and dynamic compaction can be used along with stabilization techniques to prevent soil collapse (Figure 3.7).

Gravity-driven movement of earth material can result from water saturation, slope modifications, and earthquakes. Techniques for reducing landslide and mudslide risks to structures include selecting non-hillside or stable slope sites; constructing channels, drainage systems, retention structures, and deflection walls; planting groundcover; and soil reinforcement using geosynthetic materials, and avoiding cut-and-fill building sites (Figure 3.8).

As residential developments expand into forested areas, people and property are increasingly at risk from wildfires. Fire is a natural process in any forested area and serves an important purpose; however, if ground cover is burned away, erosion, landslide, mudflow, and flood hazards can be exacerbated. A cleared safety zone of at least 10 m (30 m in pine forests) should be maintained between structures and combustible vegetation, and fire-resistant ground cover, shrubs, and trees should be used for landscaping (e.g., hardwood trees are less flammable than pines, evergreens, eucalyptus, or firs). Only fire-resistant or noncombustible materials should be used on roofs and exterior surfaces. Roofs and gutters should be regularly cleaned and chimneys should be equipped with spark arrestors. Vents, louvers, and other openings should be covered with wire mesh to prevent embers and flaming debris from entering. Overhangs, eaves, porches, and balconies can trap heat and burning embers

FIGURE 3.8 Wildfire in British Columbia.

and should also be avoided or minimized and protected with wire mesh. Windows allow radiated heat to pass through and ignite combustible materials inside, but dual- or triple-pane thermal glass, fire-resistant shutters or drapes, and non-combustible awnings can help reduce this risk.

A tsunami is a series of ocean waves generated by sudden displacements in the sea floor, landslides, or volcanic activity. In the deep ocean, the tsunami wave may only be a few inches high. The tsunami wave may come gently ashore or may increase in height to become a fast-moving wall of turbulent water several meters high. Although a tsunami cannot be prevented, the impact of a tsunami can be mitigated through urban and/or land use planning, restricting development away from shorelines, community preparedness, timely warnings, and effective response.

An area of refuge is a location designed to protect facility occupants during an emergency, when evacuation may not be safe or possible due to area contamination, obstruction, or other hazard. For example, during Hurricane Katrina, the Superdome in New Orleans served as an area of refugee for many displaced residents. Occupants can wait there until given further instructions or rescued by first responders. Areas of refuge can be safe havens, shelters, secure rooms, or protected spaces, each intended for protection from specified hazards and for different durations of occupancy. A safe haven is designed for temporary protection from various types of attacks in which building occupants may take refuge when forewarned of an imminent or ongoing attack. A shelter is an expedient retreat that provides "sheltering in place" for a relatively short duration, of hours, not days or weeks, and does not normally involve air filtration. A secure room is constructed to meet specified hardening criteria based on the design basis threat. Protected areas are readily available locations that afford safety to people by their shielding characteristics, such as a firewall. Areas of refuge may be freestanding or integrated in a building.

A risk assessment to identify likely threats is a prerequisite for validating the need and design criteria for an area of refuge. Events that might warrant an area of refuge include but are not limited to natural events such as tornados; public disturbances such as riots or demonstrations; a terrorist event involving weapons of mass destruction (WMD), firearms, or other weapons; life-threatening accidents such as the release of chemical or biological contaminants; or situations in which there is a potential for loss of life or injury due to their immobility, such as a hospital-intensive care unit. A collection of factors to be considered in the risk assessment process could include the type of hazard event, the probability of event occurrence, the severity of the event, the probable consequences of a hazard, and a cost/benefit analysis of options (see Chapter 2).

Planning, rehearsal, and preparatory procedures are paramount to a successful area of refuge. Well-written standard operating procedures (SOPs) should outline who makes decisions, when and how to evacuate, responsibilities and operating procedures, and when and how to leave the area of refuge. Depending on the function, planning may include but not be limited to personnel accountability protocol, communications for notification and instructions for comfort provisions such as seating and bedding, life-support systems such as fire apparatus, dedicated ventilation/filtration, plumbing, emergency power/fuel supply, lighting (including emergency lighting), decontamination, and maintenance. In addition to storage provisions of food, water, and first aid, other items/supplies may include current emergency contact numbers, dust masks, flashlights, vests or other identification/apparel, wind-up or battery-operated radios, toiletries, stationary for emergency administration, an audible sounding device that continuously charges or operates without a power source to signal rescue workers if refuge egress is blocked, and tools and maintenance manuals.

Key design considerations include reaction time (travel distance to refuge and activation time for emergency support systems), duration of occupancy, privacy, security (secure storage, doors, locks, windows, or view ports), isolation areas for ill or contaminated occupants or equipment, and hardening criteria to mitigate specified threats. The refuge size is based on the maximum intended occupancy. The location should be readily accessible to all people who are to be protected. Travel time is especially important when refuge users have disabilities that impair their mobility and may need assistance from others to reach the refuge. Alternate accessible routes should be considered for use in the event that the primary route is blocked or inoperable. If the refuge is integrated into a building, a centrally located interior room/space is preferable. If the refuge has windows, they should be capable of being sealed. Stand-alone refuges should be sited away from potential hazards, such as debris, excessive glazing, flooding or storm surge, power lines, or nearby structures that may be toppled or become airborne. In some situations, multiple small shelters rather than one large shelter may best suit the situation. The location should not be selected based on height above ground level if it would increase the travel time to the refuge. Signage and warning signals are critical for users to know when and where to seek refuge. Critical support systems located outside of the area of refuge area should be afforded the same protection criteria as the refuge.

During an emergency, preparatory actions may be required by designated team members outside the area of refuge, such as the turning off of fans, air-conditioning, or forced heat systems, the closing of windows or blinds, and the shutting of doors. To close all doors quickly in a large building, there must be a notification system such as a public address system. Also, occupant furniture may have to be relocated to allow the maximum number of occupants in the room or building.

Unsustainable development is a major factor in the rising damage costs as a result of natural disasters. Many design strategies and technologies not only prevent or reduce disaster losses but they also improve overall community sustainability. For example, erosion control measures designed to mitigate flood, mudslide, rainstorm, and other damage to a building's foundation may also improve the quality of runoff water entering streams and lakes. Similarly, land use regulations prohibiting new development in flood-prone areas may also help to preserve the natural and beneficial functions of the floodplains. By applying the all-hazard mitigation approach in the postdisaster recovery effort and by taking into consideration all issues related to the built environment, psychosocial recovery, economic redevelopment, and preservation or restoration of the natural environment, the net result for an impacted community is long-term disaster resilience.

Ongoing changes in climate patterns around the world may alter the behavior of hydrometeorological phenomena over the next 50 to 100 years. The frequency and severity of weather-related disasters is expected to increase, as is the risk from other weather-induced hazards such as wildfires. Buildings should be designed to incorporate strategies that both mitigate climate change such as reducing the amount of greenhouse gas emissions and adapt to changing environmental conditions by leveraging traditional hazard mitigation strategies.

The 2013 Boston Marathon bombings, the 2001 terrorist attacks on New York City's World Trade Center and the Pentagon, the 1995 bombing of Oklahoma City's Alfred P. Murrah Federal Office Building, and the 1996 bombing at Atlanta's Centennial Park made people aware of the need for better ways to protect occupants, assets, and buildings from human aggressors, which may include disgruntled employees, criminals, vandals, a lone active shooter, and terrorists. The 2001 terrorist attacks demonstrated the people's vulnerability to a wider range of threats and heightened public concern for the safety of workers and occupants in all building types. Many federal agencies responding to these concerns have adopted an overarching philosophy to provide appropriate and cost-effective protection for building occupants (Figure 3.9).

FIGURE 3.9   Ample lighting for building physical security.

The basic components of the physical security measures to address an explosive threat considers the establishment of a protected perimeter, the prevention of progressive collapse, the design of a debris mitigating facade, the isolation of internal explosive threats that may evade detection through the screening stations or may enter the public spaces prior to screening, and the protection of the emergency evacuation, rescue, and recovery systems. These protective measures are generally achieved through principles of structural dynamics, nonlinear material response, and ductile detailing.

There are a number of design standards, criteria, and guidelines in the United States for protecting government buildings; however, there are no comparable documents for commercial buildings. The American Society of Civil Engineers (ASCE) therefore undertook the task to develop a consensus-based blast standard that identifies the minimum planning, design, construction, and assessment requirements for new and existing buildings subject to the effects of accidental or malicious explosions, including principles for establishing appropriate threat parameters, levels of protection, loadings, analysis methodologies, materials, detailing, and test procedures. The document does not prescribe requirements or guidelines for the mitigation of progressive collapse or other potential postblast behavior. Unlike the government guidelines and criteria, the ASCE standards are written for structural engineers with specific information pertaining to the design and detailing of blast-resistant structure and facade systems.

There are currently no universal codes or standards that apply to all public- and private-sector buildings. However, most designers agree that security issues must be addressed with other design objectives and integrated into the building design throughout the process. This will ensure a quality building with effective security. This concept is known as an all-hazard design.

Depending on the building type, acceptable levels of risk and decisions made based on recommendations from a comprehensive threat assessment, vulnerability assessment, and risk analysis, appropriate countermeasures should be implemented to protect people, assets, and mission. Some types of attack and threats to consider include the following:

- Unauthorized entry/trespass (forced and covert)
- Insider threats
- Explosive threats: stationary and moving vehicle delivered bombs, mail bombs, package bombs
- Ballistic threats: small arms, high-powered rifles, drive-by shootings, etc.
- WMD [chemical, biological, and radiological (CBR)]
- Disruptive threats (hoaxes, false reports, malicious attempts to disrupt operations)
- Cyber and information security threats
- Supervisory Control and Data Acquisition (SCADA) system threats (relevant as they relate to HVAC, mechanical/electrical systems control, and other utility systems that are required to operate many functions within a building)

Protective design is the design of structures to mitigate blast effects. This will require the involvement of protective design and security consultants at the onset of the design phase. Early and ongoing coordination between the protective design consultant, the structural engineer, and the entire planning team is critical to providing an optimal design that is both open and inviting to the public and compliant with updated security requirements.

Crime Prevention through Environmental Design (CPTED) is a proven methodology that not only enhances the performance of these security and safety measures, but also provides aesthetics and value engineering. CPTED utilizes four primary, overlapping principles: natural surveillance, natural access control, territoriality, and maintenance. Natural surveillance follows the premise that criminals do not wish to be observed; placing legitimate "eyes" on the street, such as providing window views and lighting, increases the perceived risk to offenders, reduces fear for bona fide occupants and visitors, as well as lessens reliance on only camera surveillance.

Natural access control supplements physical security and operational measures with walls, fences, ravines, or even hedges to define site boundaries, to channel legitimate occupants and visitors to designated entrances, and to reduce access points and escape routes.

Territoriality involves strategies to project a sense of ownership to spaces such that it becomes easier to identify intruders because they don't seem to belong. Clear differentiation between public, semipublic, and private spaces by using signage, fences, pavement treatment, art, and flowers are examples of ways to express ownership.

Maintenance is a key element to preserve lines of sights for surveillance, to retain the defensiveness of physical elements, and to project a sense of care and ownership. Together, the principles of CPTED increase the effectiveness of operational, technical, and physical safety methods, thereby lessening equipment and operating costs. For total design efficiency and cost-effectiveness, security, safety, and CPTED measures are best applied at the beginning of a project.

Essential to the security plan and design of a high-quality building is the implementation of appropriate countermeasures to deter, delay, detect, and deny attacks. Often the countermeasures work on the layered defense concept or a "multibarrier approach." This concept provides for increasing levels of security from the outer areas of the site or facility toward the inner, more protected areas. Some or all of the issues outlined below need consideration for effective security design and building operations.

3.2.1 Unauthorized Entry (Forced and Covert)

Protecting the facility and assets from unauthorized persons is an important part of any security system. Some items to consider include the following:

- Compound or facility access control
  - Control perimeter: fences, bollards, antiram barriers
  - Traffic control, remote-controlled gates, antiram hydraulic drop arms, hydraulic barriers, parking control systems
  - Forced-entry–ballistic-resistant (FE–BR) doors, windows, walls, and roofs

- Barrier protection for man-passable openings (>96 square inches) such as air vents, utility openings, and culverts
- Mechanical locking systems
- Elimination of hiding places
- Multiple-layer protection processes
- Perimeter intrusion detection systems
  - Clear zone
  - Video and CCTV surveillance technology
  - Alarms
  - Detection devices (motion, acoustic, infrared)
- Personnel identification systems
  - Access control, fingerprints, biometrics, ID cards
  - Credential management
  - Tailgating policies
  - Primary and secondary credential systems
- Protection of information and data
  - Acoustic shielding
  - Shielding of electronic security devices from hostile electronic environments
  - Computer screen shields
  - Secure access to equipment, networks, and hardware, for example, satellites and telephone systems

3.2.2 INSIDER THREATS

One of the most serious threats may come from persons who have authorized access to a facility. These may include disgruntled employees or persons who have gained access through normal means such as contractors and support personnel. To mitigate this threat some items to consider include the following:

- Implement personnel reliability programs and background checks
- Limit and control access to sensitive areas of the facility
- Compartmentalize within the building/campus
- Implement two-man rule for access to restricted areas

3.2.3 EXPLOSIVE THREATS: STATIONARY AND MOVING VEHICLE-DELIVERED BOMBS, MAIL BOMBS, AND PACKAGE BOMBS

Explosive threats tend to be the terrorist weapon of choice. Devices may include large amounts of explosives that require delivery by a vehicle. However, smaller amounts may be introduced into a facility through mail, packages, or simply hand-carried in an unsecured area. Normally, the best defense is to provide a defended distance between the threat location and the asset to be protected. This is typically called standoff distance. If standoff is not available or is insufficient to prevent direct contact or reduce the blast forces reaching the protected asset, structural hardening may be required. If introduced early in the design process, this may be done in an efficient and cost-effective manner. If introduced late in a design, or if retrofitting an existing facility, such a measure may prove to be economically difficult to justify. Some items to consider for design purposes include the following:

- Including qualified security and blast consulting professionals from programming forward
- Providing defended standoff for vehicle-borne weapons using rated or certified barriers such as antiram fencing or bollards, by using reinforced street furniture such as planters, plinth walls, or lighting standards, by using natural and man-made elements such as storm water elements, berms, ditches, tree masses, and so on, by site layout strategies for parking areas, roadways, loading docks, and other locations accessible by vehicles, by critical asset location strategies, and/or by security protocol through policy and procedures (e.g., vehicle inspections)
- Considering structural hardening and hazard mitigation designs such as ductile framing that is capable of withstanding abnormal loads and preventing progressive collapse, protective glazing, strengthening of walls, roofs, and other facility components
- Providing redundancy and physical separation of critical infrastructure (HVAC), utility systems (water, electricity, fuel, communications, and ventilation)
- Providing for refuge and evacuation
- Considering plans for suicide bombers. Confer with authorities who have had previous experience
- Providing defended standoff for hand-carried weapons with anticlimb fencing, barrier (thorny) plants, natural surveillance of routine occupants and unobstructed spaces, electronic surveillance, intrusion detection, territoriality using defined spaces, natural access control using exterior and interior pedestrian layout strategies, security protocol through policy and procedures (visitor management, personnel and package screening, etc.)
- Considering handling mail at alternate or remote locations not attached to the building or in a wing of the building with a dedicated HVAC system to limit contamination and damage to the main building
- Considering loading docks in structures unattached to the main building with a dedicated HVAC system to limit contamination and damage to the main building

3.2.4 BALLISTIC THREATS

These threats may range from random drive-by shootings to high-powered rifle attacks directed at specific targets within the facility. It is important to quantify the potential risk and to establish the appropriate level of protection. The most common ballistic protection rating systems include Underwriters Laboratories (UL), National Institute of Justice (NIJ), H.P. White Laboratory, and ASTM International. Materials are rated based on their ability to stop specific ammunition. Some items to consider for design purposes include the following:

- Obscuration or concealment screening using trees and hedges, berms, solid fencing, walls, and less critical buildings
- Ballistic-resistant rated materials and products
- Locating critical assets away from direct lines of sight through windows and doors
- Minimizing the number and size of windows
- Physical energy absorption screens such as solid fences, walls, earthen parapets
- Providing opaque windows or window treatments such as reflective coatings, shades, or drapes to decrease sight lines
- Avoiding sight lines to assets through vents, skylights, or other building openings
- Using foyers or other door-shielding techniques to block observation through a doorway from an outside location
- Avoiding main entrances to buildings or critical assets that face the perimeter or an uncontrolled vantage point

3.2.5   WMD: CBR

Commonly referred to as WMD, these threats generally have a low probability of occurrence but the consequences of an attack may be severe. While fully protecting a facility against such threats may not be feasible with few exceptions, there are several commonsensical and low-cost measures that can improve resistance and reduce the risks. Some items to consider for building design include the following:

- Protect ventilation pathways into the building.
- Control access to air inlets and water systems.
- Provide detection and filtration systems for HVAC systems, air intakes, and water systems.
- Provide for emergency HVAC shutoff and control.
- Segregate portions of building spaces (i.e., provide separate HVAC for the lobby, loading docks, and the core of the building).
- Consider positive pressurization to keep contaminates outside of the facility.
- Provide an emergency notification system to facilitate orderly response and evacuation.
- Avoid building locations in depressions where air could stagnate.
- Provide access control to mechanical rooms.
- Provide CBR monitoring apparatus.

3.2.6 CYBER AND INFORMATION SECURITY THREATS

Businesses rely heavily on the transmission, storage, and access to a wide range of electronic data and communication systems. Protecting these systems from attack is critical. Some items to consider include the following:

- Understand and identify the information assets you are trying to protect. These may include personal information, business information, such as proprietary designs or processes, national security information, or simply the ability of your organization to communicate via e-mail and other LAN/WAN and wireless functions.
- Protect the physical infrastructure that supports information systems. If the computer system is electronically secured but vulnerable to physical destruction it may need more protection.
- Provide software and hardware devices to detect, monitor, and prevent unauthorized access to or the destruction of sensitive information.

3.2.7 DEVELOPMENT AND TRAINING ON OCCUPANT EMERGENCY PLANS

Occupant emergency plans should be developed for all building operations staff and occupants to be able to respond to all forms of credible attacks and threats. The emergency plan needs to include clearly defined lines of communication, responsibilities, and operational procedures. These plans are an essential element of protecting life and property from attacks and threats by preparing for and carrying out activities to prevent or minimize personal injury and physical damage. Overall, safety is accomplished by preemergency planning; establishing specific functions for operational staff and occupants; training organization personnel in appropriate functions; instructing occupants of appropriate responses to emergency situations and evacuation procedures; and conducting actual drills to ensure everyone is aware of policies and procedures (see Chapter 1).

3.3 WATER SUPPLY SYSTEMS SECURITY

The security of water systems has always been a concern for the water industry. The potential for natural, accidental, and purposeful contamination and other events, which can hinder the ability of the system to provide safe drinking water, has been the subject of many studies. After the terrorist events of September 11, 2001, the water industry began to focus their attention on the vulnerability of water supplies to terrorist activities. Water supply systems are spatially diverse (Figure 3.10), and therefore are vulnerable to a variety of activities that can compromise the system's ability to deliver clean potable drinking water.

There are several areas of vulnerability as water travels to the consumer: (1) the raw water source (surface or groundwater), (2) raw water canals and pipelines, (3) raw water reservoirs, (4) treatment facilities, (5) connections to the distribution system pipes, (6) pump stations and valves, and (7) finished water tanks and

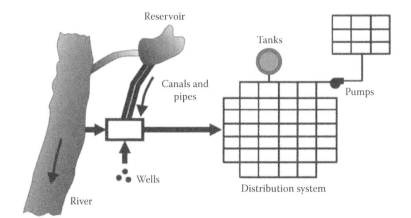

FIGURE 3.10 Major points of vulnerability in a water supply system.

reservoirs. Each of these system elements presents unique challenges to water utility in safeguarding the water supply. These challenges include the following:

- Physical disruption that prevents sufficient water flow at an acceptable pressure to all customers.
- Contamination of the water delivered to the consumer by a chemical or biological agent such that the product is not safe to use or is not of an acceptable quality to the consumer
- Loss of confidence by consumers in the ability of the water utility to deliver safe and secure drinking water

Water systems are vulnerable to natural, purposeful, or accidental events that can challenge the security of water supply systems. Examples of these types of events include the following:

- *Natural* events, such as sinkholes, floods, earthquakes, wild fires, and extreme weather, such as droughts, hurricanes, and tornadoes, which can lead to contamination of surface water and/or groundwater sources
- *Intentional* events, such as disgruntled employees, terrorist or criminal contamination, vandalism, and sabotage or destruction
- *Accidental* events, such as accidental discharges of pollutants into source waters, cross-connections between the water distribution system and waste collection system, vehicle and pipeline accidents, and *in situ* and *ex situ* explosions

The ability for a water supply system to provide clean potable drinking water to the public is dependent on several key elements. Key elements include the following:

- Raw water facilities (dams, reservoirs, pipelines, and canals)
- Treatment facilities
- Finished water elements (transmission lines and pump stations)

The physical disruption of any of these key elements can result in the disruption of service, significant economic cost, inconvenience, and loss of confidence by customers, but the direct threat to human health is generally limited. Exceptions may include (1) the destruction of a dam that causes loss of life and property in the accompanying flood wave and (2) an explosive release of chlorine gas at a treatment plant that produces a cloud injurious or lethal to nearby populations.

Water utilities need to examine their physical assets, determine their areas of vulnerability, and increase security accordingly. For example, switching from chlorine gas to liquid hypochlorite, especially in less secure locations, decreases the risk of exposure to poisonous chlorine gas. In addition, redundant system components can limit disruption of service by providing backup capability in case of accidental or purposeful damage to facilities.

The ability to deliver water of an acceptable quality can be compromised by the presence of contaminants in the raw or finished water supply. Contaminants can enter the water supply through natural causes, accidental spills or events, or purposeful terrorist or criminal acts. As illustrated in Figure 3.10, there are many potential locations where a contaminant can enter the water system. Locations include raw water sources (surface or groundwater), raw water delivery systems, the treatment facility, or the actual distribution system. Most provinces in Canada have adopted a multibarrier approach to protecting water supplies. According to Figure 3.10, the water treatment plant appears to be the primary barrier to contaminants reaching the consumer; however, this may not be as effective for some constituents found in the raw water and is not effective for contaminants that may enter the finished water through the distribution system. In turn, the province of Ontario has implemented the "source-water protection program," which is essentially the first line of defense in protecting drinking water supplies and specifically the raw water source. The purpose of this program is to identify vulnerabilities in the raw water sources and recommend safeguards or mitigation measures using all-hazards mitigation approach to offset the risks. The most vulnerable point within the distribution system is where the treated water leaves the treatment plant and enters the distribution system, usually downstream of the first valve. In turn, maintenance of a secondary disinfectant residual in a distribution system provides protection from certain bacterial agents but is not effective for all constituents.

Contamination of water supplies can date as far back as biblical times when the Nile River turned to "blood" in the first plague in the *Book of Exodus* requiring the Egyptians to turn to wells as an alternate water supply. Contamination has long been viewed as a serious potential threat to water systems. CBR agents can spread throughout the distribution system resulting in the sickness or death of people drinking the water. Deliberate contamination of water supply systems is nothing new. Such acts were reported in ancient Rome, when cyanide was used to poison the water supply. In the United States during the Civil War, farm animals were shot and their carcasses were left in ponds to poison the water supply so that advancing troops would not be able to use them. In Europe and Asia during World War II, anthrax, cholera, and raw sewage were used to contaminate water supplies. In addition, during the Kosovo crisis in the late 1990s, paints, oil, and gasoline were deliberately spilled into groundwater wells.

Accidental contamination of water systems has resulted in many fatalities and illnesses. Examples of such outbreaks include the *Escherichia coli* contamination in Walkerton, Ontario. This event is particularly significant since it led to the implementation of drinking water regulations in the province of Ontario and the establishment of the Source Water Protection Program. The event itself could also be classified as a "natural" event, since prior to the *E. coli* outbreak, a number of intense thunderstorms had passed through Southern Ontario including Walkerton, causing nuisance flooding. During these events in Walkerton, runoff from farms entered a groundwater well, which was known to be vulnerable to contamination for years. Other significant events include the *cholera* contamination in Peru, *Cryptosporidium* contamination in Milwaukee, Wisconsin, and *Salmonella* contamination in Gideon, Missouri.

The U.S. Army has compiled information on potential biological agents. Table 3.1 summarizes this information on agents that may have an impact on water supply

TABLE 3.1
Potential Threat of Biological Agents

Agent	Type	Weaponized	Water Threat	Stable in Water	Chlorine Tolerance
Anthrax	Bacteria	Yes	Yes	2 years (spores)	Spores resistant
Brucellosis	Bacteria	Yes	Probable	20–72 days	Unknown
Clostridium perfringens	Bacteria	Probable	Probable	Common in sewage	Resistant
Tularemia	Bacteria	Yes	Yes	Up to 90 days	Inactivated, 1 ppm, 5 minutes
Shigellosis	Bacteria	Unknown	Yes	2–3 days	Inactivated, 0.05 ppm, 10 minutes
Cholera	Bacteria	Unknown	Yes	Survives well	Easily killed
Salmonella	Bacteria	Unknown	Yes	8 days, fresh water	Inactivated
Q fever	Rickettsia	Yes	Possible	Unknown	Unknown
Typhus	Rickettsia	Probable	Unlikely	Unknown	Unknown
Psittacosis	Rickettsia-like	Possible	Possible	18–24 hours, seawater	Unknown
Encephalomyelitis	Virus	Probable	Unlikely	Unknown	Unknown
Hemorrhagic fever	Virus	Probable	Unlikely	Unknown	Unknown
Variola	Virus	Possible	Possible	Unknown	Unknown
Hepatitis A	Virus	Unknown	Yes	Unknown	Inactivated, 0.4 ppm, 30 minutes
Cryptosporidiosis	Protozoan	Unknown	Yes	Stable for days or more	Oocysts resistant
Botulinum toxins	Biotoxin	Yes	Yes	Stable	Inactivated, 6 ppm, 20 minutes
T-2 mycotoxin	Biotoxin	Probable	Yes	Stable	Resistant
Aflatoxin	Biotoxin	Yes	Yes	Probably stable	Probably tolerant
Ricin	Biotoxin	Yes	Yes	Unknown	Resistant at 10 ppm

(Continued)

TABLE 3.1
(Continued) Potential Threat of Biological Agents

Agent	Type	Weaponized	Water Threat	Stable in Water	Chlorine Tolerance
Staphylococcal enterotoxins	Biotoxin	Probable	Yes	Probably stable	Unknown
Microcystins	Biotoxin	Possible	Yes	Probably stable	Resistant at 100 ppm
Anatoxin A	Biotoxin	Unknown	Probable	Inactivated in days	Unknown
Tetrodotoxin	Biotoxin	Possible	Yes	Unknown	Inactivated, 50 ppm
Saxitoxin	Biotoxin	Possible	Yes	Stable	Resistant at 10 ppm

Source: Burrows, W. D. and Renner, S. E., *Environmental Health Perspectives*, 107, 975, 1998.

systems. There are several factors that should be considered when evaluating the potential threat from different chemical and biological agents. Following is a summary of some of these factors:

- Availability: Is the agent readily available or difficult to obtain?
- Monitoring response: Can the agent be detected by monitoring equipment?
- Physical appearance: Is there a telltale odor, color, or taste associated with the agent?
- Dosage/health effects: What dosage is required to have effects on human health?
- Chemical and physical stability in water: How long will the agent be stable in water?
- Tolerance to chlorine: Are chlorine or other disinfectants effective in neutralizing the agent?

In 2000, Deininger and Meier ranked various agents and compounds in terms of their relative factor of effectiveness, R, based on lethality and solubility using the following equation:

$$R = \frac{\text{Solubility in water (mg/L)}}{1000 \times \text{lethal dose (mg/human)}} \tag{3.1}$$

Table 3.2 lists the values of R for various biological agents and chemicals by decreasing level of effectiveness, that is, decreasing the degree of toxicity in the water.

Contamination of source water is of significant concern due to the number of potential locations where contaminants can enter the surface water or groundwater sources and the difficulty in providing security over an entire source water area. Both surface water and groundwater sources are subject to point source and nonpoint source pollutants.

There is an assumption that a contaminant entering a source water is subject to a number of processes, such as dilution, and by the time it enters the water distribution system its concentration has been significantly reduced. Although this may be true

TABLE 3.2
Relative Toxicity of Poisons in Water

Compound	R
Botulinum toxin A	10,000
VX	300
Sarin	100
Nicotine	20
Colchicine	12
Cyanide	9
Amiton	5
Fluoroethanol, sodium, fluoroacetate	1

Source: Deininger, R. A., and Meier, P. G. Sabotage of public water supply systems, in *Security of Public Water Supplies*, Deininger, R. A., Literathy, P., and Bartram, J. (eds.), Kluwer Academic Publishers, Boston, MA, 2000.

for large surface water bodies such as the Great Lakes, this might not necessarily be true for small basins such as rivers and streams and especially not for groundwater wells, as was in the case of the Walkerton Tragedy. Contamination of source waters may be caused by natural hydrologic processes, accidental releases of contaminants, or intentional dumping of pollutants into the water.

A mechanism for identifying and responding to a source water contamination event is an early warning system. An early warning system is a combination of equipment, institutional arrangements, and policies used in an integrated manner to achieve the goal of identifying and responding to contaminants in the source water. An early warning system may also be complemented by a source water protection plan. A source water protection plan identifies the existing and potential threats and vulnerabilities to a community's raw water sources, which may include rivers, lakes, and aquifers and makes recommendations with respect to actions to be taken to reduce or eliminate those threats.

An effective early warning system includes the following components:

- A mechanism for detecting the likely presence of a contaminant in the source water
- A means of confirming the presence of the contamination, determining the nature of the contamination event, and predicting when (and for how long) the contamination will affect the source water at the intake sites and the intensity (concentration) of the contamination at the intake
- An institutional framework generally composed of a centralized administrative unit that coordinates the efforts associated with managing the contamination event
- Communication linkages for transferring information related to the contamination event
- Various mechanisms for responding to the presence of contamination in the source water in order to mitigate its impacts on water users

The central component of an early warning system is the detection of a contaminant event and the mitigative actions to be taken if needed. There are three basic mechanisms for detecting spills and other contaminant events: monitoring, self-reporting by the facility, which is causing the spill, appropriate siting, and reporting of the event for the public and outside groups. An effective early warning system combines all three mechanisms.

A wide range of online monitoring equipment is available that can be used as part of an early warning system. Conventional sensors for online monitoring, such as dissolved oxygen, pH, conductivity, temperature, and turbidity, are inexpensive, readily available, and easily used; however, they provide little useful information in identifying the presence of transient contaminants. More advanced online monitoring equipment such as gas chromatographs and spectrophotometers are more expensive, require more expertise and maintenance, but are more effective when deployed as part of an early warning system.

Examination of online monitors around the world has shown that given sufficient resources, almost all analyses can be automated if there is a perceived need. Biomonitors that utilize living aquatic organisms (such as fish, mussels, daphnia, and bacteria) measure the stresses placed on the organisms when contaminants enter into the water. These monitors do not provide information specific to a contaminant but rather they identify that there is something unusual in the water that is affecting the organisms used in the biomonitoring. Biomonitors have been in use in many places for over 20 years; however, the field of biomonitoring still appears to be emerging. Other emerging technologies that may have future utility in early warning systems are being developed in other fields and may include electronic noses, DNA chips, flow cytometry, immunomagnetic separation techniques, and online bacteria monitors. Current state-of-the-art technologies in the identification of microbial contamination in drinking water have been summarized in the literature.

When the presence of a contaminant is identified from the source water, there are several responses that may be taken to help mitigate the impacts of a spill event: the closure of water intakes and use of alternate sources, cleanup of the spill prior to impacting water intakes, enhanced temporary chemical treatment at the treatment plant, and public notification.

Closure of the water intakes provides an absolute barrier to a contaminant entering and impacting the drinking water supply; however, this response is effective only if the presence of the contaminant is identified prior to entering the water intake, which is limited by the length of time it takes for a water utility to close the intake and provide a sufficient supply of water. Cleanup of the source water body is a practical solution; however, it is limited to a specific range of pollutants (such as petroleum products, which can be physically separated from the water body). Enhanced chemical treatment is undertaken by the addition of coagulants, carbon, disinfectant, or other chemicals. This is only effective if the chemical nature of the contamination event is known and the proper chemical dosage is determined.

If the contamination event is not identified within the source water, the only line of defense is the normal treatment processes that are online at the time of the event. Table 3.3 contains typical contaminant removal ranges for various treatment types and contaminant categories. Actual removal rates depend upon the specific

TABLE 3.3

Effectiveness of Processes for Contamination Removal

Treatment	Bacteria	Viruses	Protozoa	VOC	SOC	TOC	Taste and Odor
Aeration, air stripping	P	P	P	G–E	P–F	F	F–E
Coagulation, sediment/ filtration	G–E	G–E	G–E	G–E	P	P–G	P–G
Ion exchange	P	P	P	P	P	G–E	–
Reverse osmosis	E	E	E	F–E	F–E	G	–
Ultra filtration	E	E	E	F–E	F–E	G	–
Disinfection	E	E	E	P–G	P–G	G–E	P–E
Granular activated carbon	F	F	F	F–E	F–E	F	G–E
UV irradiation	E	E	E	G	G	G	G

Source: Hargesheimer, E., *Online Monitoring for Drinking Water Utilities*, AwwaRF, Denver, CO, 2002. E, excellent 90%–100% removal; F, fair 20%–60% removal; G good 60%–90% removal; P, poor 0%–20% removal; SOC, synthetic organic chemical; TOC, total organic carbon; VOC, volatile organic chemicals.

designs, the chemical characteristics of the raw water, and temperature. Other treatment options that provide additional barriers to the contaminants before reaching the water user include regular use of water, granular activated carbon, groundwater injection, and subsequent harvesting of water. Another mechanism—raw water storage—provides a time lag between the water intake and the water treatment, potentially allowing for additional testing prior to the use of the water.

A water distribution system is usually a closed system that is more secure than the raw water source. The implications of direct contamination into the distribution system are more severe than the water source for the following reasons: no treatment exists to provide a barrier (an exception is residual disinfectant); monitoring within a distribution system is limited; the travel time between a point of contamination in the distribution system and the consumer is short; and if a contaminant is not detected before it reaches a distribution system storage facility, contaminated water may reach many customers over an extended period of time. Potential points of contamination within the water distribution system include the water treatment plant, pump stations and valves, finished tanks and reservoirs, and hydrants and distribution system connections.

The treatment plants are the interface between the raw water source and the distribution system. They are vulnerable because the entire water supply flows through the plant. Clear wells at the end of the treatment process should receive special attention. In the clear wells, a large amount of supply is concentrated in a single location. A contaminant added at that point can impact the entire water supply leaving the plant over a period of many hours. Treatment plants have some form of security; large plants have around the clock personnel. Pump stations and valves are potential points of contamination because a large amount of flow may be passing through these points at any given time. A contaminant injected at these locations could impact many downstream consumers.

Tanks and reservoirs are vulnerable because they contain a relatively large quantity of water at a single location, are frequently located in isolated locations, are generally maintained at atmospheric pressure, and usually have access to the interior for maintenance purposes. Open reservoirs are especially vulnerable because of the open access to the contents, but they are also relatively rare. Because of the quantity of water in storage or in a tank or reservoir, a significant amount of contaminant is needed to impact the water quality. Once a sufficient amount of contaminant is added and mixed with the water in the tank, the water quality in the distribution system fed by the tank can be impacted for many days until the fill-and-draw process has sufficiently diluted the contaminant.

Hydrants are located on relatively major lines throughout water distribution systems providing water for fire protection. There are no recorded cases of individuals intentionally pumping contaminated water into the distribution system through a fire hydrant, but there have been cases of accidental backflow from tankers that were being filled directly from the hydrants. A pump that is capable of overcoming the system pressure could inject contaminants into the system affecting nearby consumers or larger areas of the distribution system depending on the location and amount of water pumped into the system.

The purposes of a vulnerability study or assessment are to examine facilities to determine their vulnerability to various events that could threaten their ability to achieve their specific operational purposes; to assess and prioritize the risks associated with the facilities and operations of the water system; and to develop a program to reduce risks and respond in case of events that could threaten the water system.

Formal vulnerability assessments have been applied to facilities that have been recognized as vulnerable to threats or hazards that could result in significant consequences. Examples of such facilities include nuclear power plants, military bases, and federal dams. Transfer of these methods to water supply systems is a recent and ongoing development. Such assessments can employ formalized information collection and analysis tools, or, more simply, a set of checklists, protocols, and procedures.

Whichever approach is adopted, the vulnerability analysis is framed in the context of managing the risks within the water supply system. The basic framework for considering risk analysis consists of the interwoven areas of risk assessment and risk management. Risk analysis recognizes that it is impossible to eliminate all risks but that risks can be identified, assessed, managed, and balanced with other resources so that the resulting risks are within an acceptable level. A risk analysis can be embedded in a series of questions through vulnerability assessment. For example, with regard to water supply systems, a typical vulnerability assessment will try to answer the following questions: What can go wrong? How can it happen? How likely is it to happen? And what are the consequences? Once the vulnerability assessment is complete, management will have their questions that will need to be answered by the risk analysis, for example, what can be done to reduce the impact of the risk described? What can be done to reduce the likelihood of the risk described? What are the trade-offs of the available options? And what is the best way to address the described risk?

In conducting a vulnerability analysis, water utilities develop a set of inspection protocols and a checklist of items. The inspection protocol includes items such as an examination of design plans, field inspection of the facilities, and review of the operational and maintenance procedures within the water utility.

Vulnerability assessments of water supply systems can involve a large number of facilities and a wide range of potential risks and solutions. There are various methods and tools available to help a water utility organize, manage, and assess such information. Examples of such tools include fault trees, event trees, decision trees, Monte Carlo simulations, and computer simulations.

Sandia National Laboratories in conjunction with the U.S. Environmental Protection Agency (EPA) and AwwaRF developed a water system vulnerability assessment training package. This stemmed from a performance-based vulnerability assessment methodology initially developed by Sandia for the nuclear security area and later applied to the security of federal dams. The central component of this analysis is a risk assessment process as illustrated in Figure 3.11. The Ministry of the Environment of Ontario, MOE, has developed specific regulations for undertaking inspections and conducting assessments at drinking water treatment facilities.

One of the tools applied in the Sandia assessment is a fault tree. Figure 3.12 illustrates an example of a generic fault tree applied to a water system. A fault tree is a top-down method of analyzing system design and performance. The top event is an undesired state of the system, in this case, defined as an occurrence that will result in the water utility being unable to meets its mission of supplying safe water to its customers. In the second level, specific credible ways that could result in not meeting

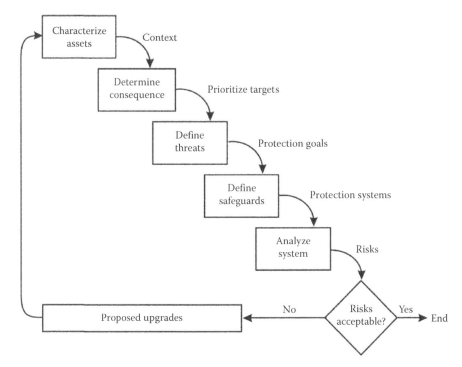

FIGURE 3.11 Risk assessment process applied to water systems (Sandia National Laboratories, http://www.awwarf.com/whatsnew.html, accessed on January 1, 2002).

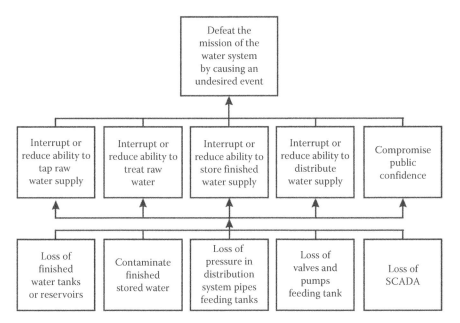

FIGURE 3.12 Risk assessment process applied to water systems (Sandia National Laboratories, http://www.awwarf.com/whatsnew.html, accessed on January 1, 2002). SCADA, supervisory control and data acquisition.

this mission are enumerated, such as interruption of the raw water supply, treatment capability, or distribution.

The fault tree continues down with events being combined through either AND or OR gates. An AND gate indicates that multiple events must occur simultaneously for the higher-level event to occur while OR gates indicate that either of two or more events could lead to the higher level event occurring. The lowest level of detail depicted in the graphical fault tree is referred to as a basic event. In the example shown, the basic events are referred to as "undeveloped" events because they may be broken down into more detail. The fault tree can be viewed as a graphical depiction of events that would lead to the overall system failure. Alternatively, it may take on more of a quantitative tool by assigning probabilities of occurrence to the basic events and then, by following the tree upward, calculating the overall probability of the top event occurring.

Monte Carlo simulation is a well-known technique for analyzing complex physical systems where probabilistic behavior is important. Monte Carlo simulation of a source water early warning system has been used to test the effectiveness of alternative designs. Spill events are represented as probabilistic occurrences, that is, probability distribution of streamflows, probabilities of spills of different substances, magnitudes, and duration of the event. The relationships of these events are embedded in the model, which is then run multiple times, with varying inputs based on the probabilities of the events. This model was applied to the Ohio River to study the effectiveness of alternative early warning system designs and on the resulting water quality of the finished water supply.

Simulation models are another tool that can be used in vulnerability studies to help a water utility understand how their system will respond to an accidental or purposeful physical or chemical event. This understanding can be used to identify the consequences of such events, to test solutions to minimize the impacts of the events, or to learn how to respond if such events occur.

Models are representations of systems that are especially effective in examining the consequences of "what if" scenarios. Within the context of water system security, some examples of "what if" scenarios may include the following: If an oil tank adjacent to a river ruptures and discharges to a river used downstream as a source of raw water, when should the utility close its water intake and for how long will they need to keep the intake closed? If a major main in the water system breaks, what happens to pressure throughout the distribution system and will there be sufficient flow and pressure to provide fire protection? If runoff contaminates a particular well, what customers would receive contaminated water and how quickly will the contaminant reach them? If a terrorist manages to dump a barrel of a particular chemical into a finished water tank, how will the chemical mix within the tank and, if the contamination is not discovered, which customers will receive the contaminated water, when will they receive it, and what will the concentration be?

In the area of water system security, computer models have been used to examine three different time frames: as a planning tool to look at what may happen in the future in order to assess the vulnerability of a system to different types of events and to plan how to respond if such an event occurs; as a real-time tool for use during an actual event to assist in formulating a response to the situation; and as a tool for investigating a past event so as to understand what happened. The characteristics of models used and the type of information that is available in these three time frames can vary significantly. There are three types of computer models for analyzing water supply systems: water distribution system models, tank and reservoir mixing models, and surface water hydraulic and water quality models.

A water distribution system model can be used to simulate flows and pressures within a distribution system and the movement and transformation of a constituent after it is introduced into the distribution system. To simulate the movement of a contaminant in a distribution system, a hydraulic extended-period simulation (EPS) model of the system is needed. The model should reflect the hydraulic conditions of concern. In using the model in a vulnerability study, it would be appropriate to select several normal operating conditions such as a typical summer day, a typical winter day, and a typical spring/autumn day. Selecting typical operating conditions provides information reflective of most situations rather than an extreme demand day that represents conditions for only a few days per year. A contaminant is represented in the model by describing the transformation characteristics of the constituent in the distribution system, where it is introduced into the system, and the time history of the amount of the constituent that is introduced. Most modern distribution system models provide options for designating this information.

In the modeling environment a constituent is represented as being a conservative substance, which means that it does not change concentration unless dilution occurs, or a nonconservative substance, which means that it follows some form of decay such as first-order exponential decay. For the purposes of a vulnerability study, it is generally assumed that the constituent will be conservative. Most modern distribution system models provide the user with several options for introducing a contaminant at a source. In each case, a time-varying pattern can be applied to the source, such as a concentration constituent, a flow-paced booster constituent, a set-point booster constituent, and a mass booster constituent. A concentration constituent source fixes the concentration of any external inflow entering the network at a node, such as flow from a reservoir or from a source placed at a junction. A flow-paced booster constituent source adds a fixed concentration to the flow resulting after the mixing of all inflow to the node from other points in the network. A set-point booster constituent source fixes the concentration of any flow leaving the source node, as long as the concentration resulting from the inflow to the node is below the set point. A mass booster constituent source adds a fixed mass flow to the flow entering the node from other points in the network.

The initial concentration in a tank may be set with the concentration changing over time due to either decay or dilution during the fill cycle. These options for specifying source concentrations allow the user to select the method that most accurately represents the physical contamination event that he or she is attempting to simulate.

A water distribution system model can be applied to a wide range of "what if" scenarios to determine the general vulnerability of the distribution system. For example, the model can be used to determine the effects of a major pipe break or the impacts of a purposeful or accidental contamination of the system. With this information in hand, a water utility is better equipped to develop an effective plan of action. An example of the results of the application of a distribution system model to represent contamination tracking in the distribution system is shown in Figure 3.13.

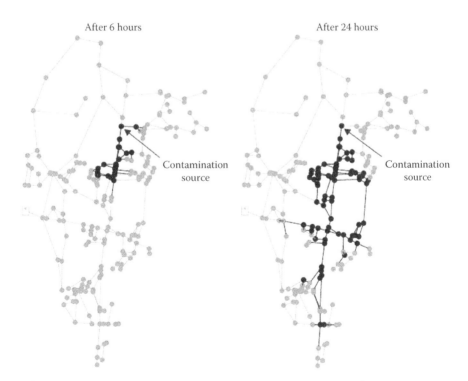

FIGURE 3.13   Risk assessment process applied to water systems (Sandia National Laboratories, http://www.awwarf.com/whatsnew.html, accessed on January 1, 2002).

This example illustrates the case where a contaminant has been pumped directly into the distribution system at a constant flow and concentration rate over a 24-hour period. The figure shows how the contaminant spreads over a significant portion of the system during the 24-hour period.

An important part of any emergency planning is the simulation of a possible emergency before it occurs. Using a model to simulate emergency drills with operators is another facet of such preparedness. An operator can be given a contamination scenario and asked to respond. The operator's response can be simulated in the model, showing where the contaminated plume would move and how the actions affected the movement of the plume and exposure for customers.

Modeling of historical contamination events is sometimes referred to as retrospective or forensic modeling using water distribution system models. The disease outbreak in Gideon, Missouri in December 1993 was identified as *salmonellosis* and a private tank was suspected as a source of the bacteria. The U.S. EPA applied a water distribution system model of the Gideon system to investigate possible scenarios by which the bacteria could propagate through the system and infect the customers. Ultimately, it was determined that it was likely that the tank, which was in bad shape, was infected by birds. A sudden drop in temperature resulted in a temperature inversion in the tank, which mixed infected bird droppings and feathers throughout the tank and resulted in taste and odor problems. An aggressive flushing program then drew the contaminated water deep into the distribution system. Similar historical modeling has been done in the investigation of *E. coli* outbreaks in Cabool, Missouri, and Walkerton, Ontario.

Improper handling and disposal of industrial chemicals led to the contamination of some groundwater in the United States during the twentieth century. In many cases, the contaminated groundwater was used as a source of drinking water and, as a result, customers were exposed to elevated levels of the contaminants. Many of these cases have resulted in legal actions and governmental studies of the impacts of the contaminants on the population exposed to the substances. Distribution system hydraulic and water quality models have played a key role in many of these cases in determining the likely movement of the contaminants through the distribution system. Some of these incidents date back to the 1950s through to the 1980s, for which the development of models and reconstruction of the operation of the water system was required. Information on the model development and reconstruction process is limited because of nondisclosure requirements associated with many legal cases.

A contamination case in Woburn, Massachusetts, has been documented in the book and movie, *A Civil Action*, and was the subject of an early modeling study in the 1980s. Industrial sources were found to have improperly discarded chemicals that seeped into the groundwater system and contaminated two wells. A series of steady-state distribution system models were used to identify areas of Woburn that received the contaminated water under different well operating patterns and demand conditions. A groundwater contamination case in Phoenix and Scottsdale, Arizona, involving the chemical trichloroethylene (TCE) was studied using EPSs of the operation of the water systems and the movement of water through those distribution systems. Most notable in this study was the use of long-term continuous simulations over multiple years of the hydraulics and water quality in the distribution system.

A recently completed detailed study of the water system in the Dover Township, New Jersey, area by the U.S. Agency for Toxic Substances and Disease Registry identified the paths between wells and customers over a multidecade period. In this study, a water distribution system model of the present-day system was first developed and calibrated. Subsequently, models of the distribution system were developed that covered the period from 1962 to 1996. The models were used to trace the percentage of water reaching nodes from the wells serving the system. In historical reconstruction modeling, the challenge is to utilize historical data to determine the characteristics and operation of the water system during the period of interest. The challenge increases as one models further back in time. One of the key challenges is reconstructing the actual operation of pumps when records are incomplete. Without any records it is difficult for the modelers to determine when the pumps were operating and for what duration, which, in turn, can impact the modeling solution in determining which customers were exposed to what sources.

In addition to historical modeling there is also real-time modeling. Real-time modeling allows the modeler to model an event as it occurs. For example, suppose that the manager of a water system receives a call from the police that an individual has been apprehended for dumping a poisonous chemical into the water system. The location, chemical, and approximate time and duration of the contamination are known. The first action the manager takes, of course, is to notify the public. Next the manager needs to determine how to operate the system to flush out the contaminant; that is, he or she needs to determine which hydrants to open and how long to keep them open. The manager has a good idea of how water moves through the system during a normal day, but he or she also knows that the flushing program could drastically change the normal flow patterns. Attempting to clean the system by trial and error would be a long, risky, and uncertain proposition. An alternate and better solution would be to model the event as it occurs in real time using a properly developed and maintained water distribution system model.

Water distribution system models have been proposed as part of a real-time or near-real-time system to assist in many aspects of the operation of a water system including energy management, water quality management, and emergency operation. The major obstacle in such use of water distribution system models is the requirement that the model be calibrated for a wide range of conditions and be ready to apply quickly and easily in an EPS mode. Information on the current state of the system must be readily available to the model through direct ties to a SCADA system. In addition, the model must be set up in an automated mode so that operation is represented by a series of logical controls that reflect the existing operating procedures. Both information requirements are feasible based on existing technology but there have only been limited demonstrations of this type of operation to date. The key to using a model as part of a real-time response lies in having the model ready to run. During an emergency, there is no time to construct a model. There is only time to make some minor adjustments to an existing model.

Distribution system models represent mixing in tanks and reservoirs using simplified, hypothetical representations such as complete and instantaneous mixing, plug flow, or by a "last in, first out" short-circuiting model. Though this has proved adequate for most planning and operational situations, a more accurate representation of how the facility mixes may be needed when planning for emergency contamination events.

Computational fluid dynamics (CFD) models use mathematical equations to simulate flow patterns, heat transfer, and chemical reactions and thus provide a much better description of the actual mixing processes in a tank. The use of CFD models has grown significantly and the technology has been applied in many planning and design studies to assess the mixing characteristics of a tank and its inlet–outlet configuration. Several commercial CFD software packages are available. Significant experience and knowledge is required to apply CFD models, and model run times of many hours, days, or even weeks are required for complex situations. In conducting a vulnerability study, knowledge of the mixing characteristics in a tank is useful in assessing the likely impacts of a contaminant being added to a storage facility.

Hydraulic and water quality models of surface water and groundwater systems can be used to study the movement of contaminants in a surface water body and the groundwater. This information is useful in assessing the vulnerability of a water intake to both point source and nonpoint source contamination. In addition, such models are also useful as part of an early warning system to predict the real-time movement of contaminants that have been detected upstream of a water intake. There are many general purpose hydraulic/water quality models and specially designed "spill models" that can be used in both vulnerability studies and real-time prediction.

Many security measures are available that can be applied to decrease the vulnerability of a distribution system to purposeful and accidental events that threaten the ability of a water utility to provide safe potable drinking water to its customers. Some of these measures can be applied quickly with minimal costs, and others may take significant time and resources to apply. The applicability of specific measures varies between water systems. In evaluating potential security measures, a water utility should balance the reduction of risks with the costs of implementing the measures. Some of these measures may include maintaining a significant disinfectant residual, an increase in security surround key facilities, installation of secure backflow preventers or check valves at key injection sites, the development of an early warning system for the raw water supply, the installation of continuous online monitoring at key locations in the distribution system, and the development of an emergency response plan.

As illustrated in Table 3.1, many of the potential biological agents are inactivated by exposure to chlorine. Actions that can be taken relative to maintaining an acceptable disinfectant residual include (1) placing continuous chlorine monitors throughout the system to report chlorine residual back to a central control center and warn of low residuals; (2) increasing the chlorine dose at the treatment plant during periods of higher alert (although this may result in undesirable higher levels of disinfectant by-products); (3) adding booster chlorination stations at locations in the distribution system that routinely experience low disinfectant residuals; and (4) modifying operating policies to reduce water age in the system, including changing the fill-and-draw patterns for tanks.

In 2000, Deininger and Meier recommended the following to increase the security of a water system: The intakes, pumping stations, treatment plants, and reservoirs should be fenced to secure them against casual vandalism. Beyond that, there should be intrusion alarms that notify the operator that an individual has entered a restricted area. An immediate response may be to shut down part of the pumping system until the appropriate authorities determine that there is no threat to the system. In underground reservoirs, the ventilation devices must be constructed in such a way as to not allow a person to pour a liquid into the reservoir. An above-ground reservoir with roof hatches should not have ladders on it that allow climbing. In addition, the hatches should be secured.

A contaminant can be injected at any connection to a water system if a pump that is capable of overcoming the system pressure is available. Backflow preventers provide an obstacle and deterrent to such action, but in order to be effective the backflow preventer must be installed so that it cannot be disengaged easily. Maintenance and cost issues associated with widespread installation of such devices should be considered when evaluating this option.

If not detected, contaminants in source waters (surface and groundwater) can pass through a treatment plant and enter the distribution system. An early warning system is a combination of equipment and institutional arrangements and policies that are used to detect and respond to contaminants in the source water.

Monitoring can provide a means of identifying the presence of unwanted contaminants in the distribution system. To be effective as a security mechanism, monitors that sample continuously or very frequently at key locations in the distribution system and are tied into a central operations center are needed.

An emergency response plan is like an insurance policy—one hopes that he or she will never have to use it but is very thankful that a plan is available if there is an emergency. Such a plan should provide detailed information on how to respond under a wide range of emergency situations (see Chapter 1). The plan should be kept up-to-date and personnel should be familiar with it so that it can be quickly implemented when needed. The procedures for developing emergency response plans have been developed by the American Water Works Association.

3.4 SECURITY FOR ENERGY FACILITIES

Protecting energy infrastructure is critical for national safety, security, and economic survival. Historically, threats to energy infrastructure have included theft, computer security, access controls, remote site vandalism, alcohol and drug-related incidents, and workplace violence. These types of threats are referred to as "low-level" or "ordinary" threats. Other threats or hazards that can directly impact energy infrastructure include aging infrastructure, poor or lack of maintenance, unintentional accidents, and extreme or severe weather events such as the 1998 ice storm in Eastern Ontario and Quebec. Hazards such as these can be further exacerbated by stressors such as climate change and unsustainable development. Aging infrastructure within the electricity sector is particularly vulnerable to both of these stressors. For this reason, out of the 10 critical infrastructure sectors within Canada and 18 within the United States, energy delivery services are considered to be the most prone to incapacity due to extreme weather events (see Chapter 2).

The electricity grid has generally been constructed and operated under a standard to maintain uninterrupted operations, even with the loss of the largest single resource on the system such as generation, a substation, or a transmission line; this standard is referred to as the $N - 1$ standard. This standard does not match current threats, for example, organized and coordinated attacks aimed at strategic elements and key resources within the energy infrastructure sector.

The September 11 attacks and the Northeast Blackout of 2003 brought about new thinking with respect to critical infrastructure and specifically the energy infrastructure sector. Energy infrastructure is considered to be the most critical out of all the different infrastructure sectors, the reason being that most elements of the energy sector are located within urban areas. People working and living within the urban centers are heavily dependent upon electricity in order to function properly. The interdependencies between the electricity sector and other critical infrastructure sectors mean that an accident could have a potentially more damaging effect than ever before. A failure in one part of the grid can cause cascading failures across a much wider area. For example, the Northeast Blackout of 2003 demonstrated that a disruption in the delivery of electricity can have a cascading and rippling effect across many different sectors in a community.

The Blackout affected approximately 10 million people in Ontario and approximately 45 million people in 8 U.S. states. The blackout's primary cause was a software bug in the alarm system at a control room of FirstEnergy Corporation in Ohio. Operators were unaware of the need to redistribute power after overloaded transmission lines hit unpruned foliage. What would have been a manageable local blackout cascaded into a widespread distress on the electric grid. Essential services remained in operation for most areas; however, in others, backup generation systems failed. Telephone networks generally remained operational, but the increased demand triggered by the blackout left many circuits overloaded. Water systems in several cities lost pressure, forcing boil-water advisories to be put into effect. Cellular service was interrupted as mobile networks were overloaded with the increase in volume of calls. Major cellular providers continued to operate on standby generator power. Television and radio stations remained on the air, with the help of backup generators, although some stations were knocked off the air for periods ranging from several hours to the length of the entire blackout (~2 days).

Since the September 11 attacks there have been significant changes with regard to infrastructure protection in Canada, the United States, and Europe. Some changes were already implemented prior to the events of September 11. For example, in the United States, the National CIP Program was already in effect, established 2 years prior by the then President Bill Clinton. However, September 11 served as a catalyst for other initiatives such as the creation of the Department of Homeland Security.

In Canada, after the September 11 attacks, there were significant changes made at the federal level including the creation of the National Critical Infrastructure Assurance Program; the adoption of an "all-hazards" approach to identifying threats, hazards, and risks for critical infrastructure; a shift in focus from an individual infrastructure sector to the overall system and networks and their associated interdependencies; the sharing of information between the public and private elements; building partnerships both domestically and internationally; mitigating damages through the Government Operations Centre; and the establishment of the Integrated Threat Assessment Centre (ITAC).

Key elements of the energy infrastructure sector include petroleum refineries, resource exploration including offshore drilling rigs, electrical generating stations, pipelines for transporting natural gas and crude oil, finished petroleum distribution, electrical substations, and electrical transmission lines. Some of these elements can cover vast distances (thousands of kilometers) and areas (thousands of kilometers), while others can be isolated unmanned facilities in remote parts of the country. The reality is that within Canada and the United States, key components of the energy sector are exposed to varying threats and hazards, and any disruption to their operations would have a significant impact. Their exposure coupled with credible threats and hazards makes it clear that the energy sector is vulnerable; however, not all elements of the infrastructure share the same levels of risk.

The level of protection afforded to each component of the energy sector will be dependent upon several factors: the type of infrastructure; its location, whether it is in a remote location or in a major urban center; function and operation; attractiveness as a target; and the level of impact due to disruption of operations. Since the level of risk varies for the various elements of the energy infrastructure, the level of protection will also vary. All facilities and components will be afforded some level of protection; the higher the risk posed by the facility or infrastructure the more protective measures will be implemented. Protective measures may include but not be limited to the following: perimeter security fencing around the facility, vehicle access control across the perimeter, personnel access control to and within the facility, surveillance and monitoring including the installation of CCTV cameras at key locations with video analytics to identify repeat visits from vehicles and individuals, protective barriers and protective housing for sensitive materials (i.e., nuclear materials), security patrols including armed emergency response, ongoing collaboration with local law enforcement, and ongoing assessments and audits of the system and operations. A nuclear plant would encompass most or all of these protective measures, whereas a remote substation or pipeline might encompass only the bare minimum, such as protective perimeter fencing.

The degree of risk toward a facility or asset will determine its level of protection. For instance, nuclear plants and large hydroelectric generating stations are considered attractive targets for terrorists and the consequences of failure can have devastating impacts. As a result, both facilities are afforded a high degree of protection. In turn, this high degree of protection also becomes a deterrent for terrorists, since the probability of mounting a successful attack is very small and the degree of difficulty in penetrating the facility is extremely high. Similarly, a pipeline or a substation in a remote location would also be considered an attractive target due to their high level of exposure and relative ease of penetration. However, the consequences of failure may not be as devastating as if they were located in an urban center, although in 2001 a single bullet hole in the Alaskan oil pipeline caused 300,000 gallons in crude oil leakage. In this case, the losses were limited to the owner including the cost for restoring the environmental damage caused by the leakage.

Where a terrorist group will strike depends largely on that group's motivation, intention, and resources. Therefore, assuming that most terrorist groups want to do serious damage in either real or symbolic terms, an attack on remote pieces of pipe or a station is less likely even though they are easier to execute. However, contrary to this presumption, history has shown that remote stations due to their high level of exposure are targeted more than often, such as in the case of the 2008–2009 British Columbia pipeline bombings. Therefore, owners and operators must remain vigilante at all times given the reality and credibility of the threat and hazards present.

Canada's oils sands are thought to hold between 175 and 300 billion recoverable barrels, based on current technologies and processes. This is over four times the total of North America's conventional crude oil. The oil sands spread across 77,000 km² of northern Alberta in the Western Canada sedimentary basin, located in four major deposits. The size of these reserves makes oil sands the largest hydrocarbons in the world. Although the major reserves are in Alberta, there are also Canadian offshore reserves. These reserves are not as vast as the oil sands; however, they are still vital to the economy and are an attractive target due to their high visibility. British Petroleum's 2010 oil disaster is an example of the vulnerability and risks posed by offshore oil drilling.

In Canada, the majority of oil and gas production takes place in the province of Alberta. Collection takes place in Edmonton, Alberta, and is delivered to refineries through three major pipelines: Enbridge, Trans Mountain, and Express. The largest is Enbridge, which stretches from Edmonton to Montreal, 37,000 km; Trans Mountain stretches from Edmonton to Vancouver, 1250 km; and Express pipeline runs from Hardisty, Alberta, to Casper, Wyoming. Canada is also a major producer of natural gas; it ranks as number 20 in the world for natural gas reserves with a proven reserve volume of 1.754 trillion cubic meters, and is the third largest producer of natural gas in the world. The majority of homes in Canada use natural gas as the primary source of heating. The major gas-producing areas in Canada are Alberta, British Columbia, and Saskatchewan, with 83% from Alberta, 13% from British Columbia, and 4% from Saskatchewan. Natural gas is collected and delivered through an interconnected series of pipelines that run through Canada and the United States (Figure 3.14).

FIGURE 3.14 Natural gas pipelines.

Oil and gas pipelines are considered to be vulnerable due to their extreme flammability, exposure, and remoteness. This level of exposure makes it a challenge to provide a reasonable degree of protection and security to the pipelines. In Canada and the United States companies are currently using real-time satellite imagery to assist in monitoring the lines. The effects of a disaster on the oil and gas sector are dependent upon the location of the disruption. The concentration of people and assets ranges from very remote unpopulated areas with a single pipeline to heavily populated urban centers with a high concentration of assets such as the City of Edmonton. The level of disruption to service delivery and public confidence that would follow is dependent on the size, location, and duration of the event. For example, an event in the City of Edmonton could conceivably cripple the Canadian oil and gas sector, disrupting service to millions and shocking the public sense of security.

Electricity plays a vital role within the national economy and is necessary for every business and household. The major source of electricity generation is hydroelectric, with Canada and the United States both being world leaders in hydroelectric power generation. Other sources of electricity generation include, coal, natural gas, oil, and nuclear. Threats to the electricity sector include extreme weather conditions, accidents, and equipment failures in addition to deliberate attacks.

Those components of the electricity sector that are most vulnerable include regional transmission systems; these are the power grids that carry high-voltage power to the consumers. These systems include high-voltage transmission lines, substations, and transformers that "step up" current for transmission and "step down" current onto the low-voltage lines necessary for distribution to the consumers. What makes these systems vulnerable is that there are very few high-voltage transformers held in excess inventory and if disrupted or damaged can take months to replace, and are difficult to transport because they can weigh in excess of 400,000 lbs.

Generation facilities such as hydroelectric dams and nuclear plants are generally not a major source of vulnerability to electricity supplies due to the number of generators and the excess generation capacity produced. Similarly, low-voltage distribution systems, although extremely exposed to attack, are not likely targets because remedying the failure of these lines is well within the industry's experience. It has been suggested that if a dam were breached or destroyed by a terrorist attack there could be significant economic damages and loss of life due to flooding, making dams and hydroelectric facilities a high-value target for terrorist groups. Historically, however, dams have been targeted only during times of conflict between states such as World War II and the Balkan Wars. In addition, dams are designed for sustained loads, whereas explosives or blast effects are instantaneous loads. Therefore, if a terrorist group wanted to breach a dam in order to cause significant damage as a result of flooding, then an excessive amount of explosive material would be required and a thorough knowledge of demolitions to carry out a successful attack.

Nuclear power is a growing and major element of energy infrastructure. There are 17 nuclear reactors operating in Canada and 104 reactors in the United States producing 12% and 19.2% of Canada's and the United States' electricity, respectively. In Canada, uranium production is roughly 10,455 tU (tonnes of uranium) and producer shipments are about 9,906 tU, one-third of the total global production. In the United States, uranium reserves are the fourth largest in the world. All uranium mining in Canada takes place in northern Saskatchewan. In Canada and the United States nuclear power plants employ numerous internal safeguards and are regulated by the Canadian Nuclear Safety Commission and U.S. Department of Energy.

Terrorists groups have indicated an interest in attacking both the U.S. and Canadian nuclear power plants. A successful attack on a reactor would be of high value, high damage, and high visibility for terrorist groups. There is a high concentration of people and assets near nuclear power generating stations as a result, and nuclear plants are relatively easy to protect against a terrorist attack. Civilian nuclear reactors are designed so that the core is contained in a heavy concrete (3.5′–6′-thick) structure, which would afford protection during a physical assault. In addition, these plants employ numerous internal safeguards including electric power pumps and redundant systems, secure perimeters, background checks of employees, and specially trained nuclear response forces.

Other systems pose a risk to the safety of these plants aside from a breach of the reactor core. These include electricity supplies, circulation pumps, the intakes for cooling water, and other piping. Problems in any of those areas could cause temperatures to rise excessively or lead to excessive steam pressure.

The concrete surrounding the reactor core should provide adequate protection against the damage caused by a large airplane flying into it; however, there is conflicting information about how well a reactor core or spent fuel storage could withstand such an attack. The Electric Power Research Institute, through computer modeling, determined that a wide-bodied aircraft striking a nuclear plant would not breach the concrete containment structure for the reactor. By way of contrast, the Nuclear Control Institute reported that a Boeing 767 could penetrate at a minimum 3 ft of reinforced concrete at a full cruising speed. The typical reactor core containment structure is between 3.5 and 6 ft thick, making a breach of the reactor core possible. In turn, civilian reactor cores were not designed with the modern threat posed by terrorism in mind.

It was mentioned earlier that electricity is used by every business and household, and therefore any disruption in electricity service could have widespread impacts. As a result, any loss in electricity would result in a loss of essential services such as health services and water supply. In addition, there is also the potential for indirect losses such as loss of confidence among the public toward utility providers and the psychosocial factor or emotional hardship on the public when service is interrupted.

The economic losses associated with energy infrastructure when service is interrupted include increased costs for security, costs of repair, consequential loss due to supply interruption, and decreased production. Politically, there would be repercussions domestically and in foreign relations. For example, if one were to examine the trading relationship between Canada and the United States, much of Canada's energy resources are shared with the United States, such as the National Electricity Grid, which is North America–wide, and the United States is heavily dependent upon Canadian energy resources, such as, oil, gas, and uranium. Therefore, any impact on Canadian energy resources would directly impact the United States. The results of this impact could be considered a "one-off" incident or it could potentially lead to a general mistrust and loss of confidence for the United States toward its largest trading partner.

In order to improve the overall resilience of a nation's energy infrastructure there needs to be greater cooperation and information sharing among government departments and industry, improved resiliency through the implementation of N-2 or N-3 standards, and investments in distributed energy resources. Increased information sharing and cooperation between private industry and government is a necessity for improving energy infrastructure resiliency. However, there are concerns on how information will be protected and stored once it is disclosed. In order for this partnership to be successful, guarantees that information will be stored and used in the most secure way need to be put in place, which involves the implementation of N-2 or N-3 resiliency standards in densely populated areas and insulating the energy grid with DC rather than AC lines. By investing in distributed energy resources, small-scale power generation sources located close to where electricity is used, such as homes and businesses, lead to increased end-user efficiency. Distributed energy resources would also include renewable energy resources such as wind systems or turbines, but for the most part they include distributed generation (any technology that produces power outside the utility grid, such as fuel cells or microturbines) and distributed power (any technology that stores power or produces power, such as batteries and fly wheels).

Some of the primary applications for distributed energy resources include premium power, backup power, peak shaving, low-cost energy, and combined heat and power. Premium power involves reduced frequency variations, voltage transients, surges, dips, or other disruptions. Backup power is used in the event of an outage, as a backup to the electric grid. Peak shaving involves the use of distributed energy resources during times when electric use and demand charges are high. Low-cost energy involves the use of distributed energy resources as base load or primary power that is less expensive to produce locally than it is to purchase from the electric utility. Combined heat and power (cogeneration) increases the efficiency of onsite power generation by using the waste heat for existing thermal process. Distributed energy resources provide the consumer with greater reliability, adequate power quality, and the possibility to participate in competitive electric power markets. They also have the potential to mitigate overloaded transmission lines, control price fluctuations, strengthen energy security, and provide greater stability to the electricity grid.

All of these investments when combined will lead to a more resilient network. The government's response to heightened threat levels has ranged from information sharing to emergency preparedness and disaster mitigation strategies. Not all elements of critical energy infrastructure share the same risks, and, therefore, the consequences of an event will vary based on size, location, and duration of the event.

3.5 FOOD AND AGRICULTURAL SECURITY

Food and agricultural security is one of the 18 critical infrastructure sectors established under the authority of Homeland Security Presidential Directive 7 (HSPD-7) in the United States. It is also one of the 10 critical infrastructure sectors established in Canada by Public Safety Canada. In the United States, the Food and Agriculture (FA) sector has the capacity to feed and clothe people well beyond the boundaries of the nation. The sector is almost entirely under private ownership and consists of an estimated 2.2 million farms, 900,000 restaurants, and more than 400,000 registered food manufacturing, processing, and storage facilities. This sector accounts for roughly one-fifth of the nation's economic activity. It is coordinated at the federal level by the U.S. Department of Agriculture (USDA) and the Department of Health and Human Services' (HHS) Food and Drug Administration (FDA).

The USDA is a diverse and complex organization with programs that touch the lives of all Americans every day. More than 100,000 employees deliver more than $75 billion in public services through USDA's more than 300 programs worldwide, leveraging an extensive network of federal, state, and local cooperators. One of USDA's key roles is to ensure that the nation's food and fiber needs are met. USDA is also responsible for ensuring that the nation's commercial supply of meat, poultry, and egg products is safe, as well as protecting and promoting U.S. agricultural health.

The FDA is responsible for the safety of 80% of the food consumed in the United States. FDA's mission is to protect and promote public health. That responsibility is shared with federal, state, and local agencies, regulated industry, academia, health providers, and consumers. FDA regulates $240 billion of domestic food and $15 billion of imported food. In addition, roughly 600,000 restaurants and institutional food service providers, an estimated 235,000 grocery stores, and other food outlets are regulated by state and local authorities that receive guidance and other technical assistance from the FDA.

In Canada, Agri-food and Agriculture is the lead federal government department with respect to food safety and security within Canada. Other agencies and departments include Health Canada and the Canadian Food Inspection Agency (CFIA). Health Canada works with governments (federal, provincial, and municipal) and government departments, industry, and consumers to establish policies, regulations, and standards related to the safety and nutritional quality of all food sold in Canada. Health Canada is responsible for assessing the CFIA's activities related to food safety. The Department also evaluates the safety of veterinary drugs used in food-producing animals. The CFIA is responsible for enforcing the food safety policies and standards that Health Canada sets. The passage of the *Safe Food for Canadians Act* in November 2012 set the stage for important changes to Canada's food safety system. The act will come into force at the beginning of 2015. The CFIA will work with consumer groups and industry to develop new regulations to support the *act*. During this period, the CFIA will also launch a number of significant food safety enhancements. The end result will be better protection for Canadian families from risks to food safety. Canada's food supply is already among the world's safest. The "Safe Food for Canadians Action Plan" focuses on continuous improvement based on science, global trends, and best practices including stronger food safety rules, more effective inspection, a commitment to service, and more information for consumers.

The food and agricultural sector depends on the water sector for clean irrigation and processed water; the transportation systems sector for movement of commodities, products, and livestock; the energy sector to power the equipment needed for agriculture production and food processing; and the banking and finance, chemical, dams, and other sectors.

Food supply is vulnerable to numerous types of threats and hazards, and because of the size and number of farms, including storage facilities, manufacturing, and processing facilities, it is very difficult to secure all elements from contamination. The cost is very prohibitive and as a result educating the public on issues of food safety and security becomes a continuous and developmental task.

Some of the more common forms of food-borne contamination include biological infection of livestock, contamination of food processing equipment and/or facilities, contamination of food storage facilities, and contamination of food retail outlets including wholesale distributors, supermarkets, grocery stores, and restaurants. Most food storage facilities have very few safeguards in place and are therefore vulnerable intentional and unintentional hazards. Other hazards that can pose a risk to food safety and security include wildfires, droughts, floods, biological infections and infestations, radiological infections, and operation and maintenance issues such as faulty irrigation systems.

Attacks on a nation's food supply can date as far back as the Roman Empire from when Roman soldiers poured salt on the Carthage Fields to Sherman's march to the sea. Attacking a country's food supply has always been a strategy of warfare. Biological infestation including the spread of disease such as the SARS pandemic outbreak is an example of how quickly a pandemic can spread globally. As a result, further education is needed to make health-care providers and first responders aware of the risks associated with the spread of viruses and the actions needed to lessen these risks.

To better protect ourselves from the risks associated with infestation and the spread of disease, and to ensure an ample and safe food supply, there needs to be increased biomonitoring and surveillance of infectious diseases and greater communication among health-care workers and food inspectors. This may include the implementation of early warning systems. Health-care workers and first responders are the first line of defense during a disease outbreak, and their health and safety is paramount to preventing the spread of disease.

The U.S. Center for Disease Control (CDC) has developed the National Electronic Database Surveillance System (NEDSS), which lays out a sort of meta-standard for both health-care information and information technology standards. All state health departments must be NEDSS-compatible; those organizations that are noncompliant cannot participate in any grant funding opportunities by the CDC. Other CDC programs include "Bio-Sense," which analyzes prediagnostic health data for indications of a disease outbreak. Bio-Sense draws information from lab tests, drug sales, and managed care hotlines, and acts as an early warning system.

Both Canada and the United States are making efforts in protecting their respective flood supplies from various threats and hazards. By having different government departments, industry, and the agricultural community work together it is possible to provide a safe environment and build resilience in our nations' food supplies. Furthermore, frontline workers including health-care providers must remain vigilant and aware of the dangers associated with infestations, thereby building resilience in our overall emergency operations and, therefore, increasing our coping capacities.

3.6   AVIATION SECURITY

There are three major threat areas with respect to aviation security; they include perimeter security, parking area security, and terminal and aircraft security. Perimeter security includes protecting aircraft from surface-to-air missiles or electromagnetic pulse attacks; protecting hangers and aircraft from sabotage or theft; and protecting passengers and aircraft from unauthorized access to terminal by way of the runway or hangers. Parking area security includes protecting the unsecured area of check-in, protecting the parking area from car bombs; and protecting passengers from attack or exposure from biological or incendiary devices. Terminal and aircraft security includes protection of aircraft and passengers from hijacking and/or air piracy, suicide bombings, and biological and/or chemical exposure.

Historically, the focus of aviation security has been on the "flight side," and the industry including its regulators has been largely reactive to the various threats and hazards facing the aviation industry. The threat of suicide bombers had been ignored in policy prior to the events of September 11, 2001. Furthermore, little has been done in keeping up with advances in technology with respect to aviation security.

After the events of September 11, several security measures were implemented to circumvent any such further incidents including in-depth background checks on all employees of the airport and airline; training and certification of airport security screeners; total access control of people within contact with the airport, baggage, food, and aircraft; in-flight security personnel and protection of the cockpit; and international response to acts of unlawful interference. In addition, there is the constant threat of vehicle-borne improvised explosive devices (IEDs) from cars, vans, and trucks. As a result, many have restricted the access of parking near the terminal. There is also a need to make further improvements to airport perimeter fencing. For example, the perimeters need to be moved back and the observation parking areas at the end of the runways need to be removed to limit immediate access to the aircraft as they land and take off, making it more difficult for an assailant to use a handheld missile or rocket.

It has been suggested that one individual with a handheld missile or rocket can stand at the end of a runway outside the perimeter fencing and take down a commercial jetliner relatively easily. Rocket-propelled grenades and shoulder-fired missiles are readily available on the black market around the world. These weapons are being sold and transported by criminal and terrorist organizations around the world. It has been suggested that commercial aircraft be equipped with the appropriate security systems to counter an attack from a surface-to-air missile. Thus far, El Al airlines has equipped all aircraft in their fleet with infrared countermeasure systems called "Flight Guard" to defend them against anti-aircraft missiles, following an attempt to shoot down an Israeli airliner in 2002. To date and as of this writing, El Al is the only airline known to have implemented such a system. Switzerland and other European countries have expressed concern that flares dropped by the Israeli system could cause fires in the vicinity of an airport. Since the September 11 attacks there has been an increase in airside security, and with more stringent security measures enforced to gain access to the flight areas, it would appear obvious that the next type of attack would be as near the secure area as possible and would most probably involve a vehicle IED.

Training and security procedures need to be standardized around the world for the aviation industry. Security audits need to be conducted and sanctions placed upon airports that fail to comply. For example, at many airports around the world there are large amounts of baggage that go uninspected; thus, it would be very easy for someone to leave a bomb in a suitcase in one of the baggage check-in areas of an airport. In addition, flight cargo by "trusted vendors" goes unchecked in most airports; this provides an opportunity for a terrorist to either steal a truck from the vendor or be hired by them and pack something inside. Access gained through vendors gives the terrorist or individual the opportunity to leave weapons, plant bombs, sabotage aircraft, or place a biological agent in the food or air-conditioning vents.

In-flight security measures include the hardening of cockpit doors, the use of in-flight security officers, also referred to as "sky marshals," and the arming of flight crews such as the "Federal Flight Deck Officer" (FFDO) program. Following the September 11 attacks in 2001, the *Arming Pilots against Terrorism Act*, part of the *Homeland Security Act* of 2002 directed the Transportation Security Administration to develop the FFDO program as an additional layer of security. Under this program, flight crew members are deputized Federal Law Enforcement Officers authorized by the Transportation Security Administration (TSA) to use firearms to defend against acts of criminal violence or air piracy undertaken to gain control of their aircraft. A flight crew member may be a pilot, flight engineer, or navigator assigned to the flight. Participants in the program are meant to remain anonymous, and, while armed, are prohibited from sharing their participation except with select personnel on a need-to-know basis. Any pilot or flight engineer employed by a commercial airline is eligible to volunteer for the FFDO program. Program size quickly exceeded TSA expectations after the program was opened for volunteers in early 2003. In December 2003, President George W. Bush signed into law legislation that expanded program eligibility to include cargo pilots and certain other flight crew. FFDOs are Sworn and Deputized Federal Law Enforcement officers commissioned by the Department of Homeland Security and Transportation Security Administration Law Enforcement Division. Officers are trained on the use of firearms, the use of force, legal issues, defensive tactics, the psychology of survival, and program SOPs. Flight crew members participating in the program are not eligible for compensation from the federal government for services provided as a FFDO. The use of armed security personnel on board an aircraft is a last line of defense, and another layer in the overall security apparatus.

Preboard screening devices and methods include X-ray machines to verify the contents of all carry-on as well as metal detectors, explosive trace detection (ETD) equipment, and random physical searches of passengers at the preboard screening points. X-ray machines, computed tomography X-ray machines, high-resolution X-rays, and ETDs are also used to scan checked bags. All checked baggage is always X-rayed at major commercial airports. Other systems that have been implemented include full body scanners and searches for randomly selected passengers. Increased vigilance in the screening of passengers is only a short-term cure. As we harden the ability to access the aircraft, the terrorist will modify how they obtain weapons by having members hired into security or maintenance positions.

Most airports were built before aviation security was such an issue. Airports were designed to give easy access and to move people quickly to their destinations. This design also makes it easy for an assailant to gain quick access. For instance, the searching of vehicles entering the airport area is almost an impossible job when attempting to accommodate the time schedules of the passengers. Figure 3.15 is an

FIGURE 3.15   Typical terminal layout.

example of a typical airport terminal layout. Although dramatic, such a structure can pose a number of safety and security concerns. Large open areas with lots of glass leave very little area for a person to take cover, with the glass functioning as projectiles. There are also few support beams, making the structure vulnerable to collapse. Most international airports have some type of rail transit system running into the basement of the terminal. These could be easily boarded by a number of suicide bombers all carrying suitcases of explosives to be detonated under the airport.

The purpose of terrorism is to instill fear; it does not take a lot to frighten the public into not flying. Even the threat of a terrorist attack on an airline causes cancellations. An aircraft that is blown out of the sky by a surface-to-air missile while on a flight plan that is proceeding toward a well-populated area could cause thousands of casualties besides shutting down the city. With the hardening of one area or target, the terrorist will simply change their focus to softer targets. If they cannot get on the aircraft, they will find other ways to bring it down by either a missile or an electromagnetic pulse weapon. Parking areas within the overall airport site layout could be used to park car bombs. They can also be used as an area to launch handheld surface-to-air missiles.

There are major construction projects going on at nearly every major airport globally to meet customer demands and to satisfy international guidelines. Construction makes it more difficult to control traffic. Construction sites provide assailants places to hide in order to commit acts of terrorism.

The fact that there have not been any major terrorist attacks at airports is not because of intense security measures. Despite all efforts, most of the security vulnerabilities are still in existence today. With governments being forced to cut spending, the cost of security is forced back onto the air carriers, which are then passed along to the consumer. Acts of terrorism can have an effect on the global or national economy. The airline industry makes itself a logical target because it is there, it has to be accessible, and it is essential to our being able to conduct business. Major improvements have been implemented in aviation security since September 11, but there is still additional work to be done. One hundred percent security in aviation is impossible as long as there is commitment to die on the part of the attacker. The best that can be achieved is to remove the easy options for an attack and then prepare our responders to counter the more difficult scenarios.

3.7 MARITIME SECURITY AND ASSET PROTECTION

Shipping and a nation's ports are vital for both the economy and its military. Nonetheless, shipping, ports, coastal facilities, and shipping containers remain highly vulnerable to terrorist attacks. As with aviation, the maritime system could be targeted directly or used as a weapon or delivery system for attacks. Maritime security is concerned with the prevention of intentional damage through sabotage, subversion, or terrorism. Maritime security is one of the three basic roles of the United States Coast Guard, which has gradually developed in response to a series of catastrophic events, which began in 1917. There are three main maritime security activities conducted by the Coast Guard: port security, vessel security, and facility security.

The following principal laws support the mission of the United States Coast Guard:

- *Espionage Act of 1917.* This act empowered the Coast Guard to make regulations to prevent damage to harbors and vessels during national security emergencies.
- *Magnuson Act, 1950.* Enacted as a result of the "Red Scare," this act provided permanent port security regulations and broad powers to search vessels in U.S. waters and control the movement of foreign vessels in U.S. ports.
- *Ports and Waterways Safety Act, 1972.* Resulting from several major groundings and oil spills, this act provided port safety authority beyond the Magnuson Act to protect the use of port transportation facilities and to enhance efforts against the degradation of the marine environment.
- *Maritime Transportation Security Act of 2002 (MTSA).* Enacted as a result of the September 11, 2001, terrorist attacks on the United States. This act provided sweeping new authorities for preventing acts of terrorism within the U.S. maritime domain.
- *The International Ship and Port Facility Security (ISPS) Code, 2002.* Adopted by the International Maritime Organization as new provisions to the International Convention for the Safety of Life at Sea to enhance maritime security.

The port security requirements found in the MTSA require security measures for U.S. ports in order to reduce the risks and to mitigate the results of an act that threatens the security of personnel, facilities, vessels, and the public. The regulations draw together assets within port boundaries to provide a framework to communicate, identify risks, and coordinate resources to mitigate threats and consequences. The captain of the port (COTP) must ensure that the total port security posture is accurately assessed and that security resources are appropriate to meet these programs. The COTP must identify critical assets within a port, develop a prioritized list of those most susceptible to acts of sabotage, and plan for adequate security measures to meet specific needs.

Both MTSA and the ISPS Code regulate vessel security. The regulations within these two documents require the owners or operators of vessels to designate security officers for vessels, develop security plans based on security assessments, implement security measures specific to the vessel's operation, and comply with current marine security levels.

A facility is defined as any structure or facility of any kind located in, on, under, or adjacent to any waters subject to the jurisdiction of the United States and used, operated, or maintained by a public or private entity, including any contiguous or adjoining property under common ownership or operation. Some examples of facilities are barge fleeting facilities, container terminals, oil storage facilities, and passenger vessel terminals.

Outer continental shelf (OCS) facilities are generally offshore fixed platforms in water depths ranging up to 1000 ft deep whose primary purpose is exploration, development, and/or production of offshore petroleum reserves. This definition also includes a novel floating design such as tension leg platforms (TLP), floating production facilities (converted MODUs), and floating production storage offloading units (FPSO).

Both MTSA and the ISPS Code regulate facility security. The regulations within these two documents require the owners or operators of facilities to designate security officers for facilities, develop security plans based on security assessments, implement security measures specific to the facility's operation, and comply with current marine security levels. Those facilities designated as OCS facilities must meet the same security requirements as those designated as waterfront facilities.

When U.S. Navy merchant vessels are in dangerous waters, security detachments are posted on the vessel. Security forces have helped deter piracy as well as terrorist attacks, such as the Maersk Alabama and the USS Cole. U.S. Navy merchant vessels normally train the deck department in firearms training, but the added Navy security detail provides for extra security. Additionally, Navy escorts might sometimes accompany the vessels, such as traveling through the Straits of Gibraltar.

Despite these efforts, maritime security currently remains wholly inadequate. Multiple vulnerabilities remain that could easily be exploited by terrorists to attack vessels, port facilities, or coastal infrastructures or to smuggle WMD or explosives into the country to be detonated at a later time. The extensive size, open accessibility, and metropolitan location of most ports ensure a free flow of trade, but these factors also make monitoring and controlling of traffic through the ports virtually impossible.

Ships, such as oil tankers, warships, cargo ships, or cruise liners, are vulnerable because they are large, slow-moving, and difficult to maneuver. They also have high strategic value. Risk assessments have shown that some of the largest ports (and port facilities) also remain vulnerable. Fences, security patrols, and access control systems are often inadequate to prevent determined attackers from entering the port perimeter. Lack of patrol boats and law enforcement officers leaves many critical port facilities largely unprotected from land or waterborne attack. Fuel and chemical storage depots, sometimes containing extremely hazardous materials, are often above ground and could be targeted by attackers from land or sea. The ports can vary in size, functions, and capabilities, and each one presents a unique set of vulnerabilities.

Critical infrastructures located on the coast or on inland waterways also remain vulnerable to waterborne attacks. Bridges, power plants, or oil refineries have few protections against approaching ships carrying explosives or other harmful materials. Similarly, coastal cities, such as Halifax, New York, and Los Angeles, have few safeguards against aggressors from the sea, except for some harbor patrols.

Despite the implementation of physical security measures, cargo containers still remain vulnerable and could be used by terrorists to introduce explosives or WMD into the country. Around 90% of the world's cargo moves by container. It is unclear exactly what percentage of these containers is actually checked upon entry into the United States or Canada, estimates range between 2% and 10%. That leaves at least 90% of all containers entering the country unchecked, and documentation requirements for containers are still easy to circumvent. Customs now receives data on shipments 96 hours before they are supposed to reach the U.S. or Canadian border, but there are no controls over what happens to the containers en route. Container routes are so complex that each individual container may pass through dozens of points before reaching its final destination. Containers are vulnerable to tampering at each of these points as most containers currently have little more than a cardboard seal to verify the integrity of the contents. Materials in the container can potentially be tampered with not only by the manufacturer or supplier of the material being shipped, but also by the carriers who are responsible for shipping the material, and the personnel who load containers onto ships, trains, and trucks. Exporters who make arrangements for shipping and loading, freight consolidators who package disparate shipments into containers, and forwarders who process the information about what is being loaded onto ships are among the many other people who interact with the cargo or have access to the records of the goods being shipped. Furthermore, the automation of the maritime transportation system has increased the vulnerability of the sector by introducing more opportunities for terrorists to access important shipping data. Terrorists can tap into inadequately protected information networks that provide real-time traffic information and purchasing data, which can be used to identify cargo in transit.

Terrorists could hijack a ship en route or they could register a ship in "flag of convenience" nations, which often ask for almost no information from shipping firms that "flag" their vessels with them, and use it for terrorist activities; or they could purchase a legitimate shipping company and its vessels to carry out acts of terrorism without coming under suspicion. These ships could be loaded with explosives and crashed into other vessels, port facilities, critical infrastructures, or population centers on the coast. Alternatively, oil tankers or vessels carrying hazardous materials could be used as terrorist weapons. The types of vessels mentioned above, major ports, coastal oil depots, power stations, harbors, or bridges, could be ideal targets for such attacks. Maritime attacks may also involve the use of small underwater craft, such as small submarines or underwater motor-propelled sleds for divers; some terrorist groups are known to have experimented with such methods.

The most frightening threat to maritime security involves terrorists smuggling explosives or WMD using cargo containers. The weapon could be detonated upon arrival at the port or at any strategic point along the container's route. Targets could include strategic transportation nodes, symbolic landmarks, or large population centers. In addition to death and destruction, any such attack using WMD would undoubtedly have a shocking effect on the psychosocial of a nation. It would also bring international cargo transports and with it the global economy to a halt until other containers could be searched for further weapons. For example, in a war game scenario in October 2002, two dirty bombs were discovered in Los Angeles and Minneapolis before they could be detonated. Despite the fact that no actual damage had been caused, every port in the United States was closed, costing the economy $58 billion. The Brookings Institution estimated that a WMD shipped by container or mail could cause damage and disruption, costing the economy as much as $1 trillion.

To reach an acceptable level of security, the maritime sector will require the development and implementation of a comprehensive security strategy, as well as sustained funding to support that strategy. A crucial first step will be to conduct comprehensive vulnerability assessments at a nation's seaports. Despite the vital role ports play in linking the different nations of the world, both economically and militarily, ports often do not understand the threats facing them because they do not have access to intelligence data. Therefore, they are unable to match threats with vulnerabilities to determine high-risk targets. Out of necessity, there will undoubtedly be a window of vulnerability, but more must be done to analyze vulnerabilities and create mechanisms to help share intelligence reports. In addition, funding must be provided to allow the ports and security units to adopt security measures. After the September 11 attacks, large amounts of funding were infused into the aviation industry, similar funding is necessary for the maritime sector. The funding would be used to purchase new technologies and equipment; improve security; hire, recruit, and/or train new and existing staff; and modernize surface fleets.

Another key area is planning and coordination. In order to prevent and deter future attacks, and to be able to respond effectively in case of a terrorist event, committees have been set up to improve the coordination of efforts among federal, state/provincial, local, and private law enforcement agencies. These agencies get together to discuss, coordinate, and rehearse potential terrorist scenarios. Training exercises and simulations help to establish who is in charge and what roles each entity will play.

Ports would be safer if potentially dangerous vessels could be intercepted before they reach a nation's coast. Therefore, coastal surveillance and the protection of ports and coastal facilities are necessary. Port perimeter security and access control measures need to be enhanced. Strong maritime security will require the smart adoption of new tools and technologies in the fight against terrorism. These would include computer software programs and databases to help sift through information on vessels, crews, and cargo in search of anomalies or to match seamen or port employees against terrorist watch lists. Information technology resources and global positioning system (GPS) technology can also help track the location of ships and cargo in transit or in near-real time. New technologies, particularly scanners and scanning tools for explosives and WMD, should be applied to the pressing issue of cargo security. As they become available, these new devices should be brought into service to protect the most strategic ports first as it will initially not be possible to scan all containers. Finally, electronic seals should be installed on all containers, thereby preventing them from being tampered with while en route. The different container security initiatives offer hope that technology can help secure containers from the moment they are filled with goods, throughout their often complex journey, to the final destination where they are unloaded.

3.8   LAND TRANSPORTATION SECURITY SYSTEMS

Land transportation consists of a diverse array of long distance passenger and freight vehicles and facilities, mass transit vehicles and facilities, and related infrastructures. These include trains, trams, subways, buses, trucks, railway lines, highways, stations, depots, control facilities, bridges, and tunnels. This complex network of systems is absolutely essential for everyday life and the national economy. The land transportation systems are the most difficult to protect. By their very nature, land transportation systems need to be open and accessible, and have thousands of entry points. Numerous key assets are distributed over a wide area, and some routes are static. All these factors mean that fully protecting land transportation is impossible. However, the recognition that a certain amount of uncertainty and risk is unavoidable should not be cause for resignation. By learning from past tactics and collecting intelligence on future threats, it is possible to apply security measures that help to detect and deter attacks; raise the likelihood that terrorists will be apprehended; and improve emergency and consequence management to limit casualties and disruptions. Dual-use security measures that reduce crime or bolster passenger safety or comfort are also attractive for land transportation.

Many land transportation and mass transit operators have improved planning and coordination efforts with law enforcement, emergency managers, first responders, and other organizations. Terrorism prevention, rapid response, and successful crisis management require cooperation between the multiple entities and jurisdictions often affected by an attack on surface transportation. Coordination with local, state, and federal authorities can also open channels of communication that can be used to share threat information. Constant training and drills, including simulations and field exercises, are extremely important to ensure readiness, test plans, and identify potential problems. Since September 11, 2001, security drills have focused increasingly on hijacking and hostage taking, chemical and biological weapons attacks, bombs, and other terrorist actions.

As with aviation and maritime security sectors, initial efforts in the land transportation sector have placed heightened emphasis on improved physical security and access control to transportation assets, such as vehicles, tracks, switching and control facilities, freight areas, and infrastructures. Due to the massive number and wide dispersal of assets, only the most critical assets are secured. Perimeter security measures include additional fences and barriers, surveillance cameras, intrusion detection systems, and increased lighting. Access to some areas, such as vehicular access to some train stations, has been restricted. In addition, employee access to critical facilities, including transit control centers, dispatch, and bus storage facilities, has also been restricted through the use of employee identification cards. As an additional security measure, all kinds of land transportation and mass transit organizations have employed additional security staff including a more visible presence at land transportation facilities to deter and prevent terrorist attacks. Law enforcement agencies have also deployed additional officers to land transportation facilities and increased patrols at stations.

To supplement security personnel, land transportation and mass transit operators have sought to train and educate staff concerning threats in order to increase awareness and vigilance. Public awareness campaigns are being employed to get travelers involved in their own protection. Passengers are being asked to be on the lookout for suspicious behavior or packages in stations and on vehicles, and to notify authorities immediately. While such campaigns extend the eyes and ears of security personnel, they may also foster anxiety among commuters and lead to false alarms.

Efforts to secure land transportation have only minimally reduced the risk of a potential attack. Vulnerabilities remain in the land transportation sector and cannot all be eliminated. Land transportation systems offer a concentration of people and a variety of targets for an assailant; in addition, the anonymity and easy escape routes that land transportation systems offer an attacker further increases their vulnerability. Different kinds of land transportation systems have different vulnerabilities, and all are linked in a complex network of interrelated services, which society and the economy dependent on. Land transportation and mass transit systems must remain accessible, convenient, and inexpensive due to their accessibility, since they cannot implement the security checks that are common in the aviation and maritime sectors.

Land transportation and mass transit systems infrastructure includes railways, highways, bridges, and tunnels. Each of these elements remains at risk because they span the entire country and cannot be completely secured. Railway lines and highways often pass through remote areas and it is impossible to totally restrict access to them. Fencing and other security measures are in place to protect critical transportation nodes, but they still remain vulnerable. Severing key rail or road arteries would probably not cause substantial loss of life, but could result in severe economic disruption. Mass transit tracks are less vulnerable because trains pass over them more frequently and they are usually in populated areas.

Numerous bridges and tunnels exist across the United States and Canada, which remain vulnerable to potential attack. There are few physical safeguards to protect these infrastructures from being damaged or destroyed; however, it is important to note that actually destroying a bridge or tunnel requires extensive planning and engineering knowledge. Key bridges and tunnels exist that have a heightened strategic or symbolic value, for example, five major bridges and one tunnel that link Ontario to Michigan and New York account for 70% of all the trade between the United States and Canada. High-risk targets such as these must be identified and offered particular protection through enhanced perimeter security, additional security personnel, and electronic monitoring systems. Attacks on transportation infrastructures could result in loss of lives and serious financial losses, and thereby weaken the morale of society. Natural disasters have demonstrated the resiliency of land transportation systems. Service on damaged parts of the transportation system is usually restored rapidly and disruptions are held to a minimum. However, the system remains vulnerable to a determined attack against several key infrastructures or nodes.

Stations and vehicles also remain vulnerable. Despite improved physical security measures and a heightened police presence, access to and movement within and between subway, bus, tram, and rail systems is free and unrestricted. Baggage and cargo undergoes little or no security checking. This makes the system vulnerable to the introduction of explosive devices, or, worse, WMD. Cars and trucks face no obstacles, freely traveling on a nation's highway network, and can enter major cities and population centers largely unchallenged. Freight and hazardous materials that are transported around the country are also vulnerable to manipulation at numerous points where they are stored, loaded and unloaded, or simply pass through.

Historically, terrorist attacks on land transportation systems have been quite common. In fact, one-third of all terrorist attacks worldwide target land transportation systems, with the weapon of choice usually being explosives. Attacks are relatively evenly split between rail systems (trains, subways, stations, and rails) and bus systems. For example, in 1995 and 1996, Algerian terrorists set off several bombs in the Paris subway system; Palestinians have carried out numerous suicide bombings of buses in Israel; Irish Republican Army (IRA) bombers have blown up various railway and subway vehicles and infrastructures during their long-running terrorist campaign in the United Kingdom; bombing of double-decker buses in London 2005, Madrid train station bombings in 2007; and Aum Shinrikyo's sarin gas attack on Tokyo's subway in 1995 raised the bar by introducing chemical weapons into the terrorist arsenal.

Thousands of hazardous material transports are conducted by rail and road daily. If not properly protected, these materials could be detonated at government buildings, at landmarks, in populated areas, or at strategically important points. News reports in the United States indicate that corrupt individuals at state motor vehicle departments have illicitly sold hazardous materials permits to people with Arabic surnames. A variety of measures may be implemented to address this threat. Hazardous material transporters could be required to develop new notification procedures to provide authorities with advance warning that chemicals or other materials will be passing through their jurisdiction. Debate has also focused on requiring background checks for individuals transporting hazardous materials and developing new identification systems for drivers, possibly using biometrics technology. Some experts have suggested removing hazardous materials warning signs from vehicles because they could help terrorists with target selection, but this may endanger first responders at regular accident sites. Other suggestions include restricting access for trucks carrying hazardous materials to particularly sensitive sites, and improving security at storage and transfer depots. Hazardous material transports could be equipped with GPS systems to monitor the location of dangerous vehicles and freight in real time. If trucks carrying hazardous materials deviate from their preassigned route, technologies could be installed that allow for the engine to be disabled remotely to prevent a catastrophe. Such technologies are already under development and in experimental use.

Better land transportation and mass transit security also requires making some changes to vehicles and stations that could minimize their vulnerability to attacks. Such measures, which have been successful in the past in other countries, include removing trash cans or replacing them with blast-proof containers, removing materials that easily fragment or give off toxic fumes from vehicles and stations, improving lighting and lines of sight, installing surveillance cameras, and locking storage and maintenance rooms. Other measures necessitate thinking ahead and developing safer stations and vehicles. An example for this would be building a subway station with good lighting, blast-resistant structures, air filtration systems, emergency evacuation routes, and open spaces that provide broad fields of vision. Better design will also produce additional benefits, such as reducing crime, enhancing efficiency, and improving customer comfort.

New tools and technologies should also be utilized to improve the security of land transportation and mass transit systems. Further research and development activity is crucial in this area. Hazardous material transports could carry GPS tracking technology and panic button and automatic engine-kill-switch technologies that could help halt truck hijackings in progress. In fact, GPS and other monitoring technologies could be used more generally to check the location of passenger and freight vehicles. Smart card or bar code technologies could be adopted for cargo containers or hazardous material transports that could be used to check that shipments are in order. This could help identify and screen high-risk transports, and prevent them from entering sensitive areas.

Due to its inherent vulnerability, the land transportation sector will have to think outside the box to protect its facilities and passengers. While elaborate security checks involving screening devices or manual searches are impossible to implement at all access points, random checks in stations or upon entering vehicles may act as a significant deterrent. Using concealed detection and scanning devices at some locations may have a similar effect. Experts have also floated different kinds of passenger screening techniques for long-distance travelers that purchase their tickets in advance. This could involve entering passenger information into a database and cross-referencing it with a variety of government databases. However, such measures could be easily circumvented. Random passenger identity checks may be more effective. Marshals have also been proposed for land transportation systems and may have a place in a comprehensive security strategy.

Preventing attacks on land transportation and mass transit systems is extremely difficult. Better physical security, vigilance, and an increased security presence will not prevent all possible attacks. Therefore, security strategies must focus on minimizing loss of life and disruption and rapid recovery. Such successful consequence management will require excellent planning, coordination, and training; sustained security funding; safer station and vehicle designs; the application of new tools and technologies; and creative thinking.

3.9 CYBERSECURITY

Cyberattacks to an organization's information technology infrastructure may cause significant damage and disruption to an organization's internal operations but they are not expected to cause any immediate disruptions to its operations. A cyberattack on the control systems and networks can have an immediate detrimental impact on an organization's operations. A prolonged disruption to the information technology infrastructure could also lead to disruptions in deliverable services.

The control systems used to connect and manage infrastructure are different than the traditional information technology systems used for business systems and are often called operational technology (OT). The National Institute of Standards and Technology (NIST) refers to these systems as industrial control systems (ICS). NIST SP 800-53 Rev 3 Appendix I defines ICS as "information systems that differ significantly from traditional administrative, mission support, and scientific data processing information systems." ICS typically have many unique characteristics including a need for real-time response and extremely high availability, predictability, and reliability.

These types of specialized systems are pervasive throughout critical infrastructure, required to meet several and often conflicting safety, operational, performance, reliability, and security requirements such as (1) minimizing risk to the health and safety of the public, (2) preventing serious damage to the environment, (3) preventing serious production stoppages or slowdowns that result in negative impact to the nation's economy and the ability to carry out critical functions, (4) protecting the critical infrastructure from cyberattacks and common human error, and (5) safeguarding against the compromise of proprietary information.

Previously, ICS had little resemblance to traditional information systems in that they were isolated systems running proprietary software and control protocols. However, as these systems have been increasingly integrated more closely into mainstream organizational information systems to promote connectivity, efficiency, and remote access capabilities, portions of these ICS have started to resemble the more traditional information systems. Increasingly, ICS use the same commercially available hardware and software components as are used in the organization's traditional information systems. While the change in ICS architecture supports new information system capabilities, it also provides significantly less isolation from the outside world for these systems, introducing many of the same vulnerabilities that exist in current networked information systems. The result is an even greater need to secure ICS.

The term "ICS" is used in its broadest sense and includes SCADA (energy, water, wastewater, pipeline, airfield lighting, locks, and dams); distributed control systems (process and manufacturing); building control systems and/or building automation systems; utility management control systems; electronic security systems; fire, life safety, and emergency management systems; exterior lighting and messaging systems; and intelligent transportation systems.

The systems and subsystems are a combination of operational technologies and information technologies. The majority of these systems were historically proprietary; analog; vendor supported; used direct serial, and/or wireless connection; and were not IP-enabled. The systems components such as remote terminal units, programmable logic controllers, physical access control, intrusion detection systems, CCTV, fire alarm systems, and utility meters have long equipment life spans, and are typically designated as OT and real property equipment.

As these systems and components became digital and IP-enabled, the interconnections to the organization's network and business systems began to expose the organization to exploits and significant vulnerabilities. Typically, there is no clear line of demarcation where one system starts and one system ends. The Department of Homeland Security Control Systems Security Program is part of the United States Computer Emergency Readiness Team (US-CERT) and provides tools, standards, training, and publications for ICS.

The top 10 vulnerabilities to any computer network are as follows: operator station logged on all the time even when the operator is not present at the workstation, thereby rendering the authentication process useless; relatively easy physical access to the ICS equipment; unprotected ICS network access from remote locations via Digital Subscriber Line (DSL) and/or dial-up modem lines; insecure wireless access points on the network; ICS networks directly or indirectly connected to the Internet; no firewall installed or one that is weak or unverified; system event logs not monitored; intrusion detection system not used; operating and ICS system software patches not routinely applied; network and/or router configuration insecure; and passwords not changed from manufacturers' default.

The following is a 21-point checklist for undertaking a vulnerability assessment of your computer network. The list was developed by Sandia National Laboratories, and was originally developed to be used for SCADA systems; however, it was applied here to include all computer networks.

1. Identify all connections to the network.
2. Remove unnecessary connections to the network.
3. Evaluate and strengthen the security of any remaining connections to the network.
4. Harden networks by removing or disabling unnecessary services.
5. Do not rely on proprietary protocols to protect your system.
6. Implement security features provided by device and system vendors.
7. Establish strong controls in any medium that is used as a back door into the network.
8. Implement internal and external intrusion detection systems and establish 24-hour incident monitoring.
9. Perform technical security audits on devices and networks and other networks.
10. Conduct physical security surveys and assess all remote sites connected to the SCADA network to evaluate their security.
11. Establish "red teams/tiger teams" to identify and evaluate possible attack scenarios.
12. Clearly define cybersecurity roles, responsibilities, and authorities for managers, system administrators, and users.
13. Document network architecture and identify systems that serve critical functions or contain sensitive information that requires several levels of protection.

14. Establish a rigorous ongoing risk management process.
15. Establish a network protection strategy based on the principle of defense in depth.
16. Clearly identify cybersecurity requirements (i.e., standards, procedures).
17. Establish effective configuration management processes.
18. Conduct routine self-assessments.
19. Establish system backups and data recovery plans.
20. Ensure that senior organizational leadership establishes expectations for cybersecurity performance and for holding individuals accountable for their performance.
21. Establish policies and conduct training to minimize the likelihood that organizational personnel will inadvertently disclose sensitive information regarding the system's design, operations, or security controls.

In evaluating an organization's computer network, there are several different attack scenarios that can be rehearsed as part of a red teaming or tiger team exercise. They are as follows: Scenario 1, "inside job," disgruntled or former employee, or employee bribed or duped into sabotaging systems or settings; Scenario 2, systems equipped with modems or Internet-based remote access points that a "war-dialer" or "port-scanning tool" can be used to automate an attack and gain access to the system; Scenario 3, wireless systems and wireless devices have a serious security flaw that compromises the wireless encryption key; Scenario 4, an employee with access to the IT/ICS systems is duped into running an innocuous application by a current or former associate, or virtually anyone on the Internet; and Scenario 5, a cyberattacker can access the network from inside or outside and run a program that starts flooding the network with useless traffic, thus jamming the links and denying service to users and ICS devices causing random disruption of service.

Some of the more popular tools used by cyberattackers to penetrate an organization's computer network include Trojan horse, worms and viruses, scanners, Windows NT hacking tools, ICQ hacking tools, mail bombs, nukes, key loggers, hacker's Swiss knife, password crackers, and BIOS crackers. Current methods for mitigating such attacks include the following: ensure physical security, disable devices not required for the operation, locate equipment in confined areas, limit and monitor physical access to the servers and ICS workstations via doors and access restriction. Access and authentication methods should include the following: change username and passwords from manufacturer's defaults, implement policies and procedures for passwords, and use strong authentication and access methods such as virtual private network (VPN) for remote access. Software improvements include the following: apply software patches routinely; upgrade software in a cyclic manner (3–5 years); install software patches on firewalls, routers, and switches; install virus protection and intrusion detection software; use various software encryption schemes; and install firewalls on all connections between IT/ICS systems, and on wireless and Internet connections. Privacy improvements include the following: consider reducing communications media vulnerabilities by introducing fiber optic or other types of more secure media and use VPN or PKI for remote access. Network topology considerations include the following: remove any network hubs, use VPN for remote access, remove any workstations with dual network interfaces or other unsecured connections to other networks, and provide UPS to all critical components of the ICS network.

What makes cybersystems such an attractive target is that a hacker can attack any system from anywhere in the world, without having to be physically present or near the actual infrastructure. Generally, cyberattacks require little more than a common personal computer, a modem, a telephone line, and a software that is generally available for free on the Internet. There have also been many cases of university and library computers being utilized to conduct attacks. In addition, the Internet serves a communication medium between hackers, thereby allowing them to conceal their identity, making it more difficult to capture them. Message boards and discussion forums can serve as drop sites.

One of the processes currently employed by hackers and criminal organizations to communicate among members and to conceal their actions is through steganography. Steganography is the process of hiding information in objects such as photos, documents, and other types of files. The photo looks the same as the original and one cannot tell that there is a message concealed within it by just looking at it. This can be done with videos, photos, or documents. For example, take the following statement, "After the theater, all clients keep a tab down at Wesley's Nook." If you were to remove the first letter in each word the message would actually read "ATTACK AT DAWN." Steganography is a way of concealing messages inside other media, such as sound files and graphic images. Many freely available steganographic utilities are available on the web. Steganography is extremely difficult, but not impossible to detect. Another method of concealing messages is through the use of cryptography or encryption tools. Encryption tools are available free throughout the Internet. Powerful encryption makes monitoring communications between two parties difficult or impossible in real time. Public key encryption means that passwords never have to be passed between hackers or criminal entities.

When responding to a cyberattack or incident there are four phases in the emergency response process: preparedness, response, mitigation, and recovery. The first phase—preparedness—is all the background work that takes place prior to an incident occurring. Preparedness would include planning, training, and routine practice. Preparedness would also include routine security audits and inspections along with a vulnerability assessment of the organization's cyber network. The next phase is the response phase. The response phase detects, identifies, and assesses the situation or incident, often referred to as triage. Once the problem has been identified the next phase is the mitigation phase. During the mitigation phase the responder(s) contain the incident or situation, collect evidence, perform analyses, and take corrective action (mitigation). The final phase is the recovery phase. During this phase the response team eradicates the problem that caused the incident in the first place, such as a virus or worm, recover system processes, and follow up to verify that the problem has been removed. During the response phase, the incident team must determine which process ICS systems are affected and to place those systems in the LOCAL/MANUAL control mode; isolate the ICS system during containment; and carry out evaluation, containment, and restoration of programmable logic controllers (PLC) and remote terminal units (RTU) programs in addition to those of ICS computers.

In order to ensure continued operations for an organization during a cyberincident, it is important to define the hazard level posed by each incident. Incidents are separated into three levels with 3 being the most severe. Level 1 are disruptions in day-to-day activities but are easily mitigated through readily available alternatives, Level 2 involves more complex solutions requiring staff time to resolve, and Level 3 are incidents directly affecting operations and may have public safety implications. The response times for taking corrective action will vary depending on the incident level. For Level 1, 1 week would be acceptable, whereas Level 3 must be mitigated as quickly as possible.

4 Targeted Violence and Violent Behavior

4.1 INTRODUCTION

Targeted violence is direct action by physical force or power against an individual, state, organization, or institution by another individual(s), state, organization, or institution outside of a battlefield. This definition of targeted violence is similar to the definition of targeted killings. Targeted killings were employed extensively by death squads in El Salvador, Nicaragua, Costa Rica, Colombia, and Haiti within the context of civil unrest and war during the 1980s and 1990s. Targeted killings have been used in Somalia and Rwanda and in the Balkans during the Yugoslav Wars. They have also been used by narcotics traffickers.

The legality of targeted killings is disputed. Some academics, military personnel, and officials describe a targeted killing as legitimate within the context of self-defense, when employed against terrorists or combatants engaged in asymmetrical warfare. They argue that drones are more humane and more accurate than manned vehicles. Others, including academics such as Gregory Johnsen and Charles Schmitz, 26 members of Congress, media sources (Jeremy Scahill, James Traub), human rights groups, and ex–Central Intelligence Agency (CIA) station chief in Islamabad, Robert Grenier, have criticized targeted killings for being similar to assassinations or extrajudicial killings, illegal within the United States and under international law.

The use of targeted killings by conventional military forces became common-place in Israel during and after the Second Intifada, when Israeli security forces used the tactic to kill Palestinian opponents. Though initially opposed by the Bush administration, targeted killings have become a frequent tactic of the U.S. government in the War on Terror. Instances of targeted killings by the United States that have received significant attention include the killing of Osama bin Laden and of U.S. citizen Anwar al-Awlaki in 2011. Under the Obama administration, the use of targeted killings has expanded, most frequently through the use of combat drones operating in Afghanistan, Pakistan, or Yemen.

Proponents of targeted killings argue that "targeted violence" is a completely different term and concept. In studies on violence, targeted violence has been defined as attacks carried out against other individuals or institutions. For the purposes of this chapter, targeted violence and violent behavior refer to acts of violence committed by terrorists and terrorist organizations. The chapter examines some key concepts and subject areas including methods and the mind-set of a terrorist, terrorism in general, organized crime, maritime piracy, incidents and indicators, suspicious activity, and the avoidance of a terrorist attack.

4.2   METHODS AND THE MIND-SET OF A TERRORIST

Thomas Hobbes once wrote that "the most fundamental human right, law of nature, is the right of self-defense," namely, the right to use force as necessary to resist and repel an attacker. Organizations use such statements to promote their own agenda. For example, terrorist organizations have filled manifestos with self-defense arguments. Statements such as the "people have the right to defend themselves against an illegitimate government" are consistently echoed and promoted among group members. In any democratic society, the underlying foundation is that the elected government or governing body represents the people. However, terrorist organizations have consistently identified themselves as the "true" representatives of the people. The question then remains: Who legitimately represents the will of the "people"?

From the terrorist's point of view, armed struggle is a legitimate means for the "people" to attain self-determination under international law. In addition, the war on terrorism is a war against the entity that can legitimately claim to represent the will of the people. Terrorist attacks against civilians are deliberate and violent in nature. To the terrorist, such acts are justified in that civilians are taxpayers who financially support the state and benefit directly from the government's illegitimate policies. Therefore, they are viewed as supporters of injustice and are not considered to be innocent; instead, they are viewed as "collaborators."

Attacks against civilians are used to intimidate people from cooperating with the state in order to undermine state control. This strategy was used in the United States in its War of Independence and in Ireland, Kenya, Algeria, and Cyprus during their independence struggles. Attacks on high-profile symbolic targets are used to incite counterterrorism by the state to polarize the population. This strategy was used by al-Qaeda in its attacks on the United States in September 2001. These attacks are also used to draw international attention to struggles that are otherwise unreported such as the Palestinian airplane hijackings in 1970 and the South Moluccan hostage crisis in the Netherlands in 1975.

Terrorism is all about the "message"; communication and violence are employed in establishing the terrorists' own state of solidarity. Most organizations issue a manifesto or declaration of war to explain their purpose of actions. The commission of a terrorist act is a demonstration that the time for negotiating with a government is over; the focus is on communicating with the "us" community or the terrorist identity group. Members must work hard to establish the identity of the group due to lack of political influence. Abraham suggests that terrorist organizations do not select terrorism for its political effectiveness. Individual terrorists tend to be motivated more by a desire for social solidarity with other members of their organization than by political platforms or strategic objectives, which are often murky and undefined.

Terrorism only occurs due to the lack of political power to advance a cause through peaceful means and the military power to move it forward by force. No one would commit terrorist acts unless they felt there was no other alternative. It is a tactic that "rational" but politically powerless people use to compel "social change."

Over the years, many people have attempted to come up with a terrorist profile to attempt to explain these individuals' actions through their psychology and social circumstances. Others, like Roderick Hindery, have sought to discern profiles in the propaganda tactics used by terrorists. Some security organizations designate these groups as *violent nonstate actors*. A 2007 study by economist Alan B. Krueger found that terrorists were less likely to come from an impoverished background (28% vs. 33%) and more likely to have at least a high-school education (47% vs. 38%). Another analysis found only 16% of terrorists came from impoverished families and over 60% had gone beyond high school, versus 15% of the populace.

To avoid detection, a terrorist will look, dress, and behave normally until the assigned mission is executed. Some claim that attempts to profile terrorists based on personality, physical, or sociological traits are not useful. The physical and behavioral description of the terrorist could describe almost any normal person. However, the majority of terrorist attacks are carried out by military-age men, aged 16–40.

Those who commit acts of terrorism can be individuals, groups, or states. According to some definitions, clandestine or semiclandestine state actors may also carry out terrorist acts outside the framework of a state of war. However, the most common image of terrorism is that it is carried out by small and secretive cells highly motivated to serve a particular cause; many of the most deadly operations in recent times, such as the September 11 attacks, the London underground bombing, and the 2002 Bali bombing, were planned and carried out by a close clique, comprising close friends, family members, and other strong social networks. These groups benefited from the free flow of information and efficient telecommunications to succeed where others had failed.

The dynamics of terrorist activity comprise the following: the terrorist group or the "us"; the sympathizers, who are considered to be part of the identity group, but do not support the cause or identify themselves as part of the community; and the nonidentity group or "them," which represents anyone outside these groups. In other words, if you do not support terrorism or the "us," you support the "them" and, therefore, you are *not* innocent.

The key to understanding terrorist interaction with their identity groups is interpreting the meaning of their public statements. Once baseline support exists with the "us" group, the establishment of influence with sympathizers begins. A number of means are used including a manifesto and actions to influence sympathizers. If the proposed act is criminal in nature, it represents the final break from societal values. Within the struggle for legitimacy, criminality in itself is lethal and, therefore, the group must prove that the cause is more legitimate than the law. There is also a risk of losing support and the cause itself at this stage. The state's ability to brand the terrorist group as criminals will also reduce support from sympathizers. For example, the purpose of a suicide bomber dying is to avoid the legal consequences of their actions.

There is a need for governments to assist in the conversion of supporters. Terrorist acts and the accompanying atrocities force governments to react. Terrorists hope that the reaction is inefficient and that it will impose hardships on the civilian population. The government has an identity group to answer to as well. Failure will result in a group looking for new leaders. The inability to trace terrorist organizations causes the government to blame the identity or community, forcing the government to treat the community with suspicion and causing animosity among "them." This is particularly true for domestic terrorism; while a democratic nation espousing civil liberties may claim a sense of higher moral ground than other regimes, an act of terrorism within such a state may cause a dilemma: whether to maintain its civil liberties and thus risk being perceived as ineffective in dealing with the problem or, alternatively, whether to restrict its civil liberties and thus risk delegitimizing its claim of supporting civil liberties. For this reason, homegrown terrorism is being perceived as a greater threat, as stated by Michael Hayden, the former CIA director. This dilemma, some social theorists would conclude, may very well play into the initial plans of the acting terrorist(s), namely, to delegitimize the state.

There has been public criticism of the mechanisms and implementation of some security initiatives to counterterrorism, with claims that they are being led by the same companies that stand to benefit the most and that they are encouraging an unnecessary culture of fear. It is argued that "the hardening of specific targets is a complete waste of time, the terrorists will simply move on to 'softer' targets." In turn, when the government is unable to stop terrorist attacks, terrorist organizations begin to have influence over the government's identity group or the "them" group. The strategy is to fool the government into destroying itself or imploding and becoming what the terrorists say it already is, that is, illegitimate, through an act of coercion.

4.3   TERRORISM

Terrorism is the systematic use of terror, often violent, especially as a means of coercion. In the international community, however, terrorism has no legally binding criminal law definition. Common definitions of terrorism refer only to those violent acts that are intended to create fear and terror; are perpetrated for a religious, political, or ideological goal; and deliberately target or disregard the safety of noncombatants such as civilians. Some definitions now include acts of unlawful violence and war. The use of similar tactics by criminal organizations for protection rackets or to enforce a code of silence is usually not labeled terrorism, though these same actions may be labeled terrorism when carried out by a politically motivated group. The writer Heinrich Böll and the scholars Raj Desai and Harry Eckstein have suggested that attempts to protect against terrorism may lead to a kind of social oppression.

The word "terrorism" is politically and emotionally charged, and this greatly compounds the difficulty of providing a precise definition. Studies have found over 100 definitions of "terrorism." The concept of terrorism may be controversial as it is often used by state authorities and individuals with access to state support to delegitimize political or other opponents and potentially legitimize the state's own use of armed force against opponents; such use of force may be described as "terror" by opponents of the state. Terrorism has been practiced by a broad array of political organizations to further their objectives. It has been practiced by right-wing and left-wing political parties, nationalistic groups, religious groups, revolutionaries, and ruling governments. An abiding characteristic is the indiscriminate use of violence against noncombatants for the purpose of gaining publicity for a group, cause, or individual. The symbolism of terrorism can leverage human fear to help achieve these goals.

The definition of terrorism has proven controversial. Various legal systems and government agencies use different definitions of terrorism in their national legislation. Moreover, the international community has been slow to formulate a universally agreed upon, legally binding definition of terrorism. These difficulties arise from the fact that the term "terrorism" is politically and emotionally charged. In this regard, Angus Martyn (2002), briefing the Australian Parliament, stated that

> [t]he international community has never succeeded in developing an accepted comprehensive definition of terrorism. During the 1970s and 1980s, the United Nations attempts to define the term floundered mainly due to differences of opinion between various members about the use of violence in the context of conflicts over national liberation and self-determination.

These divergences have made it impossible for the United Nations to conclude a Comprehensive Convention on International Terrorism that incorporates a single, all-encompassing, legally binding, and criminal law definition of terrorism. The international community has adopted a series of sectoral conventions that define and criminalize various types of terrorist activities. Since 1994, the United Nations General Assembly has repeatedly condemned terrorist acts using the following political description of terrorism:

> Criminal acts intended or calculated to provoke a state of terror in the general public, a group of persons or particular persons for political purposes are in any circumstance unjustifiable, whatever the considerations of a political, philosophical, ideological, racial, ethnic, religious or any other nature that may be invoked to justify them.

United Nations 1994

The terms "terrorism" and "terrorist" carry strong negative connotations. They are often used as political labels, to condemn violence or the threat of violence by certain actors as immoral, indiscriminate, and unjustified or to condemn an entire population segment. Those labeled "terrorists" by their opponents rarely identify themselves as such; typically the former use other terms or terms specific to their situation, such as separatist, freedom fighter, liberator, revolutionary, vigilante, militant, paramilitary, guerrilla, rebel, patriot, or any similar-meaning word in other languages and cultures. Jihadi, mujahidin, and fedayeen are similar Arabic words that have entered the English lexicon. It is common for both parties in a conflict to describe and label each other as terrorists.

Philosophers have expressed different views regarding whether particular terrorist acts, such as killing civilians, can be justified as the lesser evil in a particular circumstance: while, according to David Rodin (2006), utilitarian philosophers can (in theory) conceive of cases in which the evil of terrorism is outweighed by the good that could not be achieved in a less morally costly way, in practice the "harmful effects of undermining the convention of non-combatant immunity is thought to outweigh the good that may be achieved by particular acts of terrorism." Among the nonutilitarian philosophers, Michael Walzer (Steinfels 2003) argued that terrorism can be morally justified in only one specific case: when "a nation or community faces the extreme threat of complete destruction and the only way it can preserve itself is by intentionally targeting non-combatants, then it is morally entitled to do so."

The pejorative connotations of the word can be summed up in the aphorism— "One man's terrorist is another man's freedom fighter." This is exemplified when a group using irregular military methods is an ally of a state against a mutual enemy, but later has a fallout with the state and starts to use those very methods against its former ally. During World War II, the Malayan People's Anti-Japanese Army was allied with the British, but during the Malayan Emergency, members of its successor, the Malayan Races Liberation Army, were branded "terrorists" by the British.

During the Soviet invasion of Afghanistan, Ronald Reagan and others in the U.S. administration frequently call the Afghan Mujahideen, "freedom fighters" during the war against the Soviet Union; yet 20 years later, when a new generation of Afghan men were fighting against what they perceived to be a regime installed by foreign powers, their attacks were labeled "terrorism" by George W. Bush. Groups accused of terrorism understandably prefer terms reflecting legitimate military or ideological action. Leading terrorism researcher Professor Martin Rudner, director of the Canadian Centre of Intelligence and Security Studies at Ottawa's Carleton University, defines "terrorist acts" as attacks against civilians for political or other ideological goals, and says:

> There is the famous statement: "One man's terrorist is another man's freedom fighter." But that is grossly misleading. It assesses the validity of the cause when terrorism is an act. One can have a perfectly beautiful cause and yet if one commits terrorist acts, it is terrorism regardless.

Humphreys 2006

Some groups, when involved in a "liberation" struggle, have been called "terrorists" by the Western governments or media. Later, these same persons, as leaders of the liberated nations, are called "statesmen" by similar organizations. Two examples of this are the Nobel Peace Prize laureates Menachem Begin and Nelson Mandela.

Sometimes, states that are close allies, for reasons of history, culture, and politics, can disagree over whether or not members of a certain organization are terrorists. For instance, for many years, some branches of the U.S. government refused to label members of the Irish Republican Army (IRA) as terrorists while the IRA was engaging in acts against one of the United States' closest allies, the United Kingdom; the United Kingdom, however, branded these acts as acts of terrorism. For these and other reasons, media outlets wishing to preserve a reputation for impartiality try to be careful in their use of the term.

In early 1975, the Law Enforcement Assistance Administration in the United States formed the National Advisory Committee on Criminal Justice Standards and Goals. One of the five volumes that the committee came out with was entitled *Disorders and Terrorism* (National Advisory Committee on Criminal Justice Standards and Goals 1976), produced by the Task Force on Disorders and Terrorism under the direction of H. H. A. Cooper, Director of the Task Force staff. The Task Force classified terrorism into six categories:

1. *Civil disorder.* A form of collective violence interfering with the peace, security, and normal functioning of the community.
2. *Political terrorism.* Violent criminal behavior primarily designed to generate fear in the community, or a substantial segment of it, for political purposes.

3. *Nonpolitical terrorism.* Terrorism that does not have a political purpose but that exhibits "conscious design to create and maintain a high degree of fear for coercive purposes, but the end is individual or collective gain rather than the achievement of a political objective" (Byrnes 2009).

4. *Quasi-terrorism.* The activities incidental to the commission of crimes of violence that are similar in form and method to genuine terrorism but that, nevertheless, lack its essential ingredient. It is not the main purpose of quasi-terrorists to induce terror in the immediate victim as in the case of genuine terrorism, but the quasi-terrorist uses the modalities and techniques of the genuine terrorist and produces similar consequences and reactions. For example, the fleeing felon who takes hostages is a quasi-terrorist, whose methods are similar to those of the genuine terrorist but whose purposes are quite different.

5. *Limited political terrorism.* Genuine political terrorism is characterized by a revolutionary approach; limited political terrorism refers to "acts of terrorism which are committed for ideological or political motives but which are not part of a concerted campaign to capture control of the state" (Byrnes 2009).

6. *Official or state terrorism.* "Referring to nations whose rule is based upon fear and oppression that reach similar to terrorism or such proportions." It may also be referred to as "structural terrorism," defined broadly as terrorist acts carried out by governments in pursuit of political objectives, often as part of their foreign policy (National Advisory Committee on Criminal Justice Standards and Goals 1976).

State sponsors have constituted a major form of funding; for example, the Palestine Liberation Organization, the Democratic Front for the Liberation of Palestine, and some other terrorist groups were funded by the Soviet Union. The Stern Gang received funding from Italian Fascist officers in Beirut to undermine the British Mandate for Palestine. Pakistan has created and nurtured terrorist groups as policy for achieving tactical objectives against its neighbors, especially India. Revolutionary tax is another major form of funding and is essentially a euphemism for protection money. Revolutionary taxes are typically extorted from businesses; they also "play a secondary role as one other means of intimidating the target population." Other major sources of funding include kidnapping for ransoms, smuggling, fraud, and robbery. The Financial Action Task Force is an intergovernmental body whose mandate, since October 2001, has included combating terrorist financing.

Terrorism is a form of asymmetric warfare and is more common when direct conventional warfare cannot be effective because forces vary greatly in power. The context in which terrorist tactics are used is often a large-scale, unresolved political conflict. The type of conflict varies widely; historical examples include secession of a territory to form a new sovereign state or become part of a different state, the dominance of a territory or resources by various ethnic groups, imposition of a particular form of government, economic deprivation of a population, opposition to a domestic government or occupying army, and religious fanaticism. Terrorist attacks are often targeted to maximize fear and publicity, usually using explosives or poison. There is concern about terrorist attacks employing weapons of mass destruction. Terrorist organizations usually methodically plan attacks in advance, and may train participants, plant undercover agents, and raise money from supporters or through organized crime. Communications occur through modern telecommunications or through old-fashioned methods such as couriers.

Responses to terrorism are broad in scope and can include realignments of the political spectrum and reassessments of fundamental values. Specific types of responses include targeted laws, criminal procedures, deportations, and enhanced police powers; target hardening, such as locking doors or adding traffic barriers; preemptive or reactive military action; increased intelligence and surveillance activities; preemptive humanitarian activities; and more permissive interrogation and detention policies. The term "counterterrorism" has a narrower connotation, implying that it is directed at terrorist actors. According to a report by Dana Priest and William M. Arkin in the *Washington Post* (Priest and Arkin 2010), "some 1271 government organizations and 1931 private companies work on programs related to counterterrorism, homeland security and intelligence in about 10,000 locations across the United States."

Media exposure may be a primary goal of those carrying out terrorist activities to expose issues that would otherwise be ignored by the media. Some consider this to be manipulation and exploitation of the media. The Internet has created a new channel for groups to spread their messages. This has created a cycle of measures and countermeasures by groups in support of and in opposition to terrorist movements. The United Nations has created its own online counterterrorism resource. The mass media, on occasion, censors organizations involved in terrorism through self-restraint or regulation to discourage further terrorism. However, this may encourage organizations to perform more extreme acts of terrorism for public consumption through mass media.

4.4 ORGANIZED CRIME

Organized crime and criminal organizations are often terms that categorize transnational, national, or local groupings of highly centralized enterprises run by criminals who intend to engage in illegal activity, most commonly for monetary profit. Some criminal organizations, such as terrorist organizations, are politically motivated. Sometimes criminal organizations force people to do business with them, such as when a gang extorts money from shopkeepers for so-called protection. Gangs may become disciplined enough to be considered *organized*. An organized gang or criminal set can also be referred to as a mob.

Other organizations, such as states, the army, police, governments, and corporations, may sometimes use organized crime methods to conduct their business, but their powers derive from their status as formal social institutions. There is a tendency to distinguish organized crime from other forms of crimes, such as white-collar crimes, financial crimes, political crimes, war crimes, state crimes, and treason. This distinction is not always apparent, and the academic debate is ongoing. For example, in failed states that can no longer perform basic functions such as education, security, or governance, usually due to fractious violence or extreme poverty, organized crime, governance, and war are often complementary to each other. The term "parliamentary mafiocracy" is often attributed to democratic countries whose political, social, and economic institutions are under the control of a few families and business oligarchs.

In the United States, the Organized Crime Control Act (1970) defines organized crime as "[t]he unlawful activities of [...] a highly organized, disciplined association [...]." Criminal activity as a structured group is referred to as racketeering, and such crime is commonly referred to as the work of the *Mob*. In the United Kingdom, police estimate that organized crime involves up to 38,000 people operating in about 6,000 groups. In addition, due to the escalating violence of Mexico's drug war, the Mexican drug cartels are considered the "greatest organized crime threat to the United States" according to a report issued by the U.S. Department of Justice (National Advisory Committee on Criminal Justice Standards and Goals 1976).

Organized crime often victimizes businesses through extortion or through theft and fraud activities like hijacking cargo trucks, robbing goods, committing bankruptcy fraud, also known as "bust-out," insurance fraud, or stock fraud such as insider trading. Organized crime groups also victimize individuals by car theft either for dismantling at "chop shops" or for export, art theft, bank robbery, burglary, jewelry theft, computer hacking, credit card fraud, economic espionage, embezzlement, identity theft, and securities fraud, referred to as "pump and dump" scams. Some organized crime groups defraud national, state, or local governments by bid-rigging public projects; counterfeiting money; smuggling or manufacturing untaxed alcohol or cigarettes, which is referred to as bootlegging; and providing immigrant workers to avoid taxes.

Organized crime groups seek out corrupt public officials in executive, law enforcement, and judicial roles so that their activities can avoid, or at least receive early warnings about, investigation and prosecution. Organized crime groups also provide a range of illegal services and goods, such as loan sharking of money at very high interest rates, assassination, blackmailing, bombings, bookmaking and illegal gambling, confidence tricks, copyright infringement, counterfeiting of intellectual property, fencing, kidnapping, prostitution, smuggling, drug trafficking, arms trafficking, oil smuggling, antiquities smuggling, organ trafficking, contract killing, identity document forgery, money laundering, point shaving, price fixing, illegal dumping of toxic waste, illegal trading of nuclear materials, military equipment smuggling, nuclear weapons smuggling, passport fraud, providing illegal immigration and cheap labor, people smuggling, trading in endangered species, and trafficking in human beings. Organized crime groups also do a range of business and labor racketeering activities, such as skimming casinos, insider trading, setting up monopolies in industries such as garbage collection, construction, and cement pouring, bid rigging, getting "no-show" and "no-work" jobs, political corruption, and bullying. Many of these activities can also be associated with terrorist organizations.

The commission of violent crime may form part of a criminal organization's "tools" used to achieve their goals. For example, violent crime is considered to be threatening, authoritative, coercive, terror-inducing, or rebellious, due to psychosocial factors such as cultural conflict, aggression, rebellion against authority, access to illicit substances, countercultural dynamics, or it may, in and of itself, be a crime rationally chosen by individual criminals and the groups they form. Assaults are used for coercive measures; to "rough up" debtors, competition, or recruits; in the commission of robberies; in connection to other property offenses; and as an expression of countercultural authority. Violence is normalized within criminal organizations in direct opposition to mainstream society and the locations they control.

While the intensity of violence is dependent on the types of crime the organization is involved in as well as their organizational structure or cultural tradition, aggressive acts range on a spectrum from low-grade physical assaults to murder. Bodily harm and grievous bodily harm, within the context of organized crime, must be understood as indicators of intense social and cultural conflict, motivations contrary to the security of the public, and other psychosocial factors.

Murder has evolved from the honor and vengeance killings of the Yakuza or Sicilian, wherein significant physical and symbolic importance was placed on the act of murder, its purposes and consequences, to a much less discriminate form of expressing power, enforcing criminal authority, achieving retribution, or eliminating competition. The role of the hit man has generally been consistent throughout the history of organized crime, whether due to the efficiency or expediency of hiring a professional assassin or the need to distance oneself from the commission of murderous acts, making it harder to prove liability. This may include the assassination of notable figures including public, private, or criminal, once again depending on authority, retribution, or competition. Revenge killings, armed robberies, violent disputes over controlled territories, and offenses against members of the public must also be considered when looking at the dynamic between different criminal organizations and their at times conflicting needs.

In addition to what is considered traditional organized crime involving direct crimes of fraud, swindles, scams, racketeering, and other Racketeer Influenced and Corrupt Organizations Act (RICO) predicate acts motivated by the idea of monetary gain, there is also nontraditional organized crime that is engaged in for political or ideological gain or acceptance. Crime groups engaging in the latter activities are often labeled as terrorist organizations.

There is no universally agreed upon, legally binding, criminal law definition of terrorism. Common definitions of terrorism refer only to those violent acts that are intended to create a fear of terror; are perpetrated for a religious, political, or ideological goal; deliberately target or disregard the safety of noncombatants or civilians; and are committed by nongovernment agencies. Some definitions also include acts of unlawful violence and war, especially crimes against humanity. The use of similar tactics by criminal organizations for protection rackets or to enforce a code of silence is usually not labeled terrorism though these same actions may be labeled terrorism when done by a politically motivated group.

The alien conspiracy theory and the "queer ladder of mobility" theory state that ethnicity and "outsider" status, immigrants, or those not within the dominant ethnocentric groups and their influences dictate the prevalence of organized crime in society. The alien conspiracy theory posits that the contemporary structures of organized crime gained prominence during the 1860s in Sicily and that elements of the Sicilian population are responsible for the foundation of most European and North American organized crime, comprising Italian-dominated crime families. Bell's theory of the "queer ladder of mobility" hypothesizes that "ethnic succession" (the attainment of power and control by a more marginalized ethnic group over other less marginalized groups) occurs by promoting the perpetration of criminal activities within a disenfranchised or oppressed demographic. While early organized crime was dominated by the Sicilian Mafia, their importance has been eclipsed by the

Irish Mob (early 1900s), the Aryan Brotherhood (1960s onward), the Colombian Medellin Cartel and the Cali Cartel (mid-1970s–1990s), and more recently the Mexican Tijuana Cartel (late 1980s onward), the Russian Mafia (1988 onward), al-Qaeda (1988 onward), and the Taliban (1994 onward). Many argue this misinterprets and overstates the role of ethnicity in organized crime. A contradiction of this theory is that syndicates had developed long before large-scale Sicilian immigration in the 1860s, with these immigrants merely joining a widespread phenomenon of crime and corruption.

4.5   MARITIME PIRACY

Piracy is typically an act of robbery or criminal violence at sea. The term can include acts committed on land, in the air, or in other major bodies of water or on a shore. It does not normally include crimes committed against persons traveling on the same vessel as the perpetrator, for example, one passenger stealing from others on the same vessel. The term has been used throughout history to refer to raids across land borders by nonstate agents.

Piracy is the name of a specific crime under customary international law and also the name of a number of crimes under the municipal law of a number of states. It is distinguished from privateering, which is authorized by national authorities, and is therefore a legitimate form of war-like activity by nonstate actors. Privateering is considered commerce raiding and was outlawed by the signatories to the Peace of Westphalia (1648). Those who engage in acts of piracy are called pirates. Historically, offenders have usually been apprehended by military personnel and tried by military tribunals. In the twenty-first century, the international community is facing many problems in bringing pirates to justice.

Seaborne piracy against transport vessels remains a significant issue with estimated worldwide losses of US$13 to US$16 billion per year, particularly in the waters between the Red Sea and the Indian Ocean, off the Somali coast, and also in the Strait of Malacca and Singapore, which are used by more than 50,000 commercial ships a year. In the late 2000s, the emergence of piracy off the coast of Somalia spurred a multinational effort led by the United States to patrol the waters near the Horn of Africa. In 2011, Brazil also created an antipiracy unit on the Amazon River.

In recent years especially since 2011, shipping companies have claimed that their vessels suffer from regular pirate attacks on the Serbian and Romanian stretches of the international Danube River, for example, inside the European Union's territory. Modern pirates favor small boats, taking advantage of the small number of crew members on modern cargo vessels. They also use large vessels to supply the smaller attack/boarding vessels. Modern pirates are successful because significant international commerce occurs via shipping. Major shipping routes handle cargo ships through narrow stretches of water such as the Gulf of Aden and the Strait of Malacca, making them vulnerable to being overtaken and boarded by small motorboats. Other active areas include the South China Sea and the Niger Delta. As water traffic increases, many of these ships have to decrease cruising speeds to allow for navigation and traffic control, making them prime targets for piracy.

Also, pirates often operate in regions of developing or struggling countries with smaller navies and large trade routes. Pirates sometimes evade capture by sailing into waters controlled by their pursuer's enemies. With the end of the Cold War, navies have decreased size and patrol, and trade has increased, making organized piracy far easier. Modern pirates are sometimes linked with organized crime syndicates but are often parts of small individual groups.

The International Maritime Bureau (IMB) maintains statistics regarding pirate attacks dating back to 1995. Their records indicate that hostage taking is the dominant form of violence against seafarers. For example, in 2006, there were 239 attacks, 77 crew members were kidnapped, and 188 were taken hostage, but only 15 of the pirate attacks resulted in murder. In 2007, the attacks rose by 10% to 263 attacks. There was a 35% increase on reported attacks involving guns. Crew members that were injured numbered 64 compared to just 17 in 2006. This number does not include instances of hostage taking and kidnapping where the victims were not injured.

The number of attacks from January to September 2009 surpassed the previous year's total due to the increased pirate attacks in the Gulf of Aden and off Somalia. Between January and September the number of attacks rose to 306 from 293. The pirates boarded the vessels in 114 cases and hijacked 34 of them in 2009. Gun use in pirate attacks has gone up to 176 cases in 2009 from 76 in 2008.

Rather than targeting cargo, modern pirates target the personal belongings of the crew and the contents of the ship's safe, which potentially contains large amounts of cash needed for payroll and port fees. In other cases, the pirates force the crew off the ship and then sail it to a port to be repainted and given a new identity through false papers purchased from corrupt or complicit officials.

Modern piracy can also take place in conditions of political unrest. For example, following the U.S. withdrawal from Vietnam, Thai piracy was aimed at the many Vietnamese who took to boats to escape. Further, following the disintegration of the government of Somalia, warlords in the region attacked ships delivering the UN food aid. Environmental action groups such as Sea Shepherd have been accused of engaging in piracy and terrorism; they ram and throw butyric acid on the decks of ships engaged in commercial fishing, shark poaching and finning, seal hunting, and whaling. In two instances, they boarded a Japanese whaling vessel. The attack against the German-built cruise ship the *Seabourn Spirit* offshore of Somalia in November 2005 is an example of the sophisticated pirates mariners face. The pirates carried out their attack more than 100 miles (160 km) offshore with speedboats launched from a larger mother ship. The attackers were armed with automatic firearms and rocket-propelled grenades.

Since 2008, Somali pirates operating in the Gulf of Aden have made about US$120 million annually, reportedly costing the shipping industry between US$900 million and US$3.3 billion per year. By September 2012, the heyday of piracy in the Indian Ocean was over. Backers were reportedly reluctant to finance pirate expeditions due to the low rate of success, and pirates were no longer able to reimburse their creditors. According to the IMB, pirate attacks had dropped to a 6-year low by October 2012. Only 5 ships were captured by the end of the year, representing a decrease from 25 in 2011 and 27 in 2010, with only 1 ship attacked in the third quarter compared to 36 during the same period in 2011. However, pirate incidents off the West African seaboard increased to 34 in 2012 from 30 in 2011, and attacks off the coast of Indonesia rose from a total of 46 in 2011 to 51 in 2012. Many nations prohibit ships to enter their territorial waters or ports if the crews of the ships are armed, in an effort to restrict possible piracy. Shipping companies sometimes hire private armed security guards to thwart pirate attacks.

Modern definitions of piracy include the following acts: boarding, extortion, hostage taking, kidnapping of people for ransom, murder, robbery, sabotage resulting in the ship subsequently sinking, seizure of items or the ship, and shipwrecking done intentionally to a ship.

The Constitution of the United States has delegated power to Congress to enact penal legislation for piracy, for treason, and for offenses against the law of nations. Treason is generally making war against one's own countrymen, and violations of the law of nations can include unjust war among other nationals or by governments against their own people.

In modern times, ships and airplanes are hijacked for political reasons as well. The perpetrators of these acts could be described as pirates (for instance, the French for "plane hijacker" is *pirate de l'air*, literally "air pirate"), but in English are usually termed "hijackers." An example is the hijacking of the Italian civilian passenger ship *Achille Lauro* in 1985, which is generally regarded as an act of piracy.

Modern pirates also use a great deal of technology. It has been reported that crimes of piracy have involved the use of mobile phones, satellite phones, global positioning system (GPS), sonar systems, modern speedboats, assault rifles, shotguns, pistols, mounted machine guns, and even rocket-propelled grenades and grenade launchers. International ships equipped with helicopters patrol the waters where pirate activity has been reported, but the waters that need to be patrolled are extensive. Some ships are equipped with antipiracy weaponry such as a long range acoustic device (LRAD), a sonic device that sends a sonic wave out to a directed target, creating a sound so powerful that it bursts the eardrums and shocks pirates, causing them to become disoriented enough to drop their weapons, while the vessel being pursued picks up speed and engages in evasive maneuvering. Additional measures used to combat pirates include the deployment of unmanned aerial vehicles (UAVs) and remotely controlled boats.

Under the principle of international law known as the "universality principle," a government may "exercise jurisdiction over conduct outside its territory if that conduct is universally dangerous to states and their nationals." The rationale behind the universality principle is that states will punish certain acts "wherever they may occur as a means of protecting the global community as a whole, even absent a link between the state and the parties or the acts in question." Under this principle, the concept of "universal jurisdiction" applies to the crime of piracy. For example, the United States has a statute (section 1651 of title 18 of the United States Code) imposing a life sentence for piracy "as defined by the law of nations" committed anywhere on the high seas, regardless of the nationality of the pirates or the victims.

The goal is to "deter and disrupt" pirate activity, and pirates are often detained, interrogated, disarmed, and released. With millions of dollars at stake, pirates have little incentive to stop. In Finland, one case involves pirates who have been captured and whose boat was sunk. No prosecution of the pirates has been forthcoming, as pirates attacked a Singaporean vessel and the pirates are not EU or Finnish citizens. A further complication is that Singapore law allows the death penalty for piracy and Finland does not. Some countries have been reluctant to implement the death penalty to deter pirates.

Warships that capture pirates have no jurisdiction to try them, and the North Atlantic Treaty Organization (NATO) does not have a detention policy in place. Prosecutors have a hard time assembling witnesses and finding translators, and countries are reluctant to imprison pirates because they would be saddled with the pirates upon their release. George Mason University professor Peter Leeson has suggested that the international community appropriate Somali territorial waters and sell them, together with the international portion of the Gulf of Aden, to a private company that would then provide security from piracy by charging toll to world shipping through the Gulf.

First and foremost, the best protection against piracy is simply to avoid encountering them. This can be accomplished by using tools such as radar. In addition, while the nonwartime twentieth-century tradition has been for merchant vessels not to be armed, the U.S. government has recently changed the rules so that it is now "best practice" for vessels to embark with a team of armed private security guards. In addition, the crew themselves can be given weapons training, and warning shots, using less lethal ammunition, can be fired legally in international waters. Also, remote weapon systems can be used.

Other measures vessels can take to combat piracy are implementing a high freewall and vessel boarding protection systems, for instance, hot water wall, electricity-charged water wall, automated fire monitor, and slippery foam. Ships can also attempt to protect themselves using automatic identification systems (AISs). Every ship weighing over 300 tons carries a transponder supplying both information about the ship itself and its movements. Any unexpected change in this information can attract attention. Previously this data could only be picked up if there was a nearby ship, thus rendering single ships vulnerable. However, recently specialized satellites have been launched to detect and retransmit this data. Large ships cannot therefore be hijacked without being detected. This can act as a deterrent to attempts to either hijack the entire ship or steal large portions of cargo with another ship since an escort ship can be sent quickly.

Finally, in an emergency, warships can be called upon. In some waters such as near Somalia, naval vessels from different nations that are able to intercept vessels attacking merchant vessels are present. For patrolling dangerous coastal waters and/or keeping expenses down, robotic or remote-controlled unmanned surveillance vehicles (USVs) are also used sometimes. Also, the U.S. Army currently uses both shore-launched and vessel-launched UAVs; similar equipment can also be used to detect incoming pirate attacks.

4.6   INCIDENTS AND INDICATORS

First responders should be aware of the warning signs that indicate criminal activity, because some incidents will involve criminal acts. It is important that all first responders recognize the chief indicators of a crime scene, identify appropriate responder activities and considerations at a crime scene, differentiate between the purpose of threat assessment and risk assessment, and identify outward warning signs and indicators of the five generic agents, such as biological, nuclear, incendiary, explosive, and chemical including nerve agents.

Law enforcement officers are well versed in crime scene investigations; however, the majority of first responders such as fire, emergency medical services, and emergency management personnel are not. It is critical that first responders understand the special demands placed upon them and their activities when responding to crime scenes. Firefighters may be first responders to arson scenes; emergency medical services personnel may be called upon to administer aid to victims of a violent crime; and hazardous materials teams may have to respond to sites of clandestine dumping or intentional releases of chemicals. When at the scene, first responders need to coordinate closely with other first responding personnel. It is important to remember that no evidence should be destroyed; that even though the emergency phase of the incident is over, the incident itself has not ended; and that the incident ends only when there is successful prosecution of the guilty person(s). At the crime scene, first responders must ensure that they do not impede the investigation in any way and that their actions are coordinated with law enforcement personnel, specifically, the onsite police commander.

Three ways for solving a crime include confession of the perpetrator(s), statements by witnesses or victims, and incriminating information obtained through physical evidence. Only physical evidence provides incontestable, impartial facts. It can overcome the conflicting and confusing statements of witnesses, and it may be crucial to connect the perpetrator to the scene. The recognition, collection, and preservation of physical evidence may be the only means to identify, prosecute, and hold accountable those responsible. If a first responder is involved in an incident they essentially become part of the crime scene. As a result, first responders may be called into court to testify as to what they saw and did and did not do. The actions of the first responders can have an influence on the outcome of the case. For example, if the actions of the first responders were found to be inappropriate, then this could have a detrimental outcome.

When responding to a crime scene there is the potential for being exposed to hazardous materials. For example, there is a risk of exposure to a biological and/or chemical agent for those entering a chemical or biological incident scene. In addition, the presence of fire or collapsed building sections may intensify thermal and mechanical risks. Any risks can be reduced and incidents avoided by entering the area carefully, moving cautiously, and wearing the appropriate personal protective equipment (PPE). However, if hazardous substances are present in the area, it is best to allow only qualified personnel to secure the scene. Therefore, hazardous materials teams should have the necessary skills and equipment to be able to define the hazard and secure the area. Any appropriate response to the site of a mass chemical, biological, or radiological (CBR) attack would require decontamination of all equipment. Part of the response would be to determine the number of entry personnel, the number of survivors, and the number of casualties. In addition, the decontamination process may be the single most important task that the public safety community can perform during a CBR incident.

During an incident the first responder's role is to pay attention to hazardous conditions and be aware of the possible presence of a secondary device intended to injure or kill first responders. Upon arrival at the scene of a highly damaged structure, be aware of the structural conditions causing the unsafe buildings to collapse, the types of injuries resulting from these incidents, and any specialized precautions that need to be taken when entering the scene. At the scene note whether persons are entering or leaving on foot or by vehicle, write down their license plate number, and provide a brief description of their vehicle. Try to encourage witnesses and bystanders to remain at the scene until investigators have interviewed them. And make notes of any unusual circumstances. When making notes provide as many details as possible and supplement them with photographs and videotapes to highlight the "big picture" of the crime scene. Use rough sketches to pinpoint the location of victims and their wounds, as well as the locations of potential evidence. Take notes based on observations, organize them, and provide them to investigators as soon as possible after the response. It is critical that the scene remain undisturbed. If it is necessary to move something, make note of where it was originally, and its orientation and condition, and, if possible, photograph the object before moving it. Make notes on any holes, breaks, or scratches that may have been caused by your actions. Investigators must be able to differentiate between the results of the crime and what responders might have done to those results. Following your response, you will need to write up an after-action report summarizing your activities and observations during the incident. It is important to document the report thoroughly using your notes from a crime scene; remember your report can be used in court during a criminal proceeding.

There are a number of indicators and warning signs that first responders can use to identify the type of incident prior to arriving at the scene of an incident or event. These include indicators for biological, nuclear, incendiary, chemical, and explosive incidents. Biological incidents present themselves in two ways: the first could be a community public health emergency and the second could be a focused response to an incident, such as that involving a toxin. In the case of a biological incident, the onset of some symptoms may take days to weeks. On the other hand, some effects may be very rapid, that is, for as short as 4–6 hours. Some of the indicators for a biological event include unusual numbers of sick or dying people or animals; dissemination of unscheduled and unusual sprays, especially outdoors and/or at night; and abandoned spray devices with no distinct odors. During a biological event any number of symptoms may occur. First responders should call their local hospital to see if they have admitted casualties with similar symptoms. It is important to note that casualties may occur within minutes or they may not occur until weeks after an incident has occurred; the biological agent used determines when symptoms will manifest.

Short of an actual detonation or obvious accident involving radiological materials, there are a couple of ways to be certain that radiation is present. The first is to observe the Department of Transportation (DOT) placards and labels. The second is to use monitoring equipment that most fire department hazardous materials teams now carry. For incendiary incidents, multiple fires may indicate the use of accelerants such as gasoline, rags, or other incendiary devices. In addition, the remains of incendiary device components, the odors of accelerants, usually involving heavy burning, or fire volume are also key indicators. Signs of explosive incidents may include large-scale damage to a building, blown-out windows, and widely scattered debris. Victims may exhibit the effects of the blast, such as obvious shrapnel-induced trauma, severe burns, the appearance of shock-like symptoms, and/or damage to their eardrums.

For a chemical incident such as the release of a toxic nerve agent, the most significant sign is the rapid onset of similar symptoms among a large group of people. Dermal exposure (clammy skin) and pinpoint pupils (miosis) are the best symptomatic indications of nerve agent use. Nerve agents are so lethal that mass fatalities without other signs of trauma are the most common indicator. Other outward signs of nerve agent release include exposed individuals reporting unusual odors or tastes; explosions dispersing liquids, mists, or gases; explosions involving a package or bomb device; unscheduled dissemination of an unusual spray; numerous dead animals, fish, and birds; abandoned spray devices; absence of insect life in a warm climate; mass casualties without obvious trauma; distinct pattern of casualties and common symptoms; and civilian panic in potential target areas, such as government buildings, public assemblies, subway systems, and tourist attractions.

The responsibility of the first responder is to help ensure the preservation of evidence. The wisest course of action might be to delay entry and await the arrival of more highly trained personnel. Specific steps that can be taken by the first responder at the awareness level are to isolate the scene, deny entry, notify additional resources, and recognize key indicators of a potential incident.

4.7 SUSPICIOUS ACTIVITY

Suspicious activities relating to terrorist behavior are largely defined along lines of what is deemed "normal" by either those defining the policies or those who are tasked to enforce those policies. As every company now has clearly established policies on reporting suspicious activities, the issue therefore becomes how each company determines the veracity of those claims. Problems occur when security personnel, and law enforcement officers, incorrectly assume certain activities are benign by nature without fully understanding the true capabilities of an aggressive organization.

Many security forces are trained to report a myriad of "suspicious activities and suspicious behaviors." Most companies even document the behaviors and activities that require that a report be submitted or local law enforcement be alerted. At sensitive facilities, sites defined as critical infrastructure, or locations deemed business essential, security policies require that incidents of photography, attempts to circumvent security, or attempts to gain access without following procedures, for example, are among the many types of behaviors that must be reported.

When looking at suspicious behavior, we must look beyond the obvious or noticeable; in other words, we must look beyond the stereotypes of what we would normally associate with one being a "terrorist" or a criminal. For example, every company receptionist or operator has received that phone call where there was a request for information. Often the company employee does not understand that information may be lost in pieces at a time. Some type of information is given out in every call if not more than the operators may be aware of. The phone call from the local college student who claims to be writing a paper for a class is all too often accepted as being true without any means to verify the story. Often when "students" are confronted they offer that story or one similar and are instructed by security personnel not to take photographs or leave the area altogether. The issue then becomes whether that field adjudication by the local security force is ever reported or if the information kept at a local level because the story is believable? Has field adjudication resulted in the failure to report an incident because it sounded reasonable?

It is important to never underestimate an adversary and to recognize their abilities. An adversary taking pictures of a facility, or conducting elicitation by phone, would have a well-developed and well-rehearsed story for his or her activities. Believability allows the intelligence collector for an organization to reduce the potential for detection, and, more importantly, to reduce the risk of the incident being reported. Even local law enforcement personnel may fall victim to believing a story because there are props in the vehicle that make the story convincing. Operators working as part of a "network" are taught field craft and how to disguise their movements and true intentions without being noticed. Often the techniques employed can be very subtle, such as special knocking procedures on doors or special signs. Such signs may include hanging out a towel, opening a curtain, or placing a cushion in a special way. The sign is an indication to those wishing to enter the premises that it is safe to do so and that it is not being monitored. The same field craft techniques used by members of a "terror network" should be studied by security personnel in order to gain an appreciation and understanding of the tactics and techniques of members, thereby enabling them to respond accordingly when identifying suspicious behavior.

When screening the vehicle of a suspicious person, documents in open view should be observed. Operatives are trained to utilize post office boxes for external communication as a security measure, with consideration being given to the location where the mail is sent to and from where. The presence of an envelope with a post office box address would not by itself be an indicator of subversive or suspicious activity, but it can add to the overall suspicions when coupled with other activities.

When suspicious activities are identified, security personnel should be trained to recognize other modus operandi and techniques. For example, operatives are instructed to disguise their voices and limit telephone calls to no more than 5 minutes when engaging in wireless communications. A well-trained security officer may be able to identify efforts by a suspicious person to conceal their true identities if the person is observed during a telephone conversation. Operators engaged in clandestine activities are trained in mobility methods to maximize their ability to blend into their surroundings, thereby increasing the believability factor if questioned by security personnel. For example, if driving by automobile, there will be a strict adherence to obeying traffic rules and speed limits. Operators also try and avoid parking along streets and prefer public parking lots to avoid suspicion and blend in. Most often they park their vehicle in a public lot or garage near a major highway or freeway that provides them with a quick exit. In addition, the color of the vehicle may be changed before or after the operation is completed, and the license may be falsified. In turn, security personnel must be cognizant of changes in the vehicle as well as be able to identify signs of fabricated license plates. Security personnel must also recognize that identifying suspicious behavior is a difficult task and that signs and indicators are not easy to discern.

During an operation or mission, an organization's main priority is the safety and security of its operators. If they are jeopardized or compromised in any way it could have major repercussions for the entire organization and could lead to the entire operation being aborted. For example, the September 11 attacks originated with Operation Bojinka, a plan that was not executed, which was conceived by Khalid Sheikh Mohammed and his nephew, Ramzi Yousef. The first stage involved the assassination of Pope John Paul II, and the second the bombing of 11 airliners bound for the United States. The third stage called for a small airplane loaded with explosives to be crashed into the CIA headquarters and possibly other buildings in major cities across the United States. The plot was discovered by the Manila police on January 6, 1995, and Abdul Hakim Murad was arrested. Ramzi Yousef was arrested in Pakistan in February 1995.

When safety is being provided, certain indicators may surface that can benefit local security personnel. An organization comprises two groups of operatives: overt and covert. Overt operatives are instructed to refrain from showing any signs of curiosity or being talkative. In an effort to accomplish this, it is possible that the member may increase his or her level of suspicion. Covert operatives are trained not only in their appearance but also in their habits to ensure their activities are related to either being normal activities or being counter to the perceived threat behavior. Members operating in covert mode are instructed to have an unassuming appearance that allows them to blend in without any distinguishing features such as beards, mustaches, glasses, tattoos, ethnic clothing, or religious items. Members are instructed to avoid locations of religious or ideological significance, such as meeting places or gatherings with sympathizers to their cause, and to ensure their activities appear routine to anyone who might be observing them. The lesson for security personnel again is to always look beyond the obvious.

Anyone questioned for suspicious activities normally exhibits some type of nervousness, and security personnel must be alert to those types of reactions. Organizations recognize the importance of open source information from such places as newspapers, books, magazines and through the media. Organizations recruit operatives who are trained researchers, such as college graduates, who have experience in examining primary sources of information. This raises the question of how well trained our security personnel are to identify and prevent elicitation over the telephone. While little can be done by security personnel to prevent open source collection, the presence of multiple newspapers, magazines, and books in a suspect's possession can provide reasons for further suspicion.

Covertly collected information is categorized as information concerning either individuals and their families or strategic facilities. The information collected is of paramount importance for launching a successful attack against an individual or at a facility. During the collection phase, tradecraft and field craft skills are critical. Operatives are instructed to utilize special signals between operators to exchange orders and instructions, not to carry weapons, and to preferably have a camera. Local security personnel must be cognizant of not only what is present, but also what is missing.

Operatives are instructed to ensure that the vehicle they use is of a common make or model so that no attention is drawn to it, thus making it difficult to trace or track.

In addition, the vehicle should be in relatively good condition without any distinguishing features such as dents or scratches. Operatives are also instructed to disable the vehicle's interior light so that team members cannot be identified. Multiple vehicles may also be used during the operation so security personnel must also be aware of that possibility.

During the information collection phase, operatives are instructed to gather information on traffic patterns, traffic signals, location of parks, and the intensity and location of lighting. Operatives taking photographs are instructed to begin with an overall view of the target area and then take continuous views. They are also instructed not to take photos to an outside photo processing service for development but to develop them in-house instead. The key element for security personnel is the identification of the photographic equipment as well as the identification of possible fields of view in which photography may have occurred. Smart phones and tablets give individuals the opportunity to take pictures or videos covertly while giving the appearance that they are either working or having a conversation. Security personnel must therefore be alert to the presence of common, everyday items such as smart phones and tablets.

They must avoid being overly paranoid but must be alert to multiple indicators and react accordingly. Security education is important for local security personnel and should be continuous with changing tactics and modus operandi to tackle aggressors. Individuals in the reporting channels must look beyond the obvious and ask questions. While most incidents of photography are benign, it is important to understand that aggressors try to make their actions look benign. Information collection by a hostile enemy is designed to reduce the potential for detection and, as with any operation, collection must fall within the realm of normal activities.

Security personnel must realize that threats may come from sources that they would never consider normally. The adversary is well aware that complacency is human nature and that it can be exploited. As humans we become accepting of the obvious, which creates vulnerabilities. Routines are an ally to our adversaries as we tend to look for comfort in them. For example, we generally accept that we may have the same delivery driver for packages every day, and we would probably question someone new. However, it may be that the routine driver that is the real threat. While we do not expect a hostile aggressor to come to our front door and announce their true intentions, we must understand that suspicious behavior may not be that obvious either. In the case of the delivery driver, we need to check for more than just a driver's license or an identification card, we may need to call the delivery company and verify who they are or simply check their references.

The use of fraudulent documents and license plates is standard operating procedure for a hostile organization, and therefore security personnel must question individuals further when there is even a hint of suspicion. The problem, however, is that many security personnel lack the understanding necessary to recognize suspicions beyond the obvious. The majority of security personnel assigned to access control duties never encounter a fraudulent identification or would never recognize it if they did. For example, the security officer conducting a perimeter patrol will recognize a vehicle parked along the facility perimeter and may even stop to ask questions. However, the vast majority would take no further action if the vehicle had the hood up or the driver was changing a tire, unless the vehicle was stopped or parked in a "no stopping" or parking zone.

Proper handling and reporting of security incidents is a function of security management personnel; they must understand that there is a difference between terrorism-related behavior and other types of criminal activity. The ability to look beyond normal criminal activity may result in the identification of a greater threat. An adversary may "shroud" their true intentions with elements of normal or criminal activities in the area. For instance, just as an adversary taking pictures may have books on architecture in their vehicle, an intrusion into a facility may also be made to look like an incident of vandalism or corporate theft. Again, the key is to follow the information to the end and not just accept all incidents and scenarios at face value.

Security education and awareness are ultimately the key to recognizing suspicious activities. Security personnel undertaking access control duties must be trained to recognize that the threat may not be so obvious and must not become too accepting of the same everyday routine. Security management must ensure that field adjudications of incidents do not also follow the path of acceptance on face value alone, thereby potentially failing to report something that may be critical. While local law enforcement personnel have a better understanding of the true nature of the threats, it is important that continued vigilance be stressed as well. Aggressors are very patient and meticulous in their collection of information and planning of operations. They exploit every possible weakness in the security apparatus, especially the weakest of all, the human factor.

4.8   AVOIDING AN ATTACK

The best way to avoid an attack is to be consistently inconsistent. The most important thing to remember is to not do anything the same way at any time, constantly change your schedule, vary your routes to and from work, and vary your activities. Most importantly, follow the advice of your corporate security personnel and department, thereby reducing your chances of becoming a victim. The mind-set of an aggressor is to cause widespread fear and panic and bring attention to their cause without being caught or noticed. However, there are steps that can be taken to make one look less of a soft target and more of a hard target. If one can give the appearance of a hard target then it is expected the aggressors will move on to softer targets—"why fight a lion when there are plenty of sheep to be had." Topics covered in this section include selecting an appropriate security team or driver, the mode of behavior when in a hostile or foreign country, areas to avoid or choke points, actions to be undertaken when kidnapped and when being released or rescued.

If you are involved with a company that uses drivers for key executives or protectees, there are several things that one needs to keep in mind. The unfortunate truth is that when drivers are chosen from a motor pool, the person with the best driving record and one who is the most careful is normally provided for the chief executive officer (CEO). The problem with choosing a driver for their safety record is when the person being protected comes under attack, the last thing one would want is an overly cautious driver. Although an impeccable driving record is a good trait in a security driver, other traits they should possess include initiative, the attitude of a team player, an aggressive attitude during emergency situations, and the ability to assess and react instinctively due to changing road conditions. A security driver's main priority should be the safety and welfare of the protectee and not being too overly cautious because of the subconscious desire to have a perfect driving record. It is almost impossible for a driver to be the only member of a security detail. Drivers need to be part of a security package, and they need to report directly to the head of security or security detail team leader. If a security team leader is present, they sit in the front passenger seat and provide directions to the driver regarding what to do and where to go. In an attack scenario, the best defense is to "punch" through the attack and keep moving. If the vehicle is run off the road, the minute the aggressors begin to exit their vehicle is the time that the security driver should "ram" them. When selecting a security detail leader, avoid anyone with an autocratic mind-set. They make easy targets because they thrive on rigid schedules and routine is the norm. They believe they are invisible due to detailed planning and often ignore the advice of their security specialists because they are used to giving orders as opposed to listening to reason.

A high-profile individual or the organization they represent can become a target of political terrorism on a visit or when business is being conducted in a foreign country. If the individual or organization they represent is wealthy or gives the appearance of wealth, they could become the victims of kidnapping for the purposes of financial gain on the part of the aggressors. The objective here is not to stand out but to blend in with the crowd. Thus, there are certain steps that can be taken to allow an individual to blend in. For example, drive a vehicle that is common for the area; if using an armored vehicle, do not use a limousine. Avoid political demonstrations and protests, dress conservatively, and avoid wearing things that make reference to a country that has current political problems with the host country. Always be security conscious and maintain an awareness of your surroundings at all times.

The final point is of paramount importance since terrorists understand that the longer they are on the street the higher the likelihood of them being recognized. Intelligence shows that operatives are trained not to be on the street for more than 20 minutes at a time. In addition, surveillance teams and kidnappers constitute different individuals. This prevents recognition of the attackers before and after the attack occurs. Hostile organizations are getting larger as more people are needed to fight for their cause.

Individuals must be aware of their surroundings and must be on the lookout for indicators. Potential target locations include their office, children's day care and school, and anything else that is part of their day-to-day schedule or routine. An area an individual cannot avoid such as a staircase represents a potential choke point or vulnerable area because the probability for carrying out a successful attack at this location is very high. In addition, locations that are visited frequently by an individual or individuals for items such as food and/or gas are considered to be the most dangerous. The reasons are that these locations are known target locations and individuals feel comfortable within these locations.

Part of being vigilant and aware of one's surroundings is noticing things that are out of the ordinary for one's environment. In this situation, it is important to talk to family, friends, or traveling companions to determine if they too have noticed anything unusual. It is important to compare notes on what they have seen or thought they have seen. At this stage there are myriad questions to be asked, for example, why are the security lights out when they should be on? Why are the tires flat on that truck? Is there a new maintenance man in the building? Why did this vendor choose this location to set up his cart today? Who visits them? And do they look, dress, and act like a street vendor?

All hostile operatives use surveillance and in many cases they have surveillance teams to gather information on a potential target or target location. They typically use new recruits for this task, since they are not well known to law enforcement officials. In turn, this can be a benefit for security and law enforcement and security personnel, since the recruits do not have a lot of experience and therefore make mistakes, allowing law enforcement and security personnel too easily pick up on their activities.

Surveillance is usually the first step before an assassination or kidnapping attempt. Knowledge of surveillance detection and countersurveillance techniques can help in reducing the risk of an attack. In addition, individual(s) and organization(s) must also be aware of electronic surveillance from covert video, wiretaps, transmitters, and electronic tracking devices. A surveillance team could be watching an individual from a specific location. Progressive surveillance reduces their chance of being observed.

When approaching a vehicle, look around and under the vehicle and vehicles next to it for anything that is out of the ordinary. Is there someone hiding behind or under the vehicle? Is there a bomb attached to or under the vehicle? And is there a van that shows movement near or beside an individual's car?

Individuals must be aware of what is happening around them at all times. For example, listen to the sounds of the street or lack thereof. Is someone watching or following you? They must ensure their secretary understands the risk and hazards associated with leaking information. They must also listen to their security advisors and follow their advice and not go to places where personal protection is limited. Individuals should also make their own travel plans, limiting the number of people with knowledge of their travel arrangements. It is also important that individuals talk to their families and ask them questions about the things that have happened around them.

When driving do not allow the vehicle to be stopped or run off the road during an ambush attempt. The goal of an ambush is to get the vehicle stopped so that the ambushers can either kill the individual(s) or kidnap them. It is important to avoid known traffic congestion points or choke points. When someone is stopped in this type of traffic, it is impossible to avoid a group of armed attackers. This method or tactic is also commonly used to rob motorists. To avoid these situations, it is important for an individual to vary their schedule each day and to vary the route they take to and from work or the job site. If their office or work location has more than one entrance, they must use it. If the individual has encountered a problem on the road while traveling, the vehicle should keep moving and not stop until it reaches its destination. In situations where multiple vehicles are used in an ambush situation, the scenario may play out as follows: The first vehicle has armed men ready to shoot the security detail and break down any barriers that may exist and the next vehicle is loaded with explosives to do the damage. Suicide bombers in conjunction with the ground assault force use rocket-propelled grenades and small arms as their weapons of choice. The second vehicle carrying explosives comes in after security and response teams have been deployed and as such more responders may be injured during the blast, causing maximum damage.

It is important to know the history of the hostage takers in the area where individuals are visiting, for example, do the hostage takers normally release or kill the hostages? If captured, there are a number of steps that can be taken to increase your chances of survival. For example, ask if it would be possible to write a letter or make a call to your spouse; this way the kidnappers will perceive you as being human. When captured, always behave like a gentleman or lady, so that kidnappers do not think you are not worthy of living. Do not become combative, argumentative, or behave inappropriately toward the kidnappers; no one likes a troublemaker. Do not give them a reason to lose their temper. Do not cry and make a lot of noise; they will not take pity on you and only perceive you to be weak and inferior. If you are being held as a political hostage, make friends, become human to them. It may make it harder for them to kill you or injure you in any way. If you are a criminal hostage, keep your head down and do not look at them; they will not want you to identify them later. Carry photos of your family, especially children; this may garner sympathy from the kidnappers toward you, if you do not have any of your own borrow one from a close family member.

During a rescue attempt, do not try to pick up a weapon or try to help. Hostage rescue teams are trained to shoot anything that is standing or is carrying a weapon. During the rescue lie face down on the road and be still, placing your hands over your head. Hostage rescue personnel treat everyone as hostile until the aftermath of the assault and after authorities have determined who is a hostage and who is a foe.

Whenever planning a trip abroad, it is important to do your due diligence and research the country you are visiting; the U.S. Department of State and Foreign Affairs Canada provide up-to-date in-country information for travel abroad including the political climate and the risk level. One should also be sensitive to anniversary dates of significant events around the world, such as September 11. Finally, it is important that individuals keep things in perspective, "you are about twice as likely to be hit by lightning as to become a victim of a terrorist attack." As with anything in life, one needs to be aware of their soundings and what is going on around them.

5 Protection Strategies

5.1 PHYSICAL SECURITY

The goal of any physical security program is to

- Prevent unauthorized access to equipment, installations, materials, and documents.
- Safeguard against espionage, sabotage, damage, and theft.
- Safeguard personnel.

Physical security systems for protected facilities are generally intended to deter potential intruders through the use of warning signs and perimeter markings; distinguish authorized from unauthorized people using keycards and/or access badges; delay, frustrate, and ideally prevent intrusion attempts through the use of physical barriers; detect intrusions and monitor and record intruders through the use of surveillance systems; and trigger appropriate incident responses by security guards and the police.

It is up to security designers, architects, and analysts to balance security controls against risks, by taking into account the costs of specifying, developing, testing, implementing, using, managing, monitoring, and maintaining the controls, along with broader issues such as aesthetics, human rights, health and safety, and societal norms or conventions. Physical access security measures that are appropriate for a high-security prison or a military base may be inappropriate in an office, a home, or a vehicle, although the principles are similar.

Facilities are best protected by taking an approach to physical security that uses multiple barriers to prevent unauthorized access to the facility, referred to here as the "multibarrier approach." The first barrier in safeguarding a facility is perimeter protection and standoff distance. Perimeter protection includes both inner and outer protection controls, such as fences, walls, speed control devices, and vehicle barriers, and acts as the outermost layer of security, the purpose of which is to prevent and/or delay any potential attacks and also act as a psychological deterrent by defining the perimeter of the facility and making intrusions seem more difficult.

The standoff distance is the minimum distance required between the facility and where an explosive device is allowed to detonate without causing any significant damage to the target facility. In addition to the standoff distance, a facility may also have areas designated as exclusive and nonexclusive standoff zones. The exclusive standoff zone is the area where vehicles are not allowed within the perimeter fencing of a facility unless they have been searched and cleared by security personnel. The nonexclusive standoff zone is an area that is established when a facility or location permits both trucks and cars on its premises (Figure 5.1).

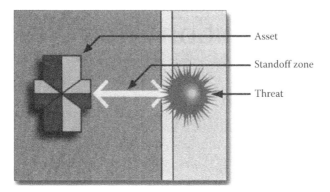

FIGURE 5.1   Standoff distance.

Speed control devices deny entry and/or exit of suspect vehicles; prohibit high-speed vehicular access through checkpoints; and control the speed of vehicles that may be used for concealing explosive devices. Vehicle barriers protect entry and/or exit to the perimeter, restricted areas, or vehicle control points; and deny entry of vehicles that may carry explosive devices. Perimeters should also be protected against standoff weapons such as rocket-propelled grenades, rifles, shotguns, and pistols. The primary defense is to obstruct the line of sight (LOS) from vantage points outside the perimeter. Aggressors will typically remain outside of controlled areas and try to gather information and assess a facility's vulnerability. Designers must eliminate or control vantage points from which aggressors can observe or eavesdrop on assets or operations. Features such as trees, bushes, fences, walls, and other buildings may aid in preventing outside observation from aggressors.

Security lighting discourages or deters unauthorized access by intruders, as intruders are less likely to enter well-lit areas for fear of being seen. Security lighting may consist of different types such as continuous, standby, movable, or in combination. Entrance and/or exit control points, command, control and communications centers, parking areas, ammunition storage areas, motor pools, and other sensitive and/or restricted areas should be well lit to allow close observation of people near a facility or within its perimeter. Lighting greatly assists security personnel to inspect people, vehicles, and cargo at the perimeter during hours of darkness. When lighting the grounds of a facility, widely distributed low-intensity lighting is generally superior to small patches of high-intensity lighting, because the latter can have a tendency to create blind spots and glare for security personnel and close circuit television (CCTV) cameras. It is important to place lighting in a manner that makes it difficult to tamper with, such as by suspending lights from tall poles, and to ensure that there is a backup power supply so that security lights will not go out in case of a blackout.

Security personnel play a central role in all layers of security. All of the technological systems that are employed to enhance physical security are useless without a security force that is trained in their use and maintenance, and which knows how to properly respond to breaches in security. Security personnel perform many functions: as patrols and at checkpoints, to administer electronic access control, to respond to alarms, and to monitor and analyze video.

A sensitive compartment and information facility (SCIF) is an accredited area, room, group of rooms, buildings, or installation where sensitive compartmented information (SCI) may be stored, used, discussed, and/or processed. Access to an SCIF is highly restricted and should be protected by multiple layers of control. The vault must have the following minimum specifications: compressive strength of 3000 psi after 28 days of aging for Class A; 5/8-inch diameter steel reinforcing bars laid 6 inches on centers in seismic areas, with 6 inches or more of reinforced concrete being required. Secured areas are required to meet the following minimum construction specifications:

- For areas designated as confidential and secret, the walls, floor, and ceiling must be constructed of either reinforced concrete over 4 inches thick or solid masonry such as stone or brick.
- For areas designated as top secret, the walls, floor, and ceiling must be at least 8 inches thick, constructed of reinforced concrete.
- Secured areas should be equipped with a GSA Class 6 vault door.
- Secured areas should preferably be constructed without windows. If accessible windows are required, they should be secured with bars and installed as specified in the construction security requirements. There are three different types of security requirements:
  - Type A: Requires a steel frame with steel bars welded on it to be bolted to the inside of the facility window frame.
  - Type B: Requires embedding the ends of steel bars in the masonry window frame of the facility.
  - Type C: Requires a grillwork of steel bars to be embedded in the masonry walls immediately adjacent to the facility window frame.
- SCIF walls, windows, floor, and ceiling, including all openings, should provide sufficient sound attenuation to preclude inadvertent disclosure of conversation and must meet the following criteria: Executive Suite, 45+ dB; briefing rooms, 45+ dB; auditoriums, 50+ dB.
- Telephone cables and wires that penetrate a facility's perimeter should enter the facility through one opening and be placed under control at the interior face of the perimeter. The number of telephone instruments servicing an SCIF should be limited to those operationally necessary.

Alarm systems can be installed to alert security personnel when unauthorized access is attempted. Alarm systems work in tandem with physical barriers, mechanical systems, and security guards, serving to trigger a response when these other forms of security have been breached. They consist of sensors including motion sensors, contact sensors, and detectors.

Different types of sensors are used to detect different types of motions. Structural vibration sensors detect energy due to hammering and drilling; point sensors detect close proximity to an object; passive ultrasonic sensors detect acoustical energy; volumetric motion sensors detect intruder motion within the interior of a protected volume of space; fence sensors detect penetration generated by mechanical vibrations and stresses in fence fabric and posts; and LOS sensors generate a beam of energy and detect changes in the received energy that an intruder causes by penetrating the beam.

When implementing an alarm system the following requirements should be met: perimeter doors should be equipped with high-security balanced magnetic door switches; vault doors should be equipped with heat detectors and balanced magnetic switches; the interior spaces not continually occupied by authorized personnel should be protected by motion detection alarms; vents and ducts over 6 inches must be protected by an approved alarm device; and windows positioned less than 18 ft from ground level must be protected by an approved alarm device.

Different types of motion sensors can trigger different types of alarms. For example, protected areas using ultrasonic, microwave, and other motion detection devices that detect overt body motion such as walking through the protected areas at the rate of one step per second for four seconds can trigger an alarm. Entry or exit switches detect motion of opening doors, windows, or other entry or exit passages using sensor devices protected by balanced magnetic door switches. Capacitance alarms are triggered by attempts to push hands, arms, or legs through the protected area such as air ducts or vents, or upon touch of a protected item such as a door, window, or wall, or upon movement of protected objects such as security containers. Tamper switches are used to trigger an alarm by removing the cover of any sensor, alarm control unit, light-activated (day or night) switches, or an end of the line supervision control unit, regardless of the status of the overall system.

However, alarms are only useful if there is a prompt response when they are triggered. In the reconnaissance phase prior to an actual attack, some intruders test the response time of security personnel to a deliberately tripped alarm system. By measuring the length of time it takes for a security team to arrive (if they arrive at all), the attacker can determine if an attack could succeed before authorities arrive to neutralize the threat. Loud audible alarms can also act as a psychological deterrent, by notifying intruders that their presence has been detected. In some jurisdictions, law enforcement will not respond to alarms from intrusion detection systems unless the activation has been verified by an eyewitness or video. Policies like this have been created to combat the 94%–99% rate of false alarm activation in the United States.

Facilities, grounds, and other areas designated as a "restricted area" are subject to high-level security precautions, including but not limited to video surveillance, access controls, and alarms to prevent unauthorized access. Various types of physical access control locks may be employed to further restrict access and/or protect assets. Examples include (1) preset locks and keys, (2) programmable and mechanical cipher locks, (3) electronic and/or digital keypad system locks, and

(4) combination locks. Security may be enhanced by frequently changing codes, the type of lock used, or random assignment of new number combinations, or the number of digits in code.

Other forms of physical control access include photo identification cards, wireless proximity readers, magnetic strip cards, and smart cards. Certain cards often require the use of a personal identification number (PIN), as well as insertion into a card reader, the use of a card swipe device, or a proximity detection device. The U.S. Department of Defense (DOD) has employed a common access card system using "smart cards" to establish and maintain physical access control in certain sensitive and restricted areas.

The use of biometric devices dramatically improves physical access control by matching unique human characteristics of authorized personnel to a reference data set. Biometric devices significantly enhance protection against unauthorized access to sensitive or restricted areas. Biometric devices may include (1) hand, fingerprint, or thumbprint scans, (2) retina scans, (3) facial geometry scans, or (4) voice recognition. Advantages include positive identification/verification, speed, and accuracy; it requires little or no monitoring and is cost-effective.

Use of preset verification times for various devices further enhances physical access control. Physical access control must be extended to visitors, cleaning teams, maintenance workers, civilians in work areas after normal work hours, and government contractors. Uniformed personnel must also be properly identified and possess authorization to enter sensitive or restricted areas, with or without escort. In a high-security facility, there should be routine and special unscheduled management review of personnel access rosters, level of access, and dates and times of entry and exit. There should also be an initial background and periodic rescreening of personnel located at the facility, including visitors from another facility or office. Special attention should be paid to disgruntled, terminated, and transferred employees, as well as socioeconomic conditions and recent change-of-life scenarios of employees. When escorting personnel through a high-security facility, the two-person rule should be implemented, that is, one escort for every two persons; escorts should also be controlled, with visitors having very limited access to only those areas needed for meetings or to carry on work activities.

In addition to physical access control at a facility, there is also the issue of data security, specifically confidentiality with regard to sharing computer accessories and data storage such as diskettes, flash drives, blu rays, DVDs, memory sticks, and other similar devices. Some of the issues surrounding sharing of these devices without the proper controls in place include compromising data integrity, user error, and inadvertently downloading malicious code and data version control.

The physical security of computing the distribution office to house the servers, both hardware and software, will have its own access control and protection control measures. Such measures would include area controls, such as limited personnel access to the actual servers themselves, entry office controls, layout property controls, electronic media controls, and climate controls for the servers such as heat and humidity adjustments. In addition, for personnel reporting to the computing distribution office, a "clean-desk" policy and space protection devices should also be implemented and adhered to.

For securing stand-alone systems and peripheral devices, a physical control cable lock using a vinyl-covered steel cable should be used to anchor the personal computer or peripheral device to a fixed object. Other security devices include port controls, switch controls, peripheral switch controls, and electronic security boards. Port controls should be used to secure data ports, such as USB ports in order to prevent their use. Switch controls are used as a cover for the on/off switch and to prevent a user from switching off the system's power. Peripheral switch controls are a lockable switch that prevents a keyboard or other peripheral device from being used. Electronic security boards are a special card inserted into a PC expansion slot that forces the user to enter a password when the unit is being booted.

Environmental and safety controls should be implemented and monitored constantly to ensure the safety and security of the facility and its occupants. Environmental and safety controls include electric power RFI, EMI implementation, humidity controls, with less than 40% of humidity increasing static electricity, initial and residual damage potential, emergency power off controls, voltage monitoring or recording, and electric power surge protection controls. In case of a blackout, power backups should be initiated. The order of precedence for electric power sources are as follows: primary electric power sources, secondary electric power sources (including distributed power sources, see Chapter 3), backup feeders, and emergency power generators. Temperature controls should also be implemented and monitored since excess temperatures can cause malfunction and damage to electronic devices and paper files. Damage occurs to paper products at 350°F, computer equipments at 175°F, disks at 150°F, and other magnetic media at 100°F.

For the purposes of fire protection, the following devices and measures can be implemented for detecting fires and excessive heat and temperatures in facility: heat-sensing devices, flame-actuated devices, smoke-actuated devices, and automatic dial-up fire alarms. In addition, fire extinguishing systems can also be implemented such as portable extinguishers, water, dry chemical, and carbon dioxide types used for small areas; overhead fire extinguisher systems use wet pipe or dry pipe mediums to suppress fire by saturation and contained area fire extinguisher systems that introduce halon or carbon IV oxide gas into an area to remove oxygen and extinguish flames. Both halon and carbon IV oxide are potentially dangerous to people and, therefore, these types of systems are used for unattended facilities or areas such as vaults and equipment cabinets.

Information system centers are important for maintaining large quantities of information, some of which can be confidential and/or personal. A number of factors are taken into consideration when designing physical security systems for these centers including site selection; low visibility to casual observation; solid construction; low natural disaster threat; and easy access to external and emergency services such as police, fire, ambulance, and hospitals. The infrastructure in information system centers such as servers, switches, and routers should be placed in secured racks and locked rooms. Wiring and cables should be routed through walls and floors to avoid potential tampering. Uninterrupted power supply should exist for the computing facility including primary, secondary, and tertiary backup power sources and systems.

A device or system in general is said to be tamper-resistant if it is difficult to modify or subvert, even for an assailant who has physical access to the system. Specialized materials such as the use of one-way screws, epoxy encapsulation, lock bolts, self-tapping screws, torque screws, and similar products to secure easily movable items are used to make tampering difficult. Many devices also have tampering detection sensors embedded into them, which give them the ability to sense whether they are under physical attack, these include switches to detect opening of device covers, sensors to detect changes in light or pressure within the device, and barriers to detect drilling or penetration of physical boundary, such as tape, paint, and other similar materials. In addition, these same devices may also have tamper responses built into them. Tamper response is a countermeasure taken upon the detection of tampering such as erase memory, shutdown and/or disable device, and enable keystroke logging. This is especially very important in the case of lost or stolen cryptographic keys. Computational errors introduced into a smart card can deduce the values of cryptographic keys hidden in the smart card. Layers of a chip can be uncovered by etching, discerning chip behavior, advanced infrared probing, and reverse-engineering chip logic.

In addition to the various protection controls implemented at a facility or high-security site, personnel reporting to these facilities must exercise operation security (OPSEC). OPSEC is an analytic process used to deny an adversary from obtaining information, generally unclassified, and potentially using it for unauthorized purposes. Personnel can apply OPSEC in his or her daily lives. It trains people on the proper methods of handling, storing, and protecting information. The question people should always be asking is "what could someone learn from watching me do this?"

5.2   IMPROVISED EXPLOSIVE DEVICES AWARENESS

An improvised explosive device (IED) is defined as an explosive device used for other than commercial or military use designed to kill personnel and/or destroy property. It may have either an explosive or an incendiary effect on its intended target. It is considered to be the contemporary terrorist's tactic of choice. It is also relatively inexpensive to produce, since most IEDs are constructed using everyday household items, and as a result there is a low risk of the perpetrator becoming injured. In addition, the destructive nature of IEDs makes them an "attention getter."

The technology or the technical know-how on how to develop a bomb or an IED is relatively easy to obtain. For example, most bomb-makers have had previous experience with explosives and incendiary devices either from the military or from the industry, such as demolition specialists working in the mining and geological exploration industries. In addition, there are numerous literatures on the subject of explosives, booby traps, and incendiary devices, such as the *Anarchist Cookbook Series*, the *Poor Man's James Bond*, and numerous previously published military field manuals available as e-books. Furthermore, the Internet is also a source of information for constructing or building IEDs.

The first thing a bomber will do is to identify his or her target. The target will vary based on the bombers motives. For example, if the bomber is a disgruntled employee looking to get even with his or her employer, the intended target may be his or her employer's place of business. However, if the disgruntled employee has no direct issues with his or her employer but the issues are directed toward his or her supervisor, the intended target may be the supervisor's residence or vehicle, or the individual themselves. Other potential targets include commercial operations, public buildings, public safety buildings, military installations or facilities, schools, and public utilities. Motivating factors for using a bomb include malicious and destructive behavior, personal animosity toward an individual or organization, political, extremist, or radical behavior toward the establishment, civil rights, labor disputes, racketeering, and monetary gain.

Explosions can be broadly defined as the sudden and rapid escape of gases from a confined space, accompanied by high temperatures, violent shock, and loud noise. The explosion caused by an IED can be chemical, mechanical, or nuclear. The blast pressure wave has two distinct phases that exert two different types of pressure on any object in its path: positive and negative.

At the point of detonation, the pressures actually compress the atmosphere producing the shock front. The shock front, followed by a positive pressure wave, moves outward, applying a sudden shattering, hammering blow to any object in its path. This cyclone-like sudden and violent push lasts only a fraction of a second. Positive pressure pushes air away, creating a partial vacuum from the point of detonation outward. The partial vacuum causes the compressed and displayed atmosphere to reverse its movement and rush inward to fill the void. Air moving inward is compared to a strong gale wind, but lasts three times longer than the positive pressure phase, which is referred to as the negative pressure wave.

IEDs also have a fragmentation effect. This causes debris from the explosion to act as projectiles, thus becoming lethal to bystanders. The fragmentation effect of a simple pipe bomb causes metal and debris from the immediate surrounding area to travel outward in a straight line at approximately 2700 ft/s. In comparison to a high-explosive device, the fragmentation effect will not only cause components from the device itself to be projected by the explosion, but also results in projectiles because of cratering, and the destruction of buildings or the delivery device itself.

A common delivery device for an IED is a vehicle, which is referred to as a vehicle-borne IED (VBIED). Several historical bombing events had truck bombs as their delivery system including the World Trade Center in 1993 (6 deaths and 1040 injured); the Oklahoma City Federal Building in 1995 (168 deaths); the Khobar Towers in Saudi Arabia in 1996 (19 dead); and the U.S. Embassy in Kenya in 1998 (213 killed). The Bureau of Alcohol, Tobacco, Firearms and Explosives has compiled a list of vehicles, their explosive carrying capacity, the minimum evacuation distance, and their lethal range. The results are listed in Table 5.1.

TABLE 5.1

Vehicle Bomb Data, Compiled from the Bureau of Alcohol, Tobacco, Firearms and Explosives

Vehicle	Maximum Explosive Capacity	Lethal Air-Blast Range	Minimum Evacuation Distance	Falling Glass Hazard
Compact sedan	500 lb	100 ft	1500 ft	1250 ft
	227 kg (in trunk)	30 m	457 m	381 m
Full size sedan	1,000 lb	125 ft	1750 ft	1750 ft
	445 kg (in trunk)	38 m	534 m	534 m
Passenger van or	4,000 lb	200 ft	2750 ft	2750 ft
cargo van	1,818 kg	61 m	838 m	838 m
Small box van	10,000 lb	300 ft	3750 ft	3750 ft
(14-ft box)	4,545 kg	91 m	1143 m	1143 m
Box van or water/	30,000 lb	450 ft	6500 ft	6500 ft
fuel truck	13,636 kg	137 m	1982 m	1982 m
Semitrailer	60,000 lb	600 ft	7000 ft	7000 ft
	27,723 kg	183 m	2134 m	2134 m

5.3 SURVEILLANCE AND COUNTERSURVEILLANCE

Surveillance is the monitoring of the behavior, activities, or other changing information, usually of people for the purpose of influencing, managing, directing, or protecting them. This can include observation from a distance by means of electronic equipment (such as CCTV cameras), or interception of electronically transmitted information such as Internet traffic or phone calls; and it can refer to simple, relatively no- or low-technology methods such as human intelligence (HUMINT) and postal interception.

Surveillance is very useful to governments and law enforcement to maintain social control, recognize and monitor threats, and prevent or investigate criminal activity. With the advent of programs such as the Total Information Awareness program, technologies such as high-speed surveillance computers and biometrics software, and laws such as the Communications Assistance for Law Enforcement Act (CALEA), governments now possess an unprecedented ability to monitor the activities of their subjects. However, many civil rights and privacy groups, such as the Electronic Frontier Foundation and the American Civil Liberties Union, have expressed concern that by allowing continual increases in government surveillance of citizens we will end up in a mass surveillance society, with extremely limited or nonexistent political and/or personal freedoms. Fears such as this have led to numerous lawsuits such as *Hepting v. AT&T* (2009).

The vast majority of computer surveillance involves the monitoring of data and traffic on the Internet. In the United States, for example, under the CALEA, all phone calls and broadband Internet traffic such as e-mails, web traffic, and instant messaging are required to be available for unimpeded real-time monitoring by federal law enforcement agencies. There is far too much data on the Internet for human investigators to manually search through all of it. So automated Internet surveillance computers sift through the vast amount of intercepted Internet traffic and identify and report to human investigators traffic considered interesting by using certain "trigger" words or phrases, visiting certain types of websites, or communicating via e-mail or chat with suspicious individuals or groups. Billions of dollars per year are spent by agencies, such as the Information Awareness Office, the National Security Agency (NSA), and the Federal Bureau of Investigation (FBI), to develop, purchase, implement, and operate systems such as Carnivore, NarusInsight, and ECHELON to intercept and analyze all of this data, and extract only the information that is useful to law enforcement and intelligence agencies.

Computers can be a surveillance target because of the personal data stored on them. If someone is able to install software, such as the FBI's Magic Lantern and CIPAV, on a computer system, they can easily gain unauthorized access to this data. Such software could be installed physically or remotely. Another form of computer surveillance, known as van Eck phreaking, involves reading electromagnetic emanations from computing devices in order to extract data from them at distances of hundreds of meters. The NSA runs a database known as "Pinwale," which stores and indexes large numbers of e-mails of U.S. citizens and foreigners.

The official and unofficial tapping of telephone lines is widespread. In the United States, for instance, the CALEA requires that all telephone communications be available for real-time wiretapping by federal law enforcement and intelligence agencies. Two major telecommunications companies in the United States—AT&T and Verizon—have contracts with the FBI, requiring them to keep their phone call records easily searchable and accessible for federal agencies, in return for $1.8 million per year. Between 2003 and 2005, the FBI sent out more than 140,000 "National Security Letters" ordering phone companies to hand over information about their customers' calling and Internet histories. About half of these letters requested information on U.S. citizens. Human agents are not required to monitor most calls. Speech-to-text software creates machine-readable text from intercepted audio, which is then processed by automated call-analysis programs, such as those developed by agencies such as the Information Awareness Office.

Law enforcement and intelligence services in the United Kingdom and the United States possess technology to activate the microphones in cell phones remotely, by accessing the phones' diagnostic or maintenance features in order to listen to conversations that take place near the person who holds the phone.

Mobile phones are also commonly used to collect location data. The geographical location of a mobile phone and thus the person carrying it can be determined easily even when the phone is not being used, using a technique known as multilateration to calculate the differences in time for a signal to travel from the cell phone to each of several cell towers near the owner of the phone. The legality of such techniques has been questioned in the United States, in particular whether a court warrant is required. Records for one carrier alone (*Sprint*) showed that in a given year federal law enforcement agencies requested customer location data 8 million times.

Surveillance cameras are video cameras used for the purpose of observing an area. They are often connected to a recording device or IP network, and may be watched by a security guard or law enforcement officer. Cameras and recording equipment used to be relatively expensive and required human personnel to monitor camera footage, but analysis of footage has been made easier by automated software that organizes digital video footage into a searchable database and by video analysis software. The amount of footage is also reduced by motion sensors that only record when motion is detected. With less expensive production techniques, surveillance cameras are simple and inexpensive enough to be used in home security systems and for everyday surveillance.

In the United States, the Department of Homeland Security (DHS) awards billions of dollars per year in Homeland Security grants for local, state, and federal agencies to install modern video surveillance equipment. For example, the city of Chicago, Illinois, used a $5.1 million Homeland Security grant to install 250 surveillance cameras, and connect them to a centralized monitoring center, along with its preexisting network of more than 2000 cameras, in a program known as Operation Virtual Shield. Speaking in 2009, Chicago Mayor Richard Daley announced that Chicago would have a surveillance camera on every street corner by the year 2016.

As part of China's Golden Shield Project, several U.S. corporations, including IBM, General Electric, and Honeywell, have been working closely with the Chinese government to install millions of surveillance cameras throughout China, along with advanced video analytics and facial recognition software, which will identify and track individuals everywhere they go. They will be connected to a centralized database and monitoring station, which will, upon completion of the project, contain a picture of the face of every person in China: over 1.3 billion people. Lin Jiang Huai, the head of China's Information Security Technology office, which is in charge of the project, credits the surveillance systems in the United States and the United Kingdom as the inspiration for what he is doing with the Golden Shield Project.

The Defense Advanced Research Projects Agency (DARPA) is funding a research project called "Combat Zones That See" that will link up cameras across a city to a centralized monitoring station, identify and track individuals and vehicles as they move through the city, and report "suspicious" activity such as waving arms, looking side to side, and standing in a group. At Super Bowl XXXV in January 2001, police in Tampa, Florida, used Identix's facial recognition software, "FaceIt," to scan the crowd for potential criminals and terrorists in attendance at the event; it found 19 people with pending arrest warrants.

Governments initially claim that cameras are meant to be used for traffic control, but many of them end up using them for general surveillance. For example, Washington, DC, had 5000 "traffic" cameras installed under this premise, and then after they were all in place, networked them all together and then granted access to the Metropolitan Police Department, so they could perform "day-to-day monitoring." The development of centralized networks of CCTV cameras watching public areas and linked to computer databases of people's pictures and identity (biometric data) that are able to track people's movements throughout the city and identify whom people have been with has been argued by some to present a risk to civil liberties.

One common form of surveillance is to create maps of social networks based on the data from social networking sites as well as from traffic analysis information from phone call records such as those in the NSA call database and others. These social network "maps" are then data mined to extract useful information such as personal interests, friendships and affiliations, wants, beliefs, thoughts, and activities. Many U.S. government agencies such as the DARPA, the NSA, and the DHS are investing heavily in research involving social network analysis. The intelligence community believes that the biggest threat to U.S. power comes from decentralized, leaderless, geographically dispersed groups of terrorists, subversives, extremists, and dissidents. These types of threats are most easily countered by finding important nodes in the network and removing them. To do this requires a detailed map of the network.

AT&T developed a programming language called "Hancock," which is able to sift through enormous databases of phone call and Internet traffic records, such as the NSA call database, and extract "communities of interest" groups of people who call each other regularly, or groups that regularly visit certain sites on the Internet. AT&T originally built the system to develop "marketing leads," but the FBI has regularly requested such information from phone companies such as AT&T without a warrant, and after using the data stores all information received in its own databases, regardless of whether or not the information was ever useful in an investigation. Some people believe that the use of social networking sites is a form of "participatory surveillance," where users of these sites are essentially performing surveillance on themselves, putting detailed personal information on public websites where it can be viewed by corporations and governments. About 20% of employers have reported using social networking sites to collect personal data on prospective or current employees.

Biometric surveillance is any technology that measures and analyzes human physical and/or behavioral characteristics for authentication, identification, or screening purposes. Examples of physical characteristics include fingerprints, DNA, and facial patterns. Examples of mostly behavioral characteristics include gait, a person's manner of walking, or voice. Facial recognition is the use of the unique configuration of a person's facial features to accurately identify them, usually from surveillance video. Both the DHS and DARPA are heavily funding research into facial recognition systems. The Information Processing Technology Office ran a program known as Human Identification at a Distance, which developed technologies that are capable of identifying a person at up to 500 ft by his or her facial features. Another form of behavioral biometrics, based on affective computing, involves computers recognizing a person's emotional state based on an analysis of his or her facial expressions, how fast he or she is talking, the tone and pitch of their voice, their posture, and other behavioral traits. This might be used for instance to see if a person is acting "suspicious," looking around furtively, "tense" or "angry" facial expressions, and waving arms.

A more recent development is DNA fingerprinting, which looks at some of the major markers in the body's DNA to produce a match. The FBI is spending $1 billion to build a new biometric database, which will store DNA, facial recognition data, iris or retina (eye) data, fingerprints, palm prints, and other biometric data of people living in the United States. The computers running the database are contained in an underground facility about the size of two American football fields. The Los Angeles Police Department is installing automated facial recognition and license plate recognition devices in its squad cars, and providing handheld face scanners, which officers will use to identify people while on patrol.

Facial thermographs are in development, which allow machines to identify certain emotions in people, such as fear or stress, by measuring the temperature generated by blood flow to different parts of their face. Law enforcement officers believe that this has potential for them to identify when a suspect is nervous, which might indicate that they are hiding something, lying, or worried about something.

Aerial surveillance is the gathering of surveillance, usually visual imagery or video, from an airborne unmanned aerial vehicle (UAV), such as a helicopter, quadrocopter, or a spy plane. Military surveillance aircraft use a range of sensors such as radar to monitor the battlefield. Digital imaging technology, miniaturized computers, and numerous other technological advances over the past decade have contributed to rapid advances in aerial surveillance hardware such as microaerial vehicles with forward-looking infrared and high-resolution imagery capable of identifying objects at extremely long distances. For instance, the MQ-9 Reaper, a U.S. drone plane used for domestic operations by the DHS, carries cameras that are capable of identifying an object the size of a milk carton from altitudes of 60,000 ft, and has forward-looking infrared devices that can detect the heat from a human body at distances of up to 60 km.

The U.S. DHS is in the process of testing UAVs to patrol the skies over the United States for the purposes of critical infrastructure protection, border patrol, "transit monitoring," and general surveillance of the U.S. population. The Miami-Dade Police Department ran tests with a vertical takeoff and landing UAV from Honeywell, which is planned to be used in SWAT operations. Houston's police department has been testing fixed-wing UAVs for use in "traffic control." The United Kingdom, as well, is working on plans to build up a fleet of surveillance UAVs ranging from microaerial vehicles to full-size drones, to be used by police forces throughout the United Kingdom. In addition to their surveillance capabilities, UAVs are capable of carrying tasers for "crowd control," or weapons for killing enemy combatants. Programs such as the Heterogeneous Aerial Reconnaissance Team (HART) developed by DARPA have automated much of the aerial surveillance process. They have developed systems consisting of large team drone planes that pilot themselves automatically decide who is "suspicious" and how to go about monitoring them, coordinate their activities with other drones nearby, and notify human operators if something suspicious is occurring. This greatly increases the area that can be continuously monitored, while reducing the number of human operators required. Thus a swarm of automated, self-directing drones can automatically patrol a city and track suspicious individuals, reporting their activities back to a centralized monitoring station.

Data mining is the application of statistical techniques and programmatic algorithms to discover previously unnoticed relationships within the data. Data profiling in this context is the process of assembling information about a particular individual or group in order to generate a profile, that is, a picture of their patterns and behavior. Data profiling can be an extremely powerful tool for psychological and social network analysis. A skilled analyst can discover facts about a person that they might not even be consciously aware of themselves. Economic transactions such as credit card purchases and social ones such as telephone calls and e-mails in modern society create large amounts of stored data and records. In the past, this data was documented in paper records, leaving a "paper trail," or was simply not documented at all. Correlation of paper-based records was a laborious process that required HUMINT operators to manually dig through documents, which was time-consuming and incomplete, at best. But today many of these records are electronic, resulting in an "electronic trail." Every use of a bank machine, payment by credit card, use of a phone card, call from home, checked out library book, rented video, or otherwise complete recorded transaction generates an electronic record. Public records such as birth, court, tax, and other records are increasingly being digitized and made available online. In addition, due to laws like CALEA, web traffic and online purchases were also available for profiling. Electronic record-keeping makes data easily collectable, storable, and accessible so that high-volume, efficient aggregation and analysis is possible at significantly lower costs. Information relating too many of these individual transactions is often easily available because it is generally not guarded in isolation, since the information, such as the title of a movie a person has rented, might not seem sensitive. However, when many such transactions are aggregated they can be used to assemble a detailed profile revealing the actions, habits, beliefs, locations frequented, social connections, and preferences of the individual. This profile is then used, by programs such as analysis, dissemination, visualization, insight, and semantic enhancement (ADVISE) and threat and local observation notice (TALON), to determine whether the person is a military, criminal, or political threat.

In addition to its own aggregation and profiling tools, the government is able to access information from third parties, for example, banks, credit companies, or employers, by requesting access informally, by compelling access through the use of subpoenas or other procedures, or by purchasing data from commercial data aggregators or data brokers. The United States has spent $370 million on its 43 planned fusion centers, which are a national network of surveillance centers that are located in over 30 states. The centers will collect and analyze vast amounts of data on U.S. citizens. It will get this data by consolidating personal information from sources such as a state driver's licensing agencies, hospital records, criminal records, school records, credit bureaus, and banks and placing this information in a centralized database that can be accessed from all of the centers, as well as other federal law enforcement and intelligence agencies. Under *United States v. Miller* (1976), data held by third parties is generally not subject to Fourth Amendment warrant requirements.

Corporate surveillance is the monitoring of a person or group's behavior by a corporation. The data collected is most often used for marketing purposes or sold to other corporations, but is also regularly shared with government agencies. It can be used as a form of business intelligence, which enables the corporation to better tailor their products and/or services to be desirable by their customers. Or the data can be sold to other corporations, so that they can use it for the aforementioned purpose. In addition, it can be used for direct marketing purposes, such as the targeted advertisements on Google and Yahoo, where ads are targeted to the user of the search engine by analyzing their search history and e-mails, if they use free webmail services, which is kept in a database. For instance, Google, the world's most popular search engine, stores identifying information for each web search. An IP address and the search phrase used are stored in a database for up to 18 months. Google also scans the content of e-mails of users of its Gmail webmail service in order to create targeted advertising based on what people are talking about in their personal e-mail correspondences. Google is, by far, the largest Internet advertising agency; millions of sites place Google's advertising banners and links on their websites in order to earn money from visitors who click on the ads. Each page containing Google advertisements adds, reads, and modifies "cookies" on each visitor's computer. These cookies track the user across all of these sites and gather information about their web surfing habits, keeping track of which sites they visit and what they do when they are on these sites. This information, along with the information from their e-mail accounts and search engine histories, is stored by Google to use for building a profile of the user to deliver better-targeted advertising.

According to the American Management Association and the ePolicy Institute (2008) that undertake an annual quantitative survey about electronic monitoring and surveillance with approximately 300 U.S. companies, "more than one fourth of employers have fired workers for misusing e-mail and nearly one third have fired employees for misusing the Internet." More than 40% of the companies monitor e-mail traffic of their workers, and 66% of corporations monitor Internet connections. In addition, most companies use software to block non-work-related websites such as sexual or pornographic sites, game sites, social networking sites, entertainment sites, shopping sites, and sport sites. The American Management Association and the ePolicy Institute (2008) also stress that companies "tracking content, keystrokes, and time spent at the keyboard … store and review computer files … monitor the blogosphere to see what is being written about the company, and … monitor social networking sites." Furthermore, about 30% of the companies had also fired employees for non-work-related e-mail and Internet usage such as "inappropriate or offensive language" and "viewing, downloading, or uploading inappropriate or offensive content" (Allmer 2012).

The U.S. government often gains access to these databases, either by producing a warrant for it or by simply asking. The DHS has openly stated that it uses data collected from consumer credit and direct marketing agencies such as Google for augmenting the profiles of individuals whom it is monitoring. The FBI, DHS, and other intelligence agencies have formed an "information-sharing" partnership with over 34,000 corporations as part of their Infragard program. The U.S. federal government has gathered information from grocery store "discount card" programs, which track customers' shopping patterns and store them in databases in order to look for "terrorists" by analyzing shoppers' buying patterns.

Organizations that have enemies who wish to gather information about the groups' members or activities face the issue of infiltration. In addition to operatives infiltrating an organization, the surveillance party may exert pressure on certain members of the target organization to act as informants, for example, to disclose the information they hold on the organization and its members. Fielding operatives can be an expensive and daunting task, and for governments with wide-reaching electronic surveillance tools at their disposal the information recovered from operatives can often be obtained from less problematic forms of surveillance. Despite this, human infiltrators are still common today. For instance, in 2007 documents surfaced showing that the FBI was planning to field a total of 15,000 undercover agents and informants in response to an antiterrorism directive sent out by George W. Bush in 2004 that ordered intelligence and law enforcement agencies to increase their HUMINT capabilities.

On May 25, 2007, the U.S. Director of National Intelligence, Michael McConnell, authorized the National Applications Office (NAO) of the DHS to allow local, state, and domestic federal agencies to access imagery from military intelligence satellites and aircraft sensors, which can now be used to observe the activities of U.S. citizens. The satellites and aircraft sensors will be able to penetrate cloud cover, detect chemical traces, and identify objects in buildings and "underground bunkers," and will provide real-time video at much higher resolutions than the still images produced by programs such as Bing Maps or Google Earth. One of the simplest forms of identification is the carrying of credentials. Some nations have an identity card system to aid identification, while many, such as Britain, are considering it but face public opposition. Other documents, such as passports, driver's licenses, library cards, banking or credit cards, are also used to verify identity. If the form of the identity card is "machine-readable," usually using an encoded magnetic stripe or identification number (such as a Social Security number), it corroborates the subject's identifying data. In this case, it may create an electronic trail when it is checked and scanned, which can be used in profiling, as mentioned above.

Radio frequency identification (RFID) tagging is the use of very small electronic devices called "RFID tags," which are applied to or incorporated into a product, animal, or person for the purpose of identification and tracking using radio waves. The tags can be read from several meters away. They are extremely inexpensive, costing a few cents per piece, so they can be inserted into many types of everyday products without significantly increasing the price, and can be used to track and identify these objects for a variety of purposes. Some critics have expressed fears that people will soon be tracked and scanned everywhere they go. In Mexico, for example, 160 workers at the Attorney General's office were required to have the chip injected for identity verification and access control purposes. In a 2003 editorial, CNET News.com chief political correspondent, Declan McCullagh, speculated that, soon, every object that is purchased, and perhaps ID cards, will have RFID devices in them, which would respond with information about people as they walk past scanners such as what type of phone they have, what type of shoes they have on, which books they are carrying, and what credit cards or membership cards they have. This information could be used for identification, tracking, or targeted marketing. As of 2012, this has largely not come to pass.

In the United States, police have planted hidden GPS tracking devices in people's vehicles to monitor their movements, without a warrant. In early 2009, they were arguing in court that they have the right to do this. Several cities are running pilot projects that require parolees to wear GPS devices to track their movements when they get out of prison.

As more people use faxes and e-mail, the significance of surveillance of the postal system is decreasing, in favor of Internet and telephone surveillance. But interception of post is still an available option for law enforcement and intelligence agencies, in certain circumstances. The U.S. Central Intelligence Agency and the FBI have performed 12 separate mail-opening campaigns targeted toward U.S. citizens. In one of these programs, more than 215,000 communications were intercepted, opened, and photographed.

Some supporters of surveillance systems believe that these tools protect society from terrorists and criminals. Other supporters simply believe that there is nothing that can be done about it, and that people must become accustomed to having no privacy. As Sun Microsystems CEO Scott McNealy (Sprenger 1999) said, "You have zero privacy anyway. Get over it." Another common argument is, "If you aren't doing something wrong then you don't have anything to fear," which follows that if you are engaging in unlawful activities, in which case you do not have a legitimate justification for your privacy. However, if you are following the law the surveillance would not affect you.

Some critics state that the claim made by supporters should be modified to read "as long as we do what we're told, we have nothing to fear." For instance, a person who is part of a political group that opposes the policies of the national government might not want the government to know his or her names and what he or she has been reading, so that the government cannot easily subvert his or her organization, arrest, or kill them. Other critics state that while a person might not have anything to hide right now, the government might later implement policies that they do wish to oppose, and that opposition might then be impossible due to mass surveillance, enabling the government to identify and remove political threats. Further, other critics point to the fact that most people *do* have things to hide. For example, if a person is looking for a new job, he or she might not want his or her current employer to know this. In addition, a significant risk of private data collection stems from the fact that this risk is too much unknown to be readily assessed today. Storage is cheap enough to have data stored forever, and the models that will be employed to analyze data a decade from now cannot reasonably be foreseen. Programs such as the Total Information Awareness and laws such as the CALEA have led many groups to fear that society is moving toward a state of mass surveillance with severely limited personal, social, political freedoms, where dissenting individuals or groups will be strategically removed in COINTELPRO-like purges.

Kate Martin (Warrick 2007), of the Center for National Security Studies, said of the use of military spy satellites to monitor the activities of U.S. citizens, "they are laying the bricks one at a time for a police state." Some point to the blurring of lines between public and private places and the privatization of places traditionally seen as public such as shopping malls and industrial parks as illustrating the increasing legality of collecting personal information. Traveling through many public places such as government offices is hardly optional for most people, yet consumers have little choice but to submit to companies' surveillance practices. Surveillance techniques are not created equal; among the many biometric identification technologies, for instance, face recognition requires the least cooperation. Unlike automatic fingerprint reading, which requires an individual to press a finger against a machine, this technique is subtle and requires little to no consent.

Some critics, such as Michel Foucault, believe that in addition to its obvious function of identifying and capturing individuals who are committing undesirable acts, surveillance also functions to create in everyone a feeling of always being watched, so that they become self-policing. This allows the state to control the populace without having to resort to physical force, which is expensive and otherwise problematic.

Numerous civil rights groups and privacy groups oppose surveillance as a violation of people's right to privacy. Such groups include the Electronic Privacy Information Center, the Electronic Frontier Foundation, and the American Civil Liberties Union. There have been several lawsuits, such as by groups or individuals, opposing certain surveillance activities. Legislative proceedings such as those that took place during the Church Committee, which investigated domestic intelligence programs such as COINTELPRO, have also weighed the pros and cons of surveillance.

Countersurveillance is the practice of avoiding surveillance or making surveillance difficult. Developments in the late twentieth century have caused countersurveillance to dramatically grow in both scope and complexity, such as the Internet, increasing prevalence of electronic security systems, high-altitude and possibly armed UAVs, and large corporate and government computer databases. Inverse surveillance is the practice of the reversal of surveillance on other individuals or groups such as citizens photographing police, although this is more of a political reference, as some groups specifically aim to harass police and retaliate for their own criminal pasts, as well as ongoing criminal activity, as was the case with regard to Rodney King's continual illegal activities. This was confirmed upon his death when authorities recorded ongoing illegal drug and alcohol use. Well-known examples are George Holliday's recording of the Rodney King beating and the organization Copwatch, which attempts to monitor police officers to prevent police brutality or for other nefarious uses such as blackmailing. It is well-known that certain criminal rights groups seek to use countermethods in efforts to deter detection of criminal activities, as was the case with Rodney King's historical criminal record, and the intentional setup to use countersurveillance as a form of entrapment to record police tactics to combat crimes. Countersurveillance can also be used in applications to prevent corporate spying or to track other criminals by certain criminal entities. Moreover, it can be used to deter stalking methods used by various entities and organizations. *Sousurveillance* is inverse surveillance, involving the recording by private individuals, rather than government or corporate entities.

5.4   CONDUCTING A SITE SECURITY SURVEY

There are numerous security threats that can impact an organization including accidents involving employees and visitors, natural disasters, data loss, fraud, industrial espionage, vandalism, threats to people, physical theft, and brand and reputation attacks. The number and merging of risk is causing organizations to rethink their security risk strategy. Even those organizations that have audited their security, risk procedures, and mitigation measures may find that they are not as resilient as they initially thought. An organization may wish to revisit its security needs for a multitude of reasons such as a recent spike in criminal activity in its area of operation, a change in regulatory requirements, recent threats to the organization, such as targeting the organization and/or personnel, or a major event such as September 11, 2001.

Many security operations already collect a myriad of information on security-related incidents as well as linking the information to specific programs. For example, information that would be collected or logged would include the number of incidents that have occurred within a specified time frame, such as six months; the types of incidents that have occurred; the primary methods and likely sources of these incidents; and their impact on the organization. Many organizations have planning tools, policies, procedures, and technologies in place; as a result, part of the review process would be to determine their effectiveness in dealing with a multitude of threats affecting the organization. Policies and procedures on security operations must be in line with the business and/or operational objectives of the organization. Having confidence in security is essential for any organization. The reason is that depending on the results of the security survey and risk assessment there may be a need for additional spending on security products and services.

Each potential threat to an organization must be explored thoroughly; solutions and recommendations must conform to regulatory standards and match the objectives of the organization. Certain threats such as those that are human induced can be prevented through an effective security and safety program, while others such as natural disasters are outside of their control. However, organizations can increase their coping capacity and resilience to these events by reducing their vulnerability and overall risk, thereby allowing them to continue their operations even under direst circumstances.

Organizations need to develop a plan and course of action to address the various threats that can impact their business and/or operations. By focusing on the right actions, an organization can make improvements at a relatively low cost. A comprehensive site security survey of their facilities can help meet this need. The site security survey is a tool for determining and identifying gaps in the perimeter and internal physical security of a facility, including electronic access control, intrusion detection, CCTV, and video analytics. Other systems such as environmental controls, fire alarms, and fire extinguishers also affect these concerns. In order to be effective, a site security survey should also consider the interaction of all building systems, surrounding environment, policies, and procedures.

The first stage in the security survey process is site preparation and planning. This stage will give the client an idea of what will be covered during the survey, a review of the basic principles of security, and the level of security the organization aims to achieve. The second stage is a review of the surrounding environment and area, including adjacent buildings, properties, walkways, and parks. The third stage is a review of the facilities perimeter and access control points. The fourth stage is a review of the building or facility, its function, its access points, and its integration to other buildings. The fifth stage is a review of existing systems such as the location of every security component including locking mechanisms, card readers, cameras, window covers, and intrusion detection devices. The sixth stage is a review of internal assets including information technology and communications infrastructure, environmental control, fire protection systems, and power sources. The seventh and final stage is a review of security policies and procedures, security regulatory requirements, and safety of personnel reporting to the facility.

The first step in the security survey process is site preparation and planning. This stage will allow the security survey team to look for conditions that may impede or preclude the effective use of physical security control measures to reduce the opportunities for human-induced threats. The preparation stage should include an analysis of incidents and crimes that have occurred on site and in the surrounding area. This stage should also identify any person(s) or organization(s) that have been a victim of crime in the surrounding area. This information can be obtained from local police services that maintain a database of reported crimes. In addition, this stage should also identify if there are any programs such as a Business Watch or Neighbourhood Watch scheme in the area. Furthermore, this stage should also identify potential or possible targets in the area, the risk against those targets, and what effects it will have? The survey team should consider all impacts including financial implications, insurance implications, loss of information or data, personnel morale, organizational image, and business continuity.

The second stage of the security survey process is a review of the surrounding areas including the streets, industrial estates, retail parks, and pedestrian walkways. If the surrounding area is well maintained, it will give the impression to an assailant that it is cared for and that security measures are currently in place. In addition, the security team should make note of any features that may assist an assailant in committing a crime such as trees, walls, and bushes to hide behind. However, features such as trees could also be used as a security control measure. This, however, will depend on the location of the facility and its function. For example, a government building located in a downtown city core could have a row of trees in front of it to act as a vehicle barrier providing additional exterior protection. Typically, a well-maintained area will act as a deterrent to criminal activity since it increases the effort required to committing a crime and the risk of being caught.

The third stage in the security review process is a review of the facility's perimeter and access control points. The security survey should make note of weak areas along the perimeter. They should also ensure that perimeter boundaries are secure and well maintained, especially where they may not be clearly visible. All entrances should be secured even when frequent access is not required. The survey team should also make note of any areas or features an assailant may be able to use to gain unauthorized access to the facility. Such features may include bins to climb wall and/or fences, tools or materials that can be used for break-ins, hiding places and areas of poor lighting, and visibility from the outside. Bins or skips are also potential targets for arson.

The fourth stage in the security survey process is a review of the facility including its function and integration to other buildings. The survey should examine the exterior of the building, including walls, windows, doors, skylights, and roof. It should also include sheds, outside storage areas, garages, cellars, and air vents, which are all potential security risks since an assailant could use them to gain unauthorized access to the facility. In addition, it should consider any gaps where a tool could be used to prize open an entry point or window. Based on these observations, the security survey team should be able to make recommendations on measures that will deter potential assailants. For example, the longer an assailant is visible from the outside and the more noise they make, the easier it is for them to be noticed.

The fifth stage of the security survey process is a review of existing security systems. The security survey should ensure that all doors and entrance points are fitted with the appropriate locking mechanisms and that they are sufficient. In addition, all locks on doors should be tested to the appropriate standards. The team should also ensure that all unattended external doors are secure and that the fire door mechanisms are working correctly. All windows should have locks that are secured with a key. For those organizations that are in medium- to high-risk areas, additional physical security features can be added to windows such as grills, shutters, mesh and hatch covers, reinforcing bars, laminated glass, and tinted glass to reduce visibility of valuable items.

The sixth stage of the security survey process is a review of internal assets. This would include information technology and communications infrastructure, environmental controls, fire protection systems, and power sources. The survey should make note of the make, model, and serial number of all equipment. At this point, the team should also consider security marking all equipment. Environmental controls, fire protection systems, and backup power sources should be tested to determine if they are working properly. If possible, high-value equipment, or equipment essential for operations, should be secured in a separate room with controlled access. The survey should also make note of any security features on equipment such as tampering sensors and response sensors. If there is an alarm system installed, the team should test that it works and determine if it is serviced regularly. They should also note who is responsible for monitoring it and who will respond to an alarm if activated.

Once the security survey team has completed its survey it will then be able to highlight any security risks, threats, and possible targets. For the final, that is, the seventh stage, the teams' notes should be documented and reviewed to ensure nothing has been left out. At this stage, the team will also be responsible for reviewing the current organizational policies and procedures with regard to security and comparing the current program against regulatory and compliance requirements. The team will also need to consider the control of stock, if applicable, personnel procedures, personnel safety, confidentiality, and OPSEC procedures for personnel reporting to the facility. At this point, the team and members of the organizations' security staff can develop a plan with a list of action items and recommendations to improve overall security at the facility. Recommendation could potentially involve adding additional lighting, cutting back hedges, increasing security to access points, setting up computer passwords, carrying equipment in the boot of the car, removing equipment from a vehicle overnight, and making sure equipment is stored securely and doors are locked to the room containing the essential equipment.

5.5   GEOSPATIAL INTELLIGENCE

Geospatial intelligence (GEOINT) is intelligence derived from the exploitation and analysis of imagery and geospatial information that describes, assesses, and visually depicts physical features and geographically referenced activities on the earth. GEOINT consists of imagery, imagery intelligence (IMINT), and geospatial information.

GEOINT encompasses all aspects of imagery including capabilities formerly referred to as advanced GEOINT and imagery-derived intelligence (MASINT) and geospatial information and services (GI&S). GI&S was formerly referred to as mapping, charting, and geodesy. It includes, but is not limited to, data ranging from the ultraviolet through the microwave portions of the electromagnetic spectrum, as well as information derived from the analysis of accurate imagery and geospatial data. GEOINT also includes information derived from the processing, exploitation, and analysis of varying types of data. The various types of data may include spectral, spatial, temporal, radiometric, phase history, polarimetric, fused products, ancillary, which are needed for data processing and exploitation, and signature information. These types of data can be collected on stationary and moving targets by electro-optical-related sensors and nontechnical means.

GEOINT is an intelligence discipline involving the exploitation and analysis of geospatial data and information to describe, assess, and visually depict physical features, both natural and constructed, and geographically referenced activities on the earth. GEOINT data sources include imagery and mapping data, whether collected by commercial satellite, government satellite, aircraft such as UAV or reconnaissance aircraft, or by other means, such as maps and commercial databases, census information, GPS waypoints, utility schematics, or any discrete data that have locations on earth.

Geospatial data can usually be applied to the output of a collector or collection system before it is processed, for example, data that was sensed. Geospatial information is geospatial data that have been processed or had value added to it by a human or machine process. Geospatial knowledge is structuring of geospatial information, accompanied by an interpretation or analysis. Simplified definition of GEOINT is data, information, and knowledge gathered about enemies or potential enemies that can be referenced to a particular location on, above, or below the earth's surface. The intelligence gathering method could include imagery, signals, measurements and signatures, and human sources, for example, IMINT, signals intelligence (SIGINT), MASINT, and HUMINT, as long as a geolocation can be associated with the intelligence.

GEOINT can also be viewed as the unifying structure of the earth's natural and constructed features, whether as individual layers in a geospatial database or as a composite set into a map or chart, imagery representations of the earth, and the presentation of the existence of data, information, and knowledge derived from and analysis of various intelligence sources and disciplines. The intelligence, defense, homeland security, and natural disaster assistance communities would all benefit from this unifying structure of spatial and intelligence data presented on a globe.

This unifying aspect of geospatial intelligence can be viewed as a global extent geographic information system (GIS) to which all community members contribute by geotagging their content. It has been suggested that "GEOINT" is just a new term used to identify a broad range of outputs from intelligence organizations that use a variety of existing spatial skills and disciplines. Spatial thinking as applied to GEOINT can synthesize any intelligence or other data that can be conceptualized in a geographic spatial context. GEOINT can be derived entirely independent of any satellite or aerial imagery and can be clearly differentiated from IMINT.

GEOINT can be described as a product occurring at the point of delivery, by the extent of analysis that occurs to resolve particular problems, not by the type of data used. For example, a database containing a list of measurements of bridges obtained from imagery is "information," while the development of an output using analysis to determine those bridges that are able to be utilized for specific purposes could be termed "intelligence." Similarly, the simple measurement of river channel and floodplain cross sections is a classical geographic information–gathering activity, while the process of selecting a river reach that matches a certain profile for a specific purpose is an analytical activity, and the output could be termed an "intelligence product." In this type of application, GEOINT is used by organizations that require definitions of their outputs for descriptive and capability development purposes.

GEOINT (Bacastow 2010) analysis can be defined as "seeing what everybody has seen and thinking what nobody has thought" or as "anticipating a target's mental map." This definition confirms that creating geospatial knowledge is a cognitive process that the geospatial analyst undertakes; it is an academic exercise that arrives at a conclusion through reasoning. Geospatial reasoning creates the connection between a geospatial problem and geospatial evidence. Here one set of activities, information searching, focuses around finding information, while another set of activities, sense making, focuses on giving meaning to the information. In turn, both activities of searching and sense making in geospatial analysis have been incorporated in the structured geospatial analytic method.

5.6 TECHNICAL SURVEILLANCE AND COUNTERMEASURES

The objective of the technical surveillance and countermeasures (TSCM) program is to detect and/or deter a wide variety of technologies and techniques that can be used to obtain unauthorized access to classified and sensitive unclassified information. The activities of a TSCM program include the following:

- Detection: Detect technical devices, security hazards, or physical security weaknesses that would permit the technical or physical penetration of sensitive activities.
- Nullification: Prevent, deter, or neutralize technical devices that may be employed within a security area. Neutralize security hazards and physical security weaknesses that are detected.
- Isolation: Restrict sensitive/classified activities to special areas established by cognizant security authorities.
- Education: Educate all personnel about the Technical Surveillance Countermeasures Program and the threat of a technical device.

At a minimum, an organization must ensure that all requirements, standards, and procedures contained in the *Technical Surveillance Countermeasures Manual* are followed. The program is not meant to be used in lieu of other security measures, but to enhance and reinforce existing security programs.

A TSCM program will include a program manager, operations manager, team manager, technicians, and a TSCM officer. The TSCM program manager is a corporate employee and is responsible for developing methods, techniques, standards, and procedures for the organization's TSCM program. The TSCM operations manager is a local employee and is responsible for implementing and managing the local TSCM program. The TSCM team is made up of a manager and two or more specially trained TSCM technicians, at least one of whom is certified to perform surveys, inspection, and monitoring activities. TSCM technicians also assist in determining if adequate physical security construction and controls are in place. The TSCM officer is responsible for identifying and nominating areas for TSCM services, provides assistance to the TSCM teams, coordinates briefing support, and tracks and accounts for responses to TSCM services.

The type of activities and services a TSCM team will provide include surveys, monitoring, inspections, technical advice, awareness briefings, and special services. TSCM services are used to circumvent the activities of hostile foreign governments, industrial espionage, terrorists, activists, and disgruntled employees. A TSCM survey is a thorough electronic, physical, and visual examination to detect listening devices, security hazards, and weaknesses. Monitoring is a limited service and is typically used to supporting classified meetings. Inspections are also a limited service and they are typically undertaken for a specific area or item. TSCM teams also provide advice and assistance in determining if adequate security standards are being met during the renovation and preconstruction phases. Awareness briefings are presentations to make individuals aware of the TSCM program. Special services are services provided to nonsecure areas such as residences, vehicles, and hotels.

Personnel within an organization must make note of the nature of the technical threat and the part they can play in the TSCM program. Illegal surveillance devices may be used for purposes other than collecting classified or sensitive unclassified information. As a result, personnel need to understand their responsibilities when they encounter this threat and how to report incidents of technical attack. Figure 5.2 illustrates the many different areas where a listening or surveillance device could be planted within an office environment.

Personnel should report all threats and incidents of a technical attack including

- Any discovery of a possible or actual technical surveillance penetration.
- Any discovery of a condition that could permit the technical surveillance of an area through equipment that, by reason of its normal design, installation, operation, or component deterioration, allows transmission of information.
- Any use of electronic surveillance equipment by persons not authorized to conduct electronic surveillance.
- Any discovery of an audio or video covert or overt surveillance device primarily designed and specifically fabricated for surreptitious acquisition of nonpublic communications or activities.
- Any evidence that might indicate that a technical surveillance device has been implanted in an area.
- Any documentation or communication giving reason to believe that your program, technology, or specific area has been or is going to be targeted with technical surveillance devices.

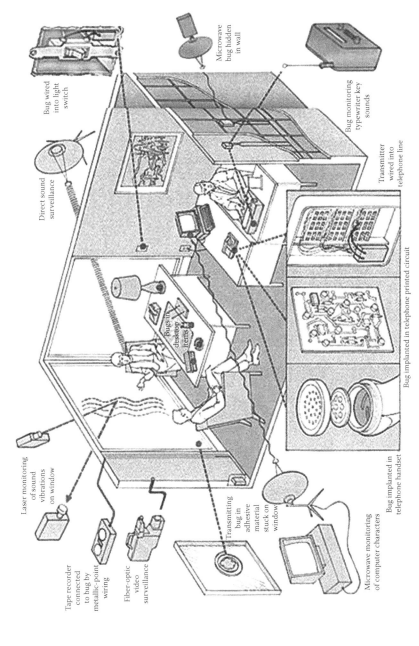

FIGURE 5.2 Potential areas where a listening or surveillance device could be planted within an office environment.

Bug wired into light switch

Microwave bug hidden in wall

Bug monitoring typewriter key sounds

Direct sound surveillance

Transmitter wired into telephone line

Bug implanted in telephone printed circuit

Laser monitoring of sound vibrations on window

Bugs in desktop items

Tape recorder connected to bug by metallic-point wiring

Fiber-optic video surveillance

Transmitting bug in adhesive material stuck on window

Microwave monitoring of computer characters

Bug implanted in telephone handset

Due to the classified nature of TSCM activities, all initial reporting of TSCM penetration or hazard investigations should be directed to the security manager, security director, or TSCM program personnel through procedures established within the organization. Upon discovery of a device, personnel should cease all classified discussions in the area, but make every effort to continue normal, routine activities. Do not mention what they have found, since it is still not clear as to who actually planted the device. They also must maintain continuous surveillance over the device and notify their technical surveillance countermeasures officer or technical surveillance countermeasures operations manager immediately by secure means.

Methods in which an organization can prevent or reduce the number of technical surveillance attacks include detection, nullification, isolation, and education. Detection is routine scheduled and nonscheduled surveys, inspections, monitoring, and special services. Once an organization has established a TSCM program, routine surveys and inspections should become the "norm." Nullification is the effort taken to negate the use of technical surveillance devices or make the emplacement of those devices more difficult. This is where the TSCM team and the facility managers for an organization can sit down and agree upon specific countermeasures (preconstruction and renovation) that can negate the use of such devices. Classified or sensitive activities should be isolated and undertaken in special security areas. Education, both formal and informal, should be ongoing; this includes regular technical surveillance countermeasures briefings and awareness of individual roles in the TSCM program.

5.7   PROTECTION AGAINST EXPLOSIVES AND BLAST EFFECTS

Seismic- and blast-resistant designs use similar analytical methodologies for evaluating performance-based design for varying levels of damage in response to varying levels of dynamic loads. Both design approaches recognize that it is economically and logistically infeasible to provide protection against all events; however, it is possible to provide an appropriate level of protection against plausible events at a reasonable cost. Both design approaches benefit from a risk assessment that evaluates the functionality, criticality, occupancy, site conditions, and design features of a building.

Natural hazards are more predictable than man-made hazards; for example, an earthquake event is more predictable than an explosive event, and earthquakes tend to affect more structures than blast events. In addition, explosive threats of the future will be very different from explosive threats of the past. As a result, owners of buildings are willing to accept varying levels of risk relative to the different events. Therefore, design limits as defined by allowable deformation, ductility, and other deformations will vary between these events.

Both seismic- and blast-resistant design methods take into account the time variance of the load. For example, for an earthquake load, the seismic motions induce forces that are proportional to the building mass, and the loads are typically applied uniformly over the foundation or other portions of the building. Whereas blast loads are not uniformly applied to a building or portions of, instead those portions of the structure that are closest to the blast will experience the highest loads. In addition, the structure's mass also contributes to its inertial resistance. Furthermore, under blast loading, in addition to the loads along the shear face of the building, there is also the ground tremor due to the blast and the vibration shear effect along the foundation of the building.

A structure's configuration can influence how loads are distributed throughout a building; characteristics such as size, shape, and location of structural elements are important for seismic and blast load designs. Structures that are designed to resist seismic forces have low height-to-base ratios, balanced resistance, symmetrical plans, uniform sections and elevations, the placement of shear walls and lateral bracing to maximize torsional resistance, short spans, direct load paths, and uniform floor heights. Although structures designed for blast resistance share the same attributes as those designed to withstand seismic loads, the reasons for doing so will differ between the two. For example, the geology of the site has a significant influence on the seismic motions that load the structure; with regard to blast loadings, the surrounding geology will influence the size of the blast crater and the reflectivity of the blast waves off the ground surface.

The similar detailed designs of structural elements to prevent deformation can be applied to structures subjected to both seismic and blast loads. The following is a listing of design recommendations for concrete columns, beams, and slabs:

- Concrete columns require lateral reinforcement to provide confinement to the core and prevent premature buckling of the rebar.
- Closely spaced ties and spiral reinforcement increase the ductility of a concrete compression element.
- Carbon fiber wraps and steel jacket retrofits provide confinement to existing structures.
- Steel column splices should be located away from regions of plastic hinging or must be detailed to develop the full moment capacity of the section.
- Closely spaced stiffeners must be used or, in the case of a blast-resistant design, the concrete encasement of the steel section is used to prevent local flange buckling.
- Reinforced concrete beam sections require resistance to positive and negative bending moments.
- Doubly reinforced sections possess greater ductility than singly reinforced counterparts.
- In order to increase the capacity of the section, steel beams should be constructed composite with the concrete deck. This increase, however, is not equally effective for both positive and negative moments. Although the composite slab may brace the top flange of the steel section, the bottom flange is vulnerable to buckling.

Buildings designed to protect its occupants and assets against explosives and other intentional threats are done so using an integrated design approach of both physical security hardening measures and progressive collapse mitigation measures. Progressive collapse mitigation measures enhance the building's performance against an intentional threat, regardless of the threat size; such design measures often lead to the selection of redundant and ductile systems.

In the United States, the various government agencies, including the U.S. DOD and the U.S. Department of State, have established antiterrorist design requirements for all government-owned facilities and assets. The design requirements will vary between the different departments, so will the level of detailed design within the technical documents. The following paragraphs discuss some of the design concepts for protecting the exterior of a building against both vehicle- and hand-delivered weapons.

When designing security measures into a building, there needs to be a balance between the security concerns of the organization and other design constraints such as accessibility, initial and life cycle costs, natural hazard mitigation, fire protection, energy efficiency, and aesthetics. The probability of a terrorist attack is very small. In addition, there is a desire for security not to interfere with daily operations of the building. However, because the effects of an attack can be catastrophic, there is a desire to incorporate measures that will save lives and minimize business interruption in the unlikely event of an attack. The countermeasures employed should be unobtrusive so that the facility presents an inviting, efficient environment and does not attract the undue attention of potential attackers. Security design needs to be a part of the overall all-hazard approach to the design, to ensure that the solution for explosion effects does not worsen the behavior of the building for other hazards. Furthermore, multiuse solutions that improve the building performance for blast and other considerations such as sustainability are to be encouraged.

The primary design objective for protecting a building against an explosive terrorist attack is to save the lives of those who visit or work in the building. The following is a list of design goals and objectives for protecting a building against a terrorist threat:

- Prevent progressive collapse that has historically caused the most fatalities resulting from terrorist incidents against buildings.
- Limit injuries to those inside the building due to the impact of flying debris and air blast during an incident, and to limit harm to innocent civilians near the building perimeter.
- Facilitate the rescue and recovery efforts for first responders by limiting the debris blocking access to the building and potential falling debris hazards that could harm rescue workers.
- Minimize disruption to building operations.

In order to achieve the above design objectives, designers must have a good understanding of the threat, loads, and damages resulting from explosions. The following paragraphs provide a detailed explanation of each.

The primary threat is a stationary vehicle weapon located along a secured perimeter line surrounding the building. The outermost perimeter line may be a public street secured using vehicle barriers and with limited secured access points. The size of the vehicle weapon considered outside the perimeter line may vary from hundreds to thousands of pounds of explosive depending on the criteria used. Weapon sizes vary depending on the specific criteria used and may be obtained from the contracting authority. The different threat scenarios against a building or facility and assumptions made are listed in Sections 5.7.1 through 5.7.4.

5.7.1 STATIONARY VEHICLE ALONG SECURED PERIMETER LINE

- Threat is to be considered on all sides of the building, with a public street or adjacent property lines along the secured perimeter line.
- It must be assumed that the highest air-blast loads are at the base of the building and decay with height; air-blast loads decay rapidly with distance.
- The required building setback may vary from tens to hundreds of feet depending on the criteria governing design.
- Design requirements given in the criteria documents are based on this setback being met; if this requirement cannot be met, the contracting authority needs to be contacted at the earliest time possible to discuss the implications.

5.7.2 STATIONARY VEHICLE IN A PARKING GARAGE OR LOADING DOCK

- It is assumed that the vehicle was able to bypass security and/or security screening.
- The amount of explosive in the vehicle is based on the maximum amount that can be carried without detection.
- A minimum separation distance between the secured surface parking areas and the building is specified.

5.7.3 MOVING VEHICLE ATTACK

- Barriers are placed along the secured perimeter with antiram capability consistent with the size of the weapon and maximum achievable velocity of 50 mph.
- For portions of the building that are parallel to adjacent streets, a maximum velocity of 30 mph is assumed.
- For street corners, or "T" intersections, a velocity of up to 50 mph is assumed.
- The weight of the vehicle may vary from a 4,000-lb car to a 15,000-lb truck depending on the criteria used.
- Landscaping features may also be used to stop a vehicle from ramming into the building including monumental stairs, permanent planters against the building, statues, concrete seating, water features, and others.
- Crime Prevention through Environmental Design (CPTED) concepts may also be effective.

5.7.4 Hand-Carried Weapon Placed against the Exterior Envelope

- The weapon is carried in a briefcase or backpack depending on the level of security screening provided outside the building.

In some circumstances, it is assumed that no weapon is able to pass through the outmost security point based on the OPSEC measures implemented. As a result, only the vehicle weapon is considered outside the secured perimeter line.

Explosive pressures used for design are much greater than the other loads considered. Explosive pressures decay extremely rapidly with time and space. The pressures generated increase linearly with the size of the weapon, measured in equivalent pounds of TNT, and decrease exponentially with the distance from the explosion. The duration of the explosion is extremely short, measured in thousandths of a second, or milliseconds. Pressures acting on the side of the building facing the explosion are amplified by factors that can be 10 times the incident pressure. This pressure is referred to as the reflected pressure, since it is not known which sides of the building the explosion will act on; all sides are designed for the worst-case scenario.

Explosive air-blast pressures have a negative or suction phase following the direct or positive pressure phase. The negative phase pressures can govern response in low-pressure regions causing windows to fail outward or sloped roof systems to fall off the building. Slender members such as exposed columns that have less surface area for the air blast to act on tend to be more sensitive to drag effects rather than direct pressure loading because the air blast tends to "wrap around" these members, lessening the time that the reflected air blast is acting. Rebound of the exterior envelope components following the explosion can pull the facade members off the building exterior. Rebound refers to the reversal of structural motion due to vibration rather than the reversal of loading direction. Since the design objective is to protect occupants, failure of the exterior envelope in the outward direction may be acceptable provided that the hazards of falling debris postevent and blocked egress points are avoided.

Immediately below the explosion, a crater will be formed that may cause damage to underground portions of the building that cause damage to the foundation and the subsurface roof and foundation walls that extend beyond the line of the structure. A proportion of the energy of the explosion is transmitted through the soil; this effect is similar to a high-intensity, short-duration earthquake that can disturb the functionality of computers and mechanical and/or electrical equipment. For above-ground explosions, ground vibration effects are negligible and are therefore neglected in design.

Damage due to explosions may be divided into direct air-blast effects and progressive collapse. Direct air-blast effects refer to damage caused by the high-intensity pressures of the air blast close to the explosive source. These may induce the failure of exterior envelope components. The severity of the damage is a function of the size of the weapon, its proximity to the exterior envelope, and the construction materials used. An example of an exterior envelope failure due to direct air-blast effects is the bombing of the Khobar Towers, the military housing complex in Daharain, Saudi Arabia, in 1996 (Figure 5.3).

FIGURE 5.3   Khobar Towers, military housing complex postblast scene, Daharain, Saudi Arabia, in 1996.

Progressive collapse refers to the spread of an initial local failure from structural element to structural element, resulting in a series of structural failures relative to impact zone. Local damage due to direct air-blast effects may or may not progress, depending on the design and construction of the building. An example of progressive collapse due to an explosive event is illustrated by the bombing of the Alfred P. Murrah Federal Building in Oklahoma City (Figure 5.4).

For a stationary vehicle weapon located outside the secured perimeter line, the building facade facing the explosion will be most affected, with the worst damage directly across from the weapon. For the design threat, there may or may not be glass breakage allowed depending on the criteria used and the level of protection assigned. There may be localized wall failure and/or damage; however, the frame will remain intact and the hazard presented by the damaged exterior envelope will be reduced. It is unlikely that this type of threat will initiate progressive collapse; however, this is dependent upon the size of the weapon used and its proximity to the building.

For hand-carried weapons placed next to the exterior envelope, the direct air-blast response will be more localized than for a vehicle weapons, but more severe, with damages and injuries extending to one or two floors. If the hand-carried weapon is strategically placed directly against a primary load bearing member such as a column, this would cause extensive damage potentially leading to progressive collapse. Critical structural members will need to be designed to resist progressive collapse either by considering the loss of the member in relation to the explosive load or by designing the member for the defined explosive loading.

Flying debris generated by nonstructural portions of the exterior envelope also has the capability to cause damage to the building envelope. An example of

FIGURE 5.4   Postblast scene, Alfred P. Murrah Federal Building, Oklahoma City, 1995.

this would be when sunshades or other lightweight materials are attached to the building exterior.

Progressive collapse can propagate vertically upward or downward from the source of the explosion, and it can propagate laterally from bay to bay. The design criteria address the issue of progressive collapse by using a variety of approaches including indirect design, direct load design, and alternate load path design approaches. The indirect design approach provides prescriptive design measures such as requiring special seismic detailing at column-spandrel connections, the design of the roof for a prescribed static loading, or using two-way systems to improve redundancy. In the direct load design approach, the member is designed to resist the loads generated by the design threat. An example of this would be the design of a column to resist the effects of a hand-carried weapon placed directly against it. The alternate load path design approach considers the effect of loss of each perimeter column or other critical load-bearing member on the stability of the building. For U.S. federal buildings, separate criteria documents are provided for progressive collapse prevention, which outline in detail what the designer needs to do to satisfy the requirements. Some solution concepts for meeting these requirements include the use of moment frames for steel buildings; doubling the columns along the perimeter; using cables embedded in concrete spandrel beams; designing load-bearing walls as deep beams to span across the damage; and designing each floor level to carry its own weight.

To reduce the hazards associated with fragments being propelled into the building interior, the envelope system is designed by keeping in mind the concepts of balanced design, ductile response, and redundancy. The use of dynamic nonlinear structural analysis methods is also beneficial when designing the exterior envelope of a building to resist air-blast effects.

Lighter systems are generally preferable for mitigating explosive effects, provided that they are designed to be ductile, redundant, balanced, and can resist the design load with the required response. Although heavier systems have added mass that can mitigate the effects of explosions, they may be more prone to brittle failure and can impart significantly larger loads into the supporting structure behind the envelope. The larger loads may cause structural failures and perhaps initiate progressive collapse. These more robust solutions have their place in high-risk buildings or in localized areas closest to the threat.

Today, the lighter, more flexible systems tend to be the preferred solution in the majority of buildings designed to resist air-blast effects. By permitting some permanent damage to the exterior envelope, which does not significantly increase the hazard to the occupants, it is possible to design lighter, more cost-effective systems that absorb energy through deformation and transmit lower forces into the connections and supporting structure, thus reducing the potential for more serious structural failures.

The design approach to be used for structural protective measures is to first design the building for conventional loads, then evaluate the response to explosive loads and augment the design, if needed, making sure that all conventional load requirements are still met. This ensures that the design meets all the requirements for gravity and natural hazards in addition to air-blast effects. However, methods used to design for explosive effects may not be appropriate for natural hazards or other loads; therefore, an iterative approach is used for structural protective measures.

As an air blast is a high-load, short-duration event, the most effective analytical technique is dynamic analysis, allowing the element to go beyond the elastic limit and into the plastic regime. Analytical methods can vary from structural handbook methods to single-degree-of-freedom (SDOF) methods to finite element methods (FEMs). Exterior envelope components such as columns, spandrels, and walls can be modeled by an SDOF system and then by solving the governing equations of motion by using numerical methods. For SDOF systems, material behavior may be modeled using idealized elastic, perfectly plastic stress-deformation functions, based on actual structural support conditions and strain rate–enhanced material properties. The model properties selected provide the same peak displacement and fundamental period as the actual structural system in flexure. Furthermore, the mass and the resistance function are multiplied by mass and load factors, which estimate the actual portion of the mass or load participating in the deflection of the member along its span. For more complex elements, the engineer must use finite element numerical time integration techniques and/or explosive testing. An SDOF approach will be used for the preliminary design, and a more sophisticated approach using finite elements and/or supported by explosive testing may be used for the final verification of the design.

A dynamic nonlinear approach is more likely to provide a section that meets the design constraints of the project compared with a static approach. Elastic static calculations are likely to give overly conservative design solutions if the peak pressure is considered without the effect of load duration. By using dynamic calculations instead of static, engineers are able to account for the very short duration of the loading. Because the pressure levels are so high, it is important to account for the short duration to mitigate response. In addition, the inertial effect included in dynamic computations greatly improves response. This is because by the time the mass is mobilized the loading is greatly diminished, enhancing response. Furthermore, by accepting that damage occurs, we are able to account for the energy absorption of ductile systems that occurs through plastic deformation. Finally, because the loading is so rapid, we are able to enhance the material strength to account for strain rate effects.

Response is evaluated by comparing the ductility, for example, the peak displacement divided by the elastic limit displacement and/or support rotation, the angle between the support and the point of peak deflection, to empirically established maximum values that have been established by the U.S. military through explosive testing. Maximum permissible values vary depending on the material and the acceptable damage level. Some criteria documents do provide the design values that need to be met; other criteria are silent on this topic or make a general reference to a source document.

The level of damage for a building caused by an explosion can be described as minor, moderate, or major depending on the peak ductility, support rotation, and collateral effects. Each is described below:

- Minor damage refers to nonstructural failure of building elements such as windows, doors, and cladding. Under this damage level, injuries may be expected, and fatalities are possible but unlikely.
- Moderate damage refers to structural damage and is confined to a localized area and is usually repairable. Structural failure is limited to secondary structural members, such as beams, slabs, and non-load-bearing walls. However, if the building has been designed for loss of primary members, localized loss of columns may be accommodated without initiating progressive collapse. Injuries and possible fatalities are expected.
- Major damage refers to loss of primary structural components such as columns or transfer girders precipitates loss of additional adjacent members that are adjacent or above the lost member. In this case, extensive fatalities are expected and the building is usually not repairable.

The four basic physical protection strategies for buildings to resist explosive threats are establishing a secure perimeter; mitigating debris hazards resulting from the damaged facade; preventing progressive collapse; and isolating internal threats from occupied spaces. Other considerations, such as the tethering of nonstructural components and the protection of emergency services, are also key design objectives.

For the most part, the size of the explosive threat will determine the effectiveness of each of these protective strategies and the extent of resources needed to protect the occupants. Therefore, a fundamental component of the design process is selecting an appropriate threat scenario and determining the design threat through careful planning and analysis (see Chapter 2).

In securing the perimeter it is important to keep the explosive as far away as possible by maximizing the standoff distance. This approach is only necessary if the threat analysis identifies the building to be at risk of attack as opposed to suffering collateral damage due to an attack on a nearby target. To guarantee the maximum keep-out distance between unscreened vehicles and the structure, antiram bollards or large planters may be placed at the curb around the perimeter of the building. The site conditions will determine the maximum speeds attainable, and thus the kinetic energy that must be resisted. Both the bollard and its foundation must be designed to resist the maximum load. If design restrictions limit the capacity of the bollard or its foundation, site restrictions will be required to limit the maximum speed attainable. If public parking is abutting the building, it must be secured or eliminated, and street parking should not be permitted adjacent to the building. Removing one lane of traffic and turning it into an extended sidewalk or plaza can gain additional standoff distance. However, this approach must be weighed against the size of the explosive; if the charge is sufficiently large, gaining additional 9 or 10 ft will have little effect in preventing damage to the building envelope.

The building's exterior is its first real defense against the effects of a bomb. The design should then focus on improving the postdamaged behavior of the facade. For instance, to prevent shattering of the windows due to blast loading, laminated glass should be used for new construction or antishatter film should be applied to existing glazing. The capacity of the frame system to resist blast loading should exceed the corresponding capacity of the glazing, referred to as the "glass fail first criteria." Factors of two to three over the nominal capacity of the glass to resist breakage should be used to design the frames. The mullions may be designed to span from floor to floor or tie into wall panels and must be capable of withstanding the reactions of a window loaded to failure. The walls to which the windows are attached must be designed to accept the reactionary forces. For new construction with low-threat criteria and limited budgets for blast protection, the weakest laminated glazing that satisfies wind and serviceability requirements should be selected, thus improving the postfailure behavior of the facade and providing the occupants with a measure of protection at a reasonable cost.

Some protective features may include insulated glazing unit with laminated inner light, glazing adhering to mullion with structural silicone sealant, and curtain-wall frame with steel backup encased in aluminum. A curtain wall is a nonbearing exterior enclosure that is supported by a building's structural steel or concrete frame and holds glass, metal, stone, or precast concrete panels. Lightweight and composed of relatively slender aluminum members, curtain-wall facades are considerably more flexible than conventional, hardened punched window systems. In a blast environment, the mullion support would absorb a portion of the blast energy and improve the performance of the glazing, allowing the glazing to sustain greater blast environments.

The design of curtain-wall systems to withstand the effects of explosive loading depends on the performance of the various elements that make up the system. While the glazing may be the most brittle component, the performance of the system and the reduction of hazard to the occupants depend on the interaction between the capacities of the various elements. This would include hardening the individual members that make up the curtain-wall system. In addition, the attachments to the floor slabs or spandrel beams must be adjustable to compensate for the fabrication tolerances and accommodate the differential interstory drifts and thermal deformations as well as be designed to transfer gravity loads, wind loads, and blast loads.

An alternative approach to blast protection takes the concept of a flexible curtain-wall system one step further by making full use of the flexibility and capacity of all the window materials to absorb and dissipate large amounts of blast energy while preventing debris from entering the occupied space. Energy-absorbing catch systems, cable protected window systems (CPWS), work in such a way that as the glass is damaged it bears against a cable catch system, which in turn deforms the window frames. Extensive explosive testing, as well as sophisticated computer simulations, has demonstrated the effectiveness of these systems.

A reinforced concrete slab when subjected to a blast load, punching shear, and softening of the moment-resisting capacity of the slab will reduce the lateral load-resisting capacity of the system. Once the moment-resisting capacity of the slabs at the columns is lost, the ability of the slab to transfer forces to the shear walls is diminished and the structure is severely weakened. In addition to the failure of the floor slab, the loss of contact between the slab and the columns may increase the unsupported column lengths, which may lead to the buckling of those columns. Furthermore, the lateral load-resisting system, which consists of the shear walls, the columns, and the slab diaphragms that transfer the lateral loads, may be weakened to such an extent that the whole building may become laterally unstable.

The exterior bays and lower floors are the most susceptible to an exterior vehicle explosive threat, and the design of the spandrel beams, which tie the structure together and enhance the response of the slab edge. Drop panels and column capitals may be used to shorten the effective slab length and improve the punching shear resistance. Shear heads embedded in the slab will improve the shear resistance and improve the ability of the slab to transfer moments to the columns. The blast pressures that enter the structure through shattered windows and failed curtain walls will load the top and bottom surfaces of the floor slabs along the height of the building. The delay in the sequence of loading and the difference in magnitude of loading will determine the net pressures acting on the slabs. Therefore, each floor will receive a net upward loading requiring the slab to be reinforced to resist loads opposing the effects of gravity.

The inclusion of beams will greatly enhance the ability of the framing system to transfer lateral loads to the shear walls. The slab–column interface should contain closed-hoop stirrup reinforcement properly anchored around flexural bars within a prescribed distance from the column face. Bottom reinforcement must be provided continuously through the column. This reinforcement serves to prevent brittle failure at the connection and provides an alternate mechanism for developing shear transfer once the concrete has punched through. The development of membrane action in the slab, once the concrete has failed at the column interface, provides a safety net for the post-damaged structure. Continuously tied reinforcement, spanning both directions, must be detailed properly to ensure that the tensile forces can be developed at the lapped splices. Anchorage of the reinforcement at the edge of the slab or at a structural discontinuity is required to guarantee the development of the tensile forces.

The reinforced concrete floor slab should be designed to prevent a punching shear failure that may in turn develop into a progressive collapse. Although research has shown that punching shear failures at interior columns are more likely to result in a progressive collapse than a failure at an exterior column, the external bay around the perimeter of the structure must be hardened at all intersecting columns for the external car bomb threat.

For blast loadings, the distance from the explosion determines the magnitude and distribution of the loading on a structure. For example, buildings located at a substantial distance from a protected perimeter, <100 ft, will be exposed to relatively low pressures, fairly uniformly distributed over the building's facade. However, buildings located at shorter distances from the curb such as in an urban environment will typically be exposed to more confined, higher-intensity blast pressures. Due to direct blast pressures, the columns of a typical building may experience severe bending deformations in addition to the axial loads they support, since the columns are designed to handle gravity loads and not ductility demands. Therefore, to enhance protection, the columns must be designed to be sufficiently ductile to sustain the combined effects of axial load and lateral displacement.

The uplift pressure on a concrete slab due to blast loading may cause tension on the concrete columns. Conventional reinforced concrete columns are not designed to resist the combined effects of bending and therefore be prone to damage under these conditions. The lower-floor columns must therefore be designed with adequate ductility and strength to resist the effects of direct lateral loading from the blast pressure and the impact of explosive debris. Reinforced concrete columns may be designed to resist the effects of an explosion by providing adequate longitudinal reinforcement, staggering the bar splices, and providing closely spaced ties at plastic hinge locations. Additional design options include sizing the columns to withstand the lateral loads; detailing splices to develop the plastic moments of the section; and encasing existing columns in a steel jacket or wrapping them up with a composite fiber to confine the concrete core, increasing shear capacity, adding mass, and preventing premature buckling of the thin flanges.

Transfer girders and the columns supporting transfer girders are particularly vulnerable to blast loading. Transfer girders typically concentrate the load-bearing system into a fewer number of structural elements, which contradicts the concept of redundancy desired in a blast environment. Typically, the transfer girder spans a large opening, such as a loading dock, or provides the means to shift the location of column lines at a particular floor. Damage to the girder may leave several lines of columns, which terminate at the girder from above, totally unsupported. Similarly, the loss of a support column from below will create a much larger span that bears critical loads. Transfer girders, therefore, create critical sections, the loss of which may result in a progressive collapse. So if a transfer girder is required and vulnerable to an explosive loading, the girder should be designed to be continuous over several supports. There should be substantial structure framing into the transfer girder to create a two-way redundancy, and thereby an alternate load path in the event of a failure. The column connections, which support the transfer girders, should be designed as Type 2 connections to provide sustained strength, despite inelastic deformations.

The conventional lateral loads including wind and seismic forces to which most buildings are designed are minimal. These minimal lateral load requirements may be resisted by a combination of shear walls, braced frames, and moment-resisting frame action. At each floor level, the slab diaphragms transfer the lateral loads to the lateral load-resisting system. Each component of the lateral load-resisting system must be checked to determine its adequacy to resist blast loads. Depending on the results of a blast analysis, the individual elements of the lateral load-resisting system may require modification.

Buildings with an irregular floor plan will induce large torsional effects on the lateral load-resisting system. Typically, symmetrical buildings behave better when subjected to blast or seismic loading. If the shear core is centrally loaded a large demand is placed on the diaphragm action of the floor slab to transmit the lateral loads from the perimeter of the floor into the central shear walls. This effect can be more critical for a blast load than for a seismic load. Seismic base motions are typically applied over the entire foundation; blast loads resulting from a close-in explosion tend to impose higher-intensity loads over a more concentrated region. Although the total base shears may be nominally the same, the lateral resisting behavior is not.

The ability of structures to resist a highly intense blast loading depends on the structural detailing of the slabs, joists, and columns that provide for the ductility of the load-resisting system. The structure has to be able to deform inelastically under extreme overload such as by dissipating large amounts of energy prior to failure. Provisions have been established for the design of structures to resist seismic forces that ensure both the ductility of the members and the capacity of the connections to undergo large rotations without failing.

In addition to providing ductile behavior, there needs to be a well-distributed lateral load-resisting mechanism in the horizontal floor plan. The use of several shear walls distributed throughout the building will improve the overall seismic as well as the blast behavior of the building. If adding more shear walls is not architecturally feasible, a combined lateral load-resisting mechanism can also be used. A central shear wall and a perimeter moment-resisting frame will provide for a balanced solution. The perimeter moment-resisting frame will require strengthening the spandrel beams and the connections to the outside columns. This will also result in better protection of the outside columns.

The walls surrounding loading docks, mailrooms, and lobbies where explosive threats, like a hand-delivered package bomb, may be introduced prior to inspection and screening must be hardened to confine the explosive shock wave and permit the resulting gas pressures to vent into the atmosphere. Specific modifications to the features of these unprotected spaces can prevent an internal explosion from causing extensive damage and injury inside the building. This hardening can be achieved by designing the slabs and erecting cast-in-place reinforced concrete walls, with the thickness and reinforcement determined relative to the appropriate threat. The isolation of occupied spaces from these vulnerable locations and any other unsecured spaces, such as basements and underground parking garages, requires both adequate levels of reinforcement and connection details capable of resisting the collected blast pressures. These structural designs must be integrated with the remainder of the structural frame to make sure they do not destabilize other portions of the system.

A variety of materials, not traditionally used in building construction, may provide alternatives to conventional blast hardening solutions. Among these alternatives, there are shock-attenuating chemically bonded ceramics (SA/CBC) and composite systems made up of carbon, aramid, and polyethylene fibers and resin 82. These materials are well-developed systems currently in use for the prevention of sympathetic detonation of explosives in munitions storage depots (SA/CBC materials) as well as in the seismic retrofit of reinforced concrete columns in highway bridges in California, such as carbon fiber wrapping. In the latter application, carbon fiber wrappings were found to have advantages over conventional steel jacketing of columns due to problems with weld seams and corrosion. Spray-on elastic polymers have been demonstrated to protect unreinforced masonry walls by providing a ductile membrane that enables these brittle elements to sustain large deformations without fragmenting and throwing hazardous debris. For retrofit construction where conventional structural treatments may be too heavy or too labor-intensive, composite materials may be attractive alternatives because of their lightweight and high tensile strength. For nonstructural building components, such as piping, ducts, lighting fixtures, and conduits, must be sufficiently tied back to structural members either below the raised floors or to the ceiling slabs, to prevent failure of the services and falling debris hazards.

5.8 RESPONDING TO CYBERATTACKS

Part of the response to cyberattacks is to ensure that an organization has the right people that will be able to conduct a thorough forensic examination of their information technology systems. An organization's forensic information technology examiners must be both qualified and certified to testify in court. As seen in many recent cases, including the Peterson case, the defense questions every detail of the forensic examiner's report and credentials. In many courts, if the individual does not have the formal education and proof of experience, his or her testimony might not be accepted as expert. An organization will want somebody who has been highly technical in the field for a long time, preferably practicing as a professional, rather than an amateur hobbyist. Ideally, they should have 5+ years in all the computer systems and operating systems involved in the incident. More recent operating systems may not have that potential depth of experience due to the relative immaturity of the field. Previous courtroom testimony and experience are key components. An organization will want their defendant to stay on trial, not their "expert's" background or credibility by opposing counsel. They should have experience with a variety of computer forensics tools; one is not enough. They need to be able to demonstrate validation, such as holding a certified information security systems professional (CISSP) credential. Technical credentials are good, too, but they must also be targeted to platform or tools in question. Professional affiliations are good to have. The ideal candidate will have either a diploma or an undergraduate degree in computer science or information technology systems.

If an organization suspects it is being "hacked" or is under cyberattack, they should assign at least two highly experienced people to initially look into the matter and determine if further action is warranted. It is important that an organization does initial "verification" of suspected activity and avoid "knee-jerk" reactions. Establish secure communications with all parties and determine if "low key" is more important than quickly finding out what happened. If a suspect is at the system when responding, immediately advise people to step away from their workstations and area; at this point, one should be prepared to have people physically removed from their workstations. If warranted, notify superiors, activate the forensic information technology team, and begin to investigate.

A forensic information technology team must have the authority and funding to act. Skill sets to address the incident can be a mixture of technical, investigative, and people skills. The team itself can be made up of internal staff, external consultants, or a mix of both. Each member of the team should have investigative experience. Team members must be calm under pressure, tenacious, and have the ability to work with others. During the investigation phase, notification of key corporate focal points is necessary during the initial, interim, and final stage of the investigation including the lessons-learned review. All communications are coordinated through the investigation command post. If necessary, a "cover story" may be implemented in order to limit poor publicity, exposure, and liability to the organization. The next question for the organization, once the investigation is complete, is whether or not to prosecute. The investigation team will need a secure room to operate from with lodging, storage, and data communications equipment. In addition, they will need access to paper shredders, computers, networks, subnets, firewalls, and routers. The team will operate on a "24-hour, 7-day a week" rotating schedule. In addition, external consultant billing and/or invoicing will need to be tracked, if an external consultant is part of the investigative team. Furthermore, team expenses will need to be tracked and reimbursements issued accordingly.

Once it has been verified that the system has come under a cyberattack, the forensics team must ensure continuous power to the system and peripherals. They should physically secure the system, attached peripherals, disks, tapes, and paperwork at the scene. They should block off logical access to the system if the system does not "hang," for example, firewall business rule, router access control list (ACL), disconnect network connection at the patch panel, but only after recording the preexisting setup of each as well as any logs. A normal reboot of the system should not be performed. If necessary, pull the power plug.

The next step would be to undertake a thorough investigation as to the root cause or origin of the attack. The first step in the investigation would be to develop a composite list of possible leads. This would include a forensic examination of all the involved equipment and media copies, interviews of all key personnel who had access to the system, a compiled list of suspects and/or attack tools, a record of all damages sustained as a result of the cyberattack, and the financial implications to the organization.

The next stage in the investigation is the collection of evidence, preservation, and control. The investigation team will work from copies of original drives and media including fully bootable drives. All the originals should be stored in a secure location. Photograph the initial state the system and/or network was found in and document the physical and logical architecture of the system. A complete inventory of configuration and files on media and/or drives should be undertaken. Tag all evidence to prevent "chain of custody" errors and maintain an evidence log book.

As part of the forensics examination, the forensics team will determine what forensic analysis tools to use: sniffers, password crackers, media-copying/analysis hardware or software, file comparison tools, and/or log analyzers. As part of the forensics examination, the team will review all system or network logs on copied drives or media. They should look for backups of those logs and note if any deletion of primary or backup logs was discovered. Compare executables and/or binaries with known vendor binaries to determine if any have been altered. Search for hidden directories or encrypted files on the system. Check to see if configuration or security files have been altered.

Ensure that prior coordination and typical attack scenario investigation types are established. Those individuals most likely involved include human resources, legal, corporate security, information technology department, chief information or chief technology officer, chief executive officer, and president. Restrict the number of people "in the know" to an absolute minimum to prevent the investigation being compromised, to reduce the likelihood of information being leaked to the media, to facilitate the accomplishment of investigative efforts, and to reduce the approval chain. At this stage, the organization may wish to obtain unique system/network technical expertise from corporate resources as needed.

The final report for review contains a section detailing the initial incident, the case status, investigative notes, interim reports, and the final report. The initial incident report along with the interim and final reports should be entered into an incident database. The Computer Emergency Response Team (CERT) and/or Computer Incident Alert Center (CIAC) should be notified if necessary.

Ultimately, it is the organization's decision if they wish to prosecute the offender. The legal officer will be the primary decision maker in the case and whether or not the organization wishes to go external. They would need to consider all options including the impact of criminally prosecuting a hacker versus a civil suit or internal administrative action against the assailant, particularly if the threat came from within the organization. The legal department will need to coordinate with the external litigation firm; state or crown attorney; and local, provincial, state, and/or federal law enforcement authorities. They might have to consider external international technical, legal, and investigative support.

In wrapping up the investigation, all evidence must be retained and returned to the evidence locker. A postinvestigation "lessons learned" meeting must be conducted with key corporate focal points including a "what happened" summary of the events that transpired. In closing the investigation, the organization must determine how to use the results to reduce the possibility that such activity does not happen again. Also, take the opportunity to review better ways to conduct future investigations. Full impact assessment should also be undertaken at this point to determine the overall damage to the organization and how to prevent another similar event. Always be proactive! Waiting for an incident to happen is not the thing to do. Work with the organization and external focal points prior to any incident and during incidents, to ensure smooth investigative operations while minimizing impact to organization's operations. Finally, use investigative results to enhance your enterprise Information Security Risk Management Program.

5.9 EXECUTIVE AND CLOSE PERSONAL PROTECTION

The executive protection or close protection officer is a type of security operative or government agent who protects a person or persons, known as the "protectee," usually a public, wealthy, or politically important figure(s) from danger: generally theft, assault, kidnapping, homicide, harassment, loss of confidential information, threats, or other criminal offences.

Most important public figures such as heads of state, heads of government, and governors are protected by several close protection agents or by a team of close protection agents from an agency, security forces, or police forces such as the U.S. Secret Service in the United States or the Royal Canadian Mounted Police Protective Policing Program in Canada. In most countries where the head of state is and has always been also the military leader, the leader's bodyguards have traditionally been Royal Guards, Republican Guards, and other elite military units. Less-important public figures, or those with lower-risk profiles, may be accompanied by a single bodyguard who doubles as a driver. A number of high-profile celebrities and chief executive officers also use close protection agents.

The role of the close protection agent is often misunderstood by the public, because the typical layperson's only exposure to bodyguarding is usually in highly dramatized action film depictions of the profession, in which bodyguards are depicted in firefights with attackers. In contrast to the exciting lifestyle depicted on the film screen, the role of a real-life close protection agent is much more ordinary: it consists mainly of planning routes, presearching rooms and buildings where the protectee will be visiting, researching the background of people that will have contact with the protectee, searching vehicles, and attentively escorting the protectee on his or her day-to-day activities. In turn, if the close protection agent can secure the protectee in event of an aggression with an escape, the agent has done his or her job. This is the life blood and birth of the protective agent and the decline of "bodyguards."

Close protection agents, formerly known as bodyguards, needed to be near the upper-class society in the middle of the crowd beside the protectee, not on the sidelines but right there prepared to stop personal attacks. Protection agents are now introduced as a professional associate and not as bodyguard. Many entertainers use their personal security like a stage prop or a status symbol wanting to show off the fact that they have a bodyguard. As a result, many professional protection companies do not take entertainers as clients. The image of a "Gorilla in a suit" is a myth and an image that should have disappeared a long time ago from the protection profession.

The professional close protection agent is well-trained, educated, articulate, and professional. A professional close protection agent puts all his or her effort and skills into never having to use a weapon or apply lethal force. It is always wiser, if possible, to be nonobtrusive and covert in the defense of the client. In some cases, this is impossible such as with the political figures that need to be in the public eye. If overt, the best defense is to make the illusion of the best protection in the world. Make the "terrorist" or assailant think it is easier to attack the next person as a failed attack is terrible for recruiting new "terrorists" and is considered bad public relations. If going overt, a show of arms can be a deterrent. It is normally considered very unprofessional to let someone see your weapon. It can also be deadly as attackers will neutralize those with weapons first during an attack. However, in a known hostile environment sometimes a large show of force and arms will deter the attack.

The role of a close protection agent depends on several factors. First, it depends on the role of a given agent in a close protection team. An agent can be a driver, a protector, one who escorts the protectee, or part of an ancillary unit that provides support such as IED detection, electronic "bug" detection, countersniper monitoring, presearches facilities, and undertakes background checks of people who will have contact with the protectee. Second, the role of a bodyguard depends on the level of risk that the client faces. An agent protecting a client at high risk of assassination will be focusing on very different roles, for example, checking cars for IED devices, bombs, and watching for potential shooters, than an agent escorting a celebrity who is being stalked by aggressive tabloid photographers. Some agents specialize in the close quarter protection of children, to protect them from kidnapping or assassination.

The close protection agent can and will have the function of being a deterrent to a violent act, not to be a trained assassin. The close protection agent's mission is to protect the protectee, in many cases as a "protection blanket" or the protectee's first layer of protection. An attack with a weapon, such as a knife, is always a very brave and determined act from the aggressor. The attacker has to come close to the protectee or protective agent and it is only a matter of a split second for the close protection agent to react or even to engage. It has been determined that an attacker with a knife can cover 21 ft quicker than the average police officer can draw his weapon. It can be very hard to defend a client if they need to be in public areas. For example, Secret Service agents were unable to prevent John Hinckley, a 19-year-old, with a history of mental disorder, armed with a .22 pistol from shooting President Reagan. The next question that plagues every protection agent is, "What if the assassin were professionally trained with covert weapons?"

The protection agents' mission is to protect the protectee: avoid the protectees' assassination, avoid their kidnapping or their being taken hostage, avoid any medical emergency that could endanger their life, and avoid any kind of embarrassment for the protectee. The protection agent must have the knowledge and skills in carrying out their assignment, which include acting as a driver, weapons and tactics, and countersniper tactics.

Close protection agents may also act as drivers. Normally, it is not sufficient for a protectee to be protected by a single driver, because this would mean that the agent would have to leave the car unattended when they escort the protectee on foot. If the car is left unattended, this can lead to several risks: an explosive device may be attached to the car; an electronic "bug" may be attached to the car; the car may be sabotaged; or city parking officials may simply tow away the vehicle or place a wheel lock on the tire. If parking services tow away or disable the car, the bodyguard cannot use the car to escape with the protectee in case there is a security threat while the protectee is at his or her meeting.

The driver should be trained in evasive driving techniques, such as executing short-radius turns to change the direction of the vehicle and high-speed cornering. The car used by the protectee will typically be a large sedan with a low center of gravity and a powerful engine, such as a BMW or Mercedes Benz. At a minimum, the vehicle should have ballistic glass, some type of armor reinforcement to protect the protectee from gunfire, a foam-filled gas tank, "run-flat" tires, and armor protection for the driver.

The car may also be equipped with an additional battery; dual foot pedal controls, such as those used by driving instruction companies, in case the driver is wounded or incapacitated; a PA system with a microphone and a megaphone mounted on the outside of the car, so that the driver can give commands to other convoy vehicles or agents who are on foot; fire extinguishers inside the vehicle in case the vehicle is struck by a Molotov cocktail bomb or other weapon; a reinforced front and rear bumper, to enable the driver to ram attacking vehicles; and additional mirrors, to give the driver a better field of view. Decoy convoys and vehicles are used to prevent tailing. In the event the convoy holding the protectee is compromised and ambushed, decoy convoys can also act as a reinforcement force that can ambush a force that is attacking the primary convoy. Some protectees rotate between residences in different cities when attending public events or meetings to prevent being tailed home or to a private location.

Depending on the laws in a agent's jurisdiction and on which type of agency or security service they are in, agents may be unarmed, armed with a less-lethal weapon such as a pepper spray, an expandable baton, a Taser, or with a lethal weapon such as a handgun, or, in the case of a government close protection agent for a Secret Service–type agency, a machine pistol. Some agents such as those protecting high-ranking government officials or those operating in high-risk environments such as war zones may carry submachine guns or assault rifles. In addition to these weapons, a close protection team may also have more specialist weapons to aid them in maintaining the safety of their principal, such as sniper rifles and antimaterial rifles, for antisniper protection, or shotguns, loaded either with buckshot as an antipersonnel weapon or with solid slugs as an antivehicle weapon.

Bodyguards that protect high-risk principals may wear body armor such as kevlar or ceramic vests. The bodyguards may also have other ballistic shields, such as kevlar-reinforced briefcases or clipboards that, while appearing innocuous, can be used to protect the principal. The principal may also wear body armor in high-risk situations.

For a close protection officer, the primary tactic against sniper attacks is defensive: avoid exposing the principal to the risk of being fired upon. This means that the principal should ideally be within an armored vehicle or a secure structure. In addition, when the principal moves between a vehicle and a building, the principal must be moved quickly to minimize the time window in which a sniper could take a shot and use a flanking escort of close protection officers to block the view of the sniper and any potential shot that the sniper may take. The use of offensive tactics against snipers will occur very rarely in a bodyguard context.

The best way to avoid a situation or problem is to minimize the cause or chances. In close protection "theory," all incidents come from a lack of security or planning. To minimize or eliminate the risk, a close protection agent or team must have a plan and execute certain assignments that are low threat and more relaxed, that is when it becomes most dangerous, since there is a risk of complacency on both the agents' and the protectees' part. In turn an agent must always make note of their surroundings and the reason for the assignment.

For a close protection officer, the primary tactic against sniper attacks is defensive: avoid exposing the principal to the risk of being fired upon. This means that the principal should ideally be within an armored vehicle or a secure structure. In addition, when the principal moves between a vehicle and a building, the principal must be moved quickly to minimize the time window in which a sniper could take a shot and use a flanking escort of close protection officers to block the view of the sniper and any potential shot that the sniper may take. The use of offensive tactics against snipers will occur very rarely in a bodyguard context.

If the protectee is under surveillance by an aggressor or enemy, it essentially means that the close protection team has not taken the proper measures to keep the security detail secure. The assailant will see that the team is more relaxed and will use the situation to launch his or her attack.

If the client asks the close protection agent to conduct duties for which the close protection agent is not retained or assigned, and it may happen, use the task to check the outer perimeter, or do a thorough search of the residence, to verify if everything is secure. Under no circumstances, no matter what the protectee asks the agent to do, should they neglect their duties as a close protection agent. To avoid any unfortunate situation or challenges with the protectee, it is important to have a legal contract in place signed by both parties and their attorneys. Close protection agents are not butlers or maids and those things have to be determined before the assignment starts in the initial interview with the protectee. Protective agents are not obligated to do certain tasks around the protectees' residence, but they still have to maintain a certain tolerance. Therefore, there must be compromise on the part of both parties. Going a little out of the way for the protectee as long as it does not jeopardize their safety will help the long-term security detail by positive cooperation from the protectee. Much of what you need to know about the VIP, his or her life, family, business, and activities should be determined in the initial interview. It is therefore important to plan ahead and to know exactly what to ask during the interview, including a prebackground check on the VIP. It is important to inform the VIP that the initial interview will provide the information needed to keep him or her safe and that they must be totally honest and upfront for the sake of their safety more than they might realize.

A close protection team protecting a high-profile politician who is at risk of attack would be based around escorting the client from a secure residence such as an embassy to the different meetings and other activities he or she has to attend during the day (whether professional or social), and then to escort the client back to the residence.

The day would begin with a meeting of the close protection team led by the team leader (TL). The team would review the different activities that the protectee plans to do during the day, and discuss how the team would undertake the different transportation, escorting, and monitoring tasks. During the day, the protectee or "principal" may have to travel by car, train, and plane and attend a variety of functions, including meetings and invitations for meals at restaurants, and do personal activities such as recreation and errands. Over the day, the protectee will be exposed to a range of risk levels, ranging from higher risk such as meeting and greeting members of the public at an outdoor rally to low-risk dining at an exclusive, gated country club with high security.

Some planning for the day would have begun on previous days. Once the itinerary is known, one or more agents would travel the route to the venues, to check the roads for unexpected changes in road work, detours, and closed lanes and to check the venue. The venue needs to be checked for bugs, and the security of the facility including exits and entrances will need to be inspected. In addition, the agents will want to know the names of the staff that will have contact with the protectee, so that a simple electronic background check can be run on these individuals. Agents learn how to examine a premises or venue before the protectee arrives, to determine where the exits and entrances are, find potential security weaknesses, and meet the staff, so that a would-be attacker cannot pose as a staff member. In addition, some agents learn how to do research to make note of potential threats to their VIP, by doing a thorough assessment of the threats facing the principal, such as a protest by a radical group or the release from custody of person who is a known threat. Close protection officers also learn how to escort a client in potentially threatening situations.

An hour prior to leaving with the protectee to his or her first appointment, the driver and another close protection agent remove the cars that will be used to transport the principal from the locked garage and inspect them. There may be only one car for a lower-risk principal. A higher-risk principal will have additional cars to form a protective convoy of vehicles that can flank the principal's vehicle. The vehicles are inspected before leaving.

Once the cars have been inspected and they are deemed to be ready for use, they are brought into position near the exit door where the principal will leave the secure building. At least one driver or close protection agent stays with the cars while waiting, because the now-searched cars cannot be left unattended. If the convoy is left unattended, an attacker could attach an IED or sabotage one or more of the vehicles. Then the close protection team flanks the principal as they move from the secure residence to the principal's car. Depending on the size of the close protection team, different foot formations will be used to escort the principal as illustrated in Figures 5.5 through 5.9.

The convoy then moves out toward the destination. The team will have chosen a route or two and in some cases it may involve three routes that are designated for travel along, which avoids the most dangerous "choke points," such as one-lane bridges or tunnels, because these routes have no way of escape and they are more vulnerable to ambush. Figures 5.5 through 5.9 illustrate the formation of the vehicles upon traveling and the position of each member of the close protection team. Each position has specific task, for example, the TL calls the checkpoints while the motorcade is moving, number 1 is usually the only shooter if needed, and the drivers (D) stay with the vehicles at all times. In some cases, if the client has to travel by train, the bodyguards will inspect the rail car they are traveling in and the other cars he or

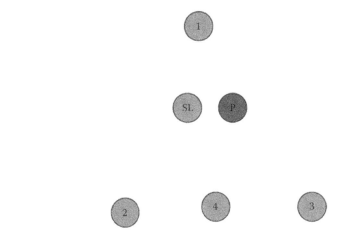

FIGURE 5.5   Five-man wedge formation.

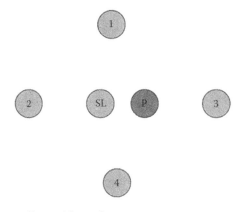

FIGURE 5.6   Five-man diamond formation.

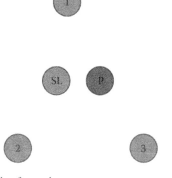

FIGURE 5.7   Four-man wedge formation.

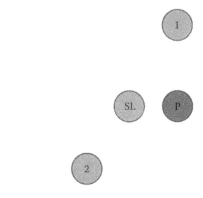

FIGURE 5.8   Three-man diagonal formation.

FIGURE 5.9   Two-man formation.

she will use. Traveling by transport other than vehicle is very dangerous because of the lack of control over the environment.

When the convoy arrives at the location, one or more agents will exit first to confirm that the location is secure and that the staff that were booked to work that day are the ones who are present. If the location is secure, these agents signal that it is safe to bring in the principal. The principal is escorted into the building using a flanking procedure. If the principal is attending a private meeting inside the building and the building itself is secure (controlled entrances), the principal will not need to have an agent escort in the building. The agents can then pull back to monitor his or her safety from a further distance. Agents could monitor entrances and exits while the driver watches the cars. If the principal is moving about in a fairly controlled environment such as a private golf course, which has limited entrances and exits, the security detail may drop down to one or two agents, with the other agents monitoring the entrances to the facility, the cars, and remaining in contact with the agents escorting the principal. Throughout the day, as the principal goes about his or her activities, the number of agents escorting the principal will increase or decrease according to the level of risk. Figures 5.10 through 5.24 illustrate the formations of escorting a client when arriving at a destination, walking in a hallway, moving into and exiting an elevator, walking a picket line, and engaging in public speaking.

Dedication is a very important aspect of being a professional protection agent. Stress is normal in the lifestyle of a professional protection agent. It is normal for anyone who has many and important responsibilities in their work. Mostly it is

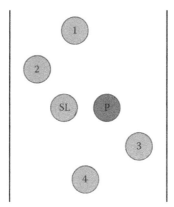

FIGURE 5.10   Five-man hallway formation.

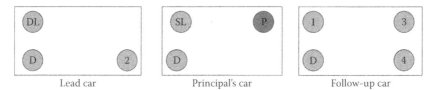

Lead car   Principal's car   Follow-up car

FIGURE 5.11   Arriving at a destination.

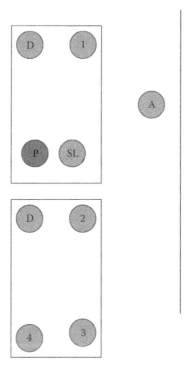

FIGURE 5.12   Arriving at a destination with attendant present.

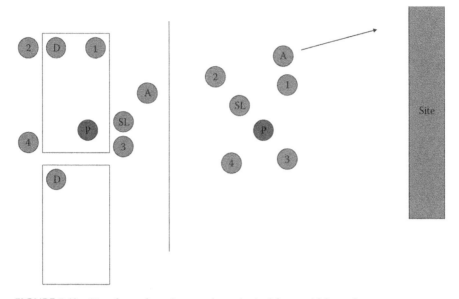

FIGURE 5.13   Foot formation when moving principal from vehicle to site.

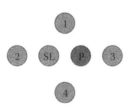

FIGURE 5.14   Foot formation approaching the picket line.

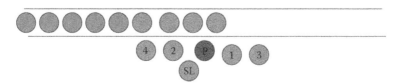

FIGURE 5.15   Foot formation walking the picket line.

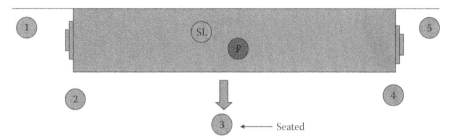

FIGURE 5.16   Stage and banquet deployment.

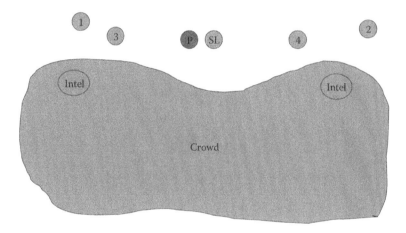

FIGURE 5.17   Formation during speaking engagement.

accompanied by the feeling and risk of danger. There are three stages of stress a close protection agent will experience. In Stage I, the close protection agent is alerted by a situation or event that places them into the first stage and all their resources are used and guided toward the situation; at this point, the agent should try and avoid tunnel vision. Their mind and body are already in the alert stage. In Stage II, resistance arrives, where the stress is controlled by their body's natural defenses, their training, and their professionalism so it does not exceed their limits. In Stage III, there is exhaustion. If the stress is not resolved or reduced at this point, the body's natural defense system crashes, and the effects of stress are no longer controlled. The consequences may include chronic diseases, psychological side effects, physical symptoms such as chest pains, and ultimately in the most severe cases death.

The stress of a professional protection agent is different than the stress of a police officer. The professional protection agent has to be proactive. They are not obligated to engage or use deadly force against the assassin. They never need to take the initiative to pursue an enemy or chase after them. They also carry weapons and equipment that they selected themselves. Since anything can happen, the

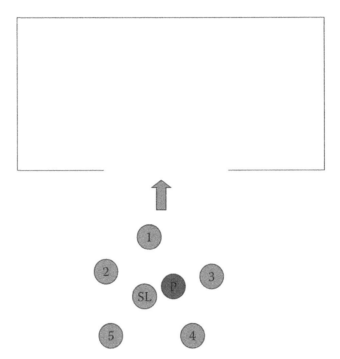

FIGURE 5.18   Entry into elevator, Phase I.

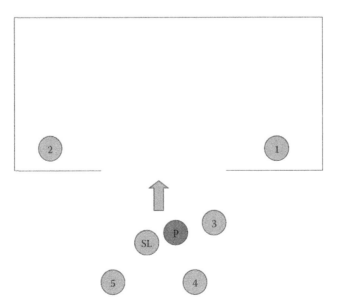

FIGURE 5.19   Entry into elevator, Phase II.

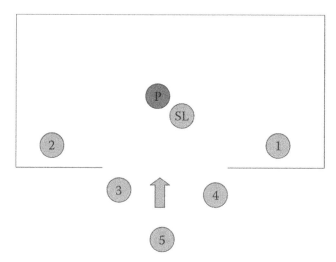

FIGURE 5.20 Entry into elevator, Phase III.

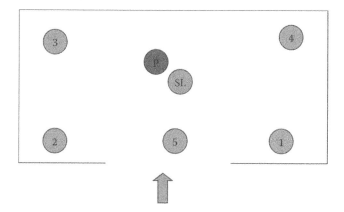

FIGURE 5.21 Entry into elevator, Phase IV.

protective agent needs to know how to handle any crisis. It is unlikely that the protectee will only be grazed by a bullet or even killed by it. The protective agent must be able to assist the protectee with injuries such as sharp objects, knives, grenade fragments, and burns from Molotov cocktails or explosions. In turn, the protection agent should be trained in the use of AED and CPR, which are necessary skills for heart attack victims.

If the principal becomes injured, the close protection agent should examine the protectee and determine what injuries they have? Use their knowledge of first aid and determine how severe the injuries are? And intervene when necessary. Upon examination the agent should determine if the protectee is conscious or unconscious? Are they currently breathing? Is there a heartbeat? They should attempt to stop all severe bleeding, start resuscitation in the case of cardiac arrest, prevent and treat shock, and call paramedics. Sometimes on a close protection detail, particularly a high-risk

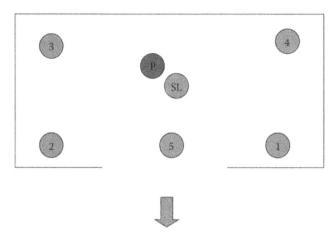

FIGURE 5.22   Exit elevator, Phase I.

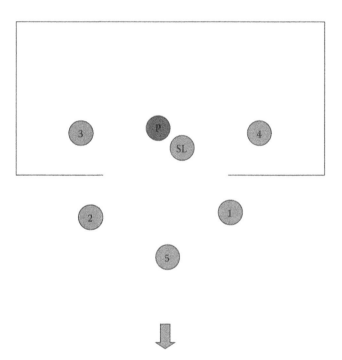

FIGURE 5.23   Exit elevator, Phase II.

detail, one member of the team should be cross-trained as a paramedic, with a full medical kit to intervene in dire circumstances.

There are different stages of a protection assignment. The close protection team must develop a threat assessment on the protectee and analyze the different risk factors toward the protectee and the potential scenarios against the protectee including harassment of protectee, kidnapping of the protectee, and an assassination attempt.

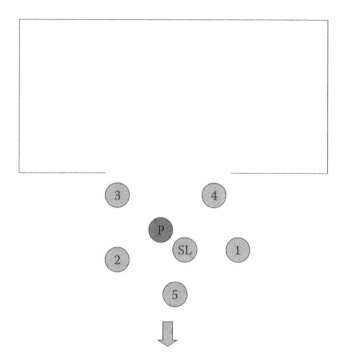

FIGURE 5.24 Exit elevator, Phase III.

At this point the team must determine what the attacker's motivations are and how high is the risk? In addition, they must also determine the attacker's interest in the protectee and why? The questions that are most often asked at this point are as follows:

- Is the protectee a dominant type, leader?
- Is the protectee a wealthy individual?
- Is the protectee a key executive in a major corporation?
- Does the protectee represent a symbol in any government institution or some other institution?
- Does the attacker want money?
- Does the attacker want to manipulate decisions?
- Does the attacker want to manipulate upcoming elections?
- In case of an assassination attempt, is the attack conducted against a corporation or a country?

A risk assessment of an individual country or city that the protectee is traveling to has to be conducted. The political situation of a country can play a key role in determining the safety and risk posed to the protectee during his or her visit. Other factors include the protectees' country of origin and the country the protectee's company is from. Upon entering a foreign country, always inform the Ministry of Foreign Affairs from the country you are departing from and from the country you are arriving. Due diligence is the key here. There are different types of risk that can be encountered including minor risks, real risks, and major risks. In turn, the protection needs to put in context the different types of threats affecting the principal, including the identity of the attackers and identifying the attacker's methods. According to these characteristics, the protection team must then evaluate the protectee's day-to-day operations/actions/routine; his or her residence, office(s), secondary residence, country clubs and gyms, and his or her surroundings including family, coworkers, friends, and servants. The protection team must also evaluate the principal's vehicles including cars, helicopters, planes, and boats. A protectee's biographical data sheet for all family and associates is also important to have.

The protection agent has to be intuitive, informed, and inquisitive. It is very important that the protective agent maintain the integrity of the detail. If the principal becomes upset with the agent and cancels the detail and is then killed, it is not good public relations for the agent's firm or agency. The following is a list of actions that a protection agent can take to reduce the probability of having future problems with his or her principal:

- The protection agent is a technical aid and advisor on security issues to the executive so as to assist them in learning to protect themselves or to make his or her job easier by them not making mistakes.
- In an attack scenario, the agent may be injured so they should constantly be teaching the executive to protect themselves and how to avoid attacks, deal with an attack, and respond if captured.
- An agent's primary mission is to protect the lives of his or her protectee and then their property.
- An agent must also realize that they must abide by the image of the organization that they represent as well as the principal's organization in how they carry out their task.
- An agent's appearance and actions must be beyond reproach.
- An agent must maintain a low-key demeanor.
- An agent must fit into the environment.
- An agent should not be making a spectacle of themselves in front of the neighbors or coworkers and family.
- An agent's weapon and badge should rarely if ever be seen by anyone.
- If the principal were familiar with weapons they would not be needed. Many people are made very uncomfortable by weapons.
- If the principal is uncomfortable with anything the agent does, it will shorten the duration of the assignment or end the detail.
- An agent should do everything in his or her power to be as nonobtrusive as possible. No one wants people in their private space.
- An agent's job is to let the client go about their lives with as little change as possible.
- An agent should always be polite but firm where security measures are concerned.

When on a stationary post the agent should try to have as little contact with the principal as possible. A protection agent's job is to be alert and attentive at all times, focusing on the job and the assignment, and avoiding distractions. If an agent observes something that is suspicious they should put themselves between the problem and the protectee or the protectee's residence. An agent can then take up a good defensive position. Observe until the assailant does something that is offensive or illegal. Use the radio to contact another unit to call the police or a manager to assist. The agent should make note of his or her location at all times. Be aware that the initial disturbance could be a distraction and the real threat may be from someone else. After stopping the threat, document the incident. Use a video camera and obtain real or physical evidence and the license number if possible. Agents have no legal right to use deadly force unless their life is in danger. Do not tell the principal any more details than is necessary at this point; an agent does not want to frighten the principal. All details and raw intelligence should be given only to the director or to the TL, then they can decide what to give out.

There a number of checks an agent should be undertaking as part of their protection detail and/or assignment; they are as follows:

- When with a principal an agent should always make a show of checking their vehicle for bombs; if under surveillance, this could discourage the assailants from placing a bomb on the vehicle.
- The car should be started by the agent with the door open and one leg on the outside; that way if the car were to explode, it could blow the agent out and not up. On a close-quarters detail, when entering the house always bring the protectee and have them in one place near the door and then go check the house for danger.
- When first entering the house do not turn on the lights; assess the situation first. Look for signs of entry and smell for gas or chemicals.
- Check any packages that come for the principal; ask them if they were expecting it. Do not open mail unless they ask.
- Rehearse an emergency plan with the principal until they are familiar with their role.
- An agent should call the director, special agent-in-charge, or TL at the start of each shift.
- Protection agents should have the proper equipment with them.
- Protection agents should have their lunch and water with them.
- Protection agents should keep an emergency go-bag at all times, in case of last-minute changes in schedule or emergencies.

When on a protection detail, there are certain rules that a protection agent must follow:

- Do not use foul language especially if it is directed toward the principal.
- Do not become argumentative with the principal, such as reminding them that your assignment does not include secretarial work and porter duties.
- Do not drink on the job.

- Do not bring children to work.
- Do not bring a spouse or significant other to work.
- Do not bring friends to work.
- Never leave a post.
- Do not talk about other work assignments; the current principal is the first priority.
- Do not talk about problems at work or at home, especially problems with pay; they should be dealt with separately with either the director or special agent-in-charge.
- Do not litter.
- Do not sleep.
- Do not bring alcohol on the job or drink on the job; also, do not drink when armed.

If the principal asks the agent to do something that they feel is not part of their duties, they should inform them that they cannot keep them safe and protect them if they are doing other things. The agent should also notify the principal that his or her supervisor prohibits them from doing any duties other than his or her job. If the problem persists, they should contact the special agent-in-charge or the director of the agency or firm. However, if an agent is conducting their security and protective services properly, the principal will see in plain sight that an agent is too busy to carry out any duties other than his or her assignment. A protective agent's task is to protect the life and belongings of his or her principal. It is also to present a professional appearance of themselves and their organization. The position of a protective agent has no room for mistakes. If an agent makes a mistake, the principal, they themselves, their teammates, or bystanders could die. The best protection for an agent is his or her training, professionalism, and vigilance.

5.10   TRAVEL SECURITY

When on a trip abroad whether for business or pleasure, it is always important for individuals to know his or her travel plans as early as possible, especially if his or her destination or the country he or she is traveling to has any political issues (terrorism), business issues (criminal and organized crime), and health issues (SARS and swine flu). Determine the travel dates ahead of time and if there are any anniversary dates of past terror acts or other significant events happening on those dates. For example, determine what political events are happening at the time of travel and avoid them at all costs. Before traveling, ensure all documentation is in order including passports, visas, and health requirements, such as travel health insurance and prescription medications. In the United States, the Overseas Advisory Council (OSAC) and the Department of State are excellent sources of information for much of the material mentioned above including travel advisories; in-country threat assessments and risk assessments; and local laws, customs and cultural awareness. Additional information on such things as the prohibition of alcohol and restrictions on prescription medications is also available. In Canada, the Department of Foreign Affairs offers similar information for Canadians traveling abroad.

When traveling abroad there a number of steps an individual can take to ensure his or her safety and security and reduce their risk. The following is a list of preventive measures an individual can take when traveling by air:

- Process through and get past security as quickly as possible; the most dangerous time at an airport is before passing through security.
- Ask for a window seat near an exit; if hijacked people in the aisle seats get struck more often as they are closest to the attackers.
- Do not work on or discuss confidential company or organizational information. People are not cognitive about the fact that others can hear and see what they are talking about or working on, particularly on an airplane, which is relatively close quarters.
- During a hijacking or other incident, avoid eye contact, and do not draw attention. Obey instructions unless it is evident that it is a September 11–type hijacking.
- For personal protection, carry a compact, pocket-sized, high-intensity, tactical-level incandescent flashlight.
- Since the events of September 11, manufactures have developed a number of survival kits when in similar situations. Some of the items in these kits include survival masks to protect against smoke inhalation, mouthwash, which can kill some biological agents such as anthrax, and hand sanitizers to protect the skin from epidermal exposure.
- If in a shooting situation, crouch or lay on the floor and do not move.
- If a recovery is attempted do not try to help. The police are trained to shoot anyone standing.
- When you are met at the airport, do not use organization or personal name on placard. This helps to avoid detection or identification, if being targeted for kidnapping.

The following is a list of preventive measures an individual can take when in the country:

- They should avoid conversations about politics or company business. People are often ignorant to the fact that they never know who is listening or who they may repeat their comments to.
- When in the country they should know how to use payphones and have the correct coins, tokens, or phone cards at all times, if possible. Prior to a trip determine if your cell phone works.
- They should contact their cell provider and ensure that they are signed up for international service, both inbound and outgoing.
- When traveling they may want to rent or buy a satellite phone before they leave or rent one at the airport.
- They should also keep close track of their briefcase or similar valuable material or carrier; there will always be items in their possession that if lost could hurt them or their organization personally or financially.
- They should always ensure that they do not leave their briefcase or suitcase alone or unattended in any area that could be a safety risk; they not only

risk the item being stolen, they risk it being confiscated by security as a potential risk.

- They must travel together with another person, or group, whenever possible.
- They must keep a map, local currency, and emergency contacts; have copies of important documents, including passports and visas on them at all times; and a copy in a hotel safe with the originals.
- They must carry money in zipped or internal pockets.
- They must consider using a money belt or holster and separate funds, so as to minimize losses if pickpocketed.
- They must not isolate themselves but must stay in well-lit, populated areas.
- They must avoid Western-named hangouts or places frequented by foreigners.
- They must use only approved taxis at the airport; they must consider hiring a car for the whole day or evening.
- If they are a witness to an accident or incident, they must withdraw and move away and not get caught up in political discussions or demonstrations.
- They must be leery of all strangers and avoid getting into a conversation; if talking with a stranger they must be extraalert to what is going on around such as if another person is coming up from behind.
- In many countries, a common method of robbery starts with a free cup of coffee. An assailant provides an unsuspecting victim with a free or "sample" cup of coffee that is laced with various types of knockout drugs. The technique is also being used by prostitutes.

When traveling abroad individuals need to be cognizant of fire safety; the following is a list of fire safety measures that can be practiced when traveling overseas:

- Carry lightweight Cyalume light sticks, which are great items to carry; in case of emergencies, they are an excellent light source.
- If leaving the room during a fire always take the room key in order to get back into the room or to get out of some stairway doors.
- If the doors or walls are hot, douse them with water using the ice bucket.
- If you remain in the room, turn off the air-conditioner, fill the bathtub with water, wet the sheets or towels and stuff them completely around the door, and block the vents emitting smoke.
- Open the window only if fresh air is available.
- Call the Fire Department giving them the room number and notify them that you are staying in the room.
- Hold wet towels over the nose and mouth to prevent smoke inhalation.

The following is a list of preventive measures an individual can take at the hotel they are staying at:

- Avoid leaving your key at the hotel desk and responding to the hotel paging you when in the restaurant or lobby.
- Do not identify appointments; avoid providing home address and use organization address.

- Avoid organization luggage labels, or anything that could give information to a would-be attacker.
- Avoid ground floors and those above the eighth floor that fire trucks cannot reach.
- Know the location of the fire alarm nearest to the hotel room and the emergency exits for fire and other evacuation purposes.
- Secure all valuable items and documents, preferably in a hotel safe, but be aware that proprietary documents, even in a hotel safe, are subject to examination.
- Keep the room locked, turn on the lights, play the television, and use the deadbolt and chain on the door when inside the room.
- Use a personal alarm and rubber door stop to jam door and to jam bathroom door when in the shower.
- Do not answer the door to unexpected visitors; if the person claims to be hotel staff and is at the door, call the desk or housekeeping and verify that a person was sent to the room.
- Do not use "clean room" signs; instead use "do not disturb" signs and be in the room when being cleaned.
- Assume all telephone calls, faxes, and probably that the room itself is "bugged."
- One common activity that is being used in hotel business centers is for someone to plug in a "Key Stroke Reader." These items can be plugged into the computer in seconds and they can capture whatever an individual types, be it passwords and logins or company data.

When traveling abroad it is important to have the necessary travel documentation, which includes the following:

- Passport; proper visa
- Airline tickets; itinerary (shared only with key people and family)
- Vaccinations/inoculations
- Traveler's checks (provided they are accepted in the countries of travel)
- Credit cards need to be kept secure and it must be ensured that they are taken in the area that you are traveling.
- Telephone numbers and addresses of company, embassy, and consulate
- Driver's license and/or international driver's license
- International medical insurance, medical evacuation insurance (may not be accepted in some counties)

Copies of all of the above items, including copies of serial numbers and make/model of any camera, computer, or other electronic equipment, should be carried separately, and one copy should be left at home. Copies should be e-mailed by self and maintained on the server to access if needed.

Money should be changed in a small amount ahead of time. Money should be split up between traveler's checks and cash. Do not change money at unofficial money changing locations; results could be counterfeit currency or police blackmail. The best exchange rates are normally found at the automated telemachines (ATM). When traveling it is a good idea to bring a basic first-aid kit and any required medicine or prescriptions, including extra contacts and an extra pair of glasses. Prescription medications should be kept in their original bottles with labels clearly visible. Prior to disembarking ensure the country traveling too allows the prescription medications; if not, it may be considered drug trafficking. Some countries carry heavy penalties for drug trafficking including the death penalty.

If one becomes ill while traveling they can always go to a U.S. military installation for treatment, even if they are a civilian. This involves charges but they will receive Western standard of care. Carry your prescription meds with you and have some extra in your bags in case your luggage is delayed or lost. In some countries, antibiotics are unavailable at any price. Be sure to wear a medical alert necklace or bracelet listing any allergies, blood type, and emergency contact.

When traveling abroad it is always a good idea to have in one's possession a travel first-aid kit and a travel emergency kit. A travel first-aid kit should consist of the following items:

- Small bag with emergency items
- Small flashlight, extra batteries, and a small radio
- Lock picks and door tool
- Medication, needle, and thread
- Signal mirror
- Band aids, large bandage
- Permanent marker to leave messages on walls for EMS or fire
- Emergency tool
- Whistle
- Bottle of water and ready-to-eat snack
- Pair of leather gloves to open doors in case of fire
- Pencil and spiral-bound notepad

An emergency travel kit should consist of the following items:

- Package of chewing gum
- Aspirin or similar pain reliever
- Dry washcloth in zip-lock bag and wet wipes and hand gel
- Toothbrush and paste in zip-lock bag and dental floss
- Extra contacts or eye glasses

When traveling abroad your appearance and demeanor should be conservative. Dress in darker colors and try to blend in if possible. Dress casually but "smartly" while traveling; dress "down" but not too "down." Wear clothes that are not too "Western," no jackets or tee-shirts with flags, no cowboy boots, cowboy hats, or belt buckles, and no expensive jewelry. If possible, avoid jeans and tennis shoes; although you want to be comfortable, you should make an effort to blend in with locals, so that you do not stand out. Do not wear baseball-style caps, or military-style clothing. Take just what you need, such as a suit; in addition, check the weather and bring a dress accordingly. Women should be cognizant of their appearance; especially when visiting religious sites they should have appropriate head covering and clothing. Similarly, men must also have appropriate clothing when visiting the same sites.

If being held hostage or in a crisis situation, maintain a low profile and be quiet. Do not let yourself be provoked. Do not act emotional. Crying and screaming only draws attention and do not be a hero. If offered something to drink or smoke, take it. Do not offend them. Under no circumstances should you talk politics, pro or con; the last thing you want to do is offend someone and get into an argument. When being rescued hit the ground and cover up; do not try and help.

If something is too good to be true, it is! Pay attention to people around you, especially those just "loitering." Avoid demonstrations and civil unrest; do not take pictures of military installations or civil unrest. Know the capabilities of police, medical and fire in the country you are visiting. Consider a cell phone that works in the area. Assume you are not going to be liked by all; do not get into political or religious discussions. Be alert to what is going on around you. Watch for surveillance. Vary routes and times. Listen to your "gut" and take appropriate action. Do not be in a "daze"; pay attention and always think!

6 Management Strategies

6.1 CRISIS AND INCIDENT MANAGEMENT

Incidents managed in a systematic way are the most successful at achieving the intended goals. The incident commander (IC) and their staff make operational decisions, some strategic, others tactical in nature, and carefully allocate resources to implement them. First responders need to understand the role of the IC as the ultimate decision maker responsible for the outcome of the incident. The Incident Command System (ICS) is the framework necessary to manage the resources, personnel, apparatus, and equipment used to mitigate the incident.

Strategic decisions identify the overall approach to the incident, and operational decisions spell out the best use of those resources. During routine emergencies, most firefighters follow a standard approach to onsite incident management including performing a size up, choosing a strategy, implementing various tactics, and conducting ongoing evaluation.

All first responders and ICs must know their standard operating procedures (SOPs): With an increased emphasis on nonroutine incidents such as hazardous materials, terrorism, and maritime piracy, other methods have been developed to address new aspects related to nonroutine situations. In these situations, it is especially critical to know exactly what steps to take and the sequence in which they must occur because of the presence of hazards other than those traditionally encountered. In addition, they must know the appropriate course of action: For example, during a bombing incident a responder may find it difficult to determine an appropriate course of action due to the nature or the magnitude of the incident. Furthermore, they may feel extreme pressure to act.

Regardless of the specific process used, responders go through a number of similar steps in dealing with their response. Five common steps include conducting size up, evaluating the situation, setting incident priorities, estimating potential incident course and harm, and choosing strategic goals and tactical objectives. Size up refers to the rapid mental evaluation of the factors that influence an incident; it is the first step in determining a course of action. For many responders, it begins even before the incident in the form of preplanning. The more information a responder has prior to an incident, the greater the chances of having a safe and successful response.

Incident factors are dynamic and must be evaluated continually. Therefore, in a sense, size up continues throughout the incident. In the same way that the military studies its enemy prior to battle and constantly evaluates its battle plans, so should the first responder with respect to an incident.

Incident situation refers to the type, the cause, and the status of the incident. The type of incident refers to whether it is one of the threats or hazards identified in Chapter 2. The cause of the incident refers to whether it is an accident, such as a system failure, or something intentional, such as a bombing. The incident status refers to whether the incident is in a somewhat controlled state or static or in an uncontrolled state, dynamic, and expanding. Incident priorities include life safety for the responders as well as the public; protecting critical systems such as the infrastructure, including transportation, public services, and communication networks; and incident stabilization.

Potential incident course and harm includes a series of predictions based upon the incident situation and available information. The responders estimate the probable course that the incident will take and the probable harm or damage that is likely to occur. For example, if faced with an explosion, a responder should be concerned about the possible presence of a secondary device that may cause harm to personnel or create additional property damage.

Strategic goals are broad, general statements of the desired outcome. An example of a strategic goal would be "to prevent loss of life for both civilians and responders." Tactical objectives are specific operations or functions to meet the goal. For example, to meet the strategic goal of preventing loss of life, you should "isolate the hazard area and deny entry into that area." Tactics are the specific steps and actions taken by the assigned personnel to meet the determined objectives. For example, to accomplish the tactical objective of isolation, you could "position apparatus in such a fashion as to block the area, and cordon off the area with banner tape." At each level there are more specifics involved. In the case of the tactical methods, using the apparatus and cordoning off the area are only two possible approaches.

Hazardous materials incidents differ from the more traditional incidents in that they have historically been the "bread and butter" of the fire service. This training is organized around five levels: awareness, operations, technician, specialist, and the incident manager. In implementing its training programs, the National Fire Academy (NFA) has followed these classifications. Furthermore, the NFA has adopted for its hazardous materials curriculum an incident analysis process called GEDAPER developed by David M. Lesak. The seven steps of GEDAPER provide the responders with the needed processes for analyzing and handling a hazardous materials incident safely and prudently. This same tool, although not the only one available, can be very helpful in dealing with a range of potential incidents. There are seven steps to this process, which are as follows:

1. Gathering information
2. Estimating course and harm
3. Determining strategic goals
4. Assessing tactical options and resources
5. Planning and implementing actions
6. Evaluating
7. Reviewing

A first responder should gather as much information as possible about the incident through observation, using their senses. However, a first responder in personal protective equipment (PPE), including positive pressure self-contained breathing apparatus (SCBA), can only use sight and hearing. Given the likely presence of hazardous materials at a hazardous materials incident, it would be in their best interest to observe from a distance, using only the senses of sight and hearing. The use of touch, taste, or smell could result in exposure. A first responder's education, training, and experience will help them evaluate this information before going any further.

Today, there are numerous information resources available in hard copy or electronic format to access information that will assist them at an incident scene. If they cannot access this information at the scene, then they should contact those who can access it for them. For instance when the term "mass casualty incident" is used to describe an incident scene, responders can relate to the situation automatically. The term triggers a mental assessment based on education, training, and experience. On top of this there are other layers, perhaps many, of technical information including data provided by other sources, such as texts, computers, preplans, and floor plans. For example, if responding to an incident involving hazardous materials, such as chemical, biological, radiological, nuclear, and explosives (CBRNE), the first responder may consult the *North American Emergency Response Guidebook* (2012) for recommendations on initial isolation and protective action distances. Information received from a dispatcher, such as type of incident, incident location, and number of reported casualties. Collectively, all of these factors could indicate a possible terrorist incident.

Using the above example, information obtained during the responder's size up may include unusual signs and symptoms, presence of dead animals or people, unexplained odors, unusual metal debris and placards or labels, or outward warning signs and detection clues. Environmental information would include information such as time of day or night; location of the incident including address, neighborhood, and occupancy; weather including temperature, wind direction, and relative humidity; topography including lay of the land, hills, and bodies of water; and exposures, which would include people, property, and the environment.

Regardless of the incident, the first step is to collect all the information as quickly as possible. Then, once a responder has made some initial decisions, they need to continue to collect information and reassess it. Estimating the course of an incident involves using the information they have gathered to make a series of predictions and to assess the potential harm. This involves damage assessment, hazard identification, vulnerability assessment, and risk determination. Damage assessment involves figuring the damage that has already occurred. Hazard identification means determining what product is involved, where it is, what it can do, and how much there is? Vulnerability assessment is figuring out whom and what is at risk, in other words, all persons and things the hazard may affect. Risk determination involves estimating the probability that the situation might get worse before it is controlled.

Initially, strategic goals and tactical options should be based on the most likely situation outcome. Strategic goals are broad, general statements of intent. Always to be included in determining strategic goals are the incident priorities of life safety of both the first responder and the civilian; protection of critical infrastructure,

infrastructure that is in place for the betterment of the community, such as public utilities and transportation and hospitals; and incident stabilization.

To meet the strategic goals, first responders and ICs need to select appropriate tactical objectives and methods. For instance, if the strategic goal is isolation, then the tactical objectives must include establishing perimeters and operational zones, denying entry into the "hot zone," and removing the public and emergency personnel far from the hot zone. Perimeters and zones represent a safety factor, or buffer, against the hazards presented by the incident. The establishment of zones, or perimeters, is critical to protect both first responders and civilians. Denial of entry includes the use of physical barriers, such as tape, rope, and barricades. Public protection involves establishing an area of safe refuge for those who are contaminated, thus reducing the chances of secondary contamination. It also involves assisting those individuals who are in harm's way to safety. Doing so will set the stage for decontamination and subsequent medical treatment.

All of these objectives require the use of resources, including personnel and equipment. The level of effort required, coupled with the amount of resources available, will determine if the goals and objectives can be attained. If the resources are adequate, or if other assistance is available, then the next step, planning and implementing actions, becomes possible.

Withdrawal is an option where the situation is too dangerous or too large for intervention. The best course of action may be to evacuate the area, deny entry, and allow the incident to run its course. The plan of action is a written document that consolidates all of the operational actions to be taken by various personnel in order to stabilize the incident.

It is important for first responders to appreciate the purposes of the written plan. It helps pinpoint the exact actions planned. SOPs or standard operating guidelines (SOPs/SOGs) are linked to the plan of action. They spell out the functions, roles, and responsibilities of personnel on the incident scene. They should be agreed upon long before the incident, and the staff must be trained in implementing them. The plan of action references SOPs/SOGs; it does not create them.

Another important planning step is to create a "site safety and health plan." If the incident involves hazardous materials, the Federal Regulations Occupational Safety and Health Act (OSHA 1910.120) requires that a plan be created. A site safety and health plan is a series of checklists used to manage an incident and to assure the safety of all involved. Like SOPs/SOGs, the checklists are developed before the incident and are implemented during the incident. The site safety and health plan identifies the health and safety hazards faced at the incident scene. It further identifies the appropriate PPE, decontamination considerations, Emergency Medical Services (EMS) concerns, and similar safety issues. When the incident involves chemical or biological hazards, it assists in fulfilling employee right-to-know requirements. The site safety and health plan helps to document the specific actions and safety procedures used. It assists in documenting whether the chosen plan of action and the specific procedures are followed. In addition, the site safety and health plan tracks activities and performances and ensures that personnel safely perform those tasks for which they received appropriate training. Someone trained only to the awareness level should not perform tasks specific to the operations or technician levels.

Included in the site safety and health plan are the location and the extent of zones, the nature of the hazards found on the scene, the types of PPE worn by personnel, and the types of decontamination procedures followed.

The goal of the evaluation process is to determine whether the plan of action is working as intended. Evaluation will help identify possible errors and allow the responders to correct them. Responders should monitor and evaluate all incident scenes, regardless of the severity or impact. If the plan is failing rapidly, an alternate plan of action that can be implemented quickly and, depending on the available resources, used to solve the problem will be needed. It is foolish to stick with a plan that is not working.

The review process involves revisiting and confirming the GEDAPER process. Review occurs either when strategic goals are accomplished or when there is an extended response period, and it is not wise to wait until the entire operation has concluded. If the entire process is managed effectively from the start, there should be no problems with the plan of action. Specifically, if the information gathered initially is thorough, comprehensive, and well managed, the estimate of course and harm should be accurate and the strategic goals and tactical objectives chosen should also be appropriate. If problems are discovered with the plan, then the existing plan should be modified to reflect the appropriate changes, or a new plan should be developed to replace the flawed one. In summary, the plan tells what should be done, the evaluation tells what was not done, and the review makes the corrections. Ongoing evaluation ensures that the plan is working or alerts the responder that the plan is failing.

First responders must be cognizant of the fact that their role is critical to the management of an incident. The actions a first responder takes and the decisions they make early in the incident have a dramatic effect on the outcome of the event. One of the first concerns they should address is their safety. Depending upon the situation they may find upon their arrival, coupled with prearrival information such as the incident location and situation as dispatched, they will need to make early decisions that will affect the incident, always keeping in mind the outward warning signs and detection clues mentioned earlier. On-scene considerations should be similar to their existing response guidelines dealing with hazardous materials. They must keep the following key points in mind:

- Their safety and that of their fellow personnel is paramount; otherwise they cannot possibly mitigate the incident.
- The initial steps of gaining control of the scene will greatly affect incident management. Simple procedures, such as staging apparatus uphill and upwind, performing isolation, and establishing perimeters, will help immensely.
- This may be all they can do prior to the arrival of additional resources, but that does not minimize its importance.

In summary, in a crisis situation or incident response, the first responders need to be proactive, not reactive. They need to stay a few steps ahead of the current situation to be better prepared for what may occur next. It is also important to remember that first responders are only human and that they can do only a limited number of tasks simultaneously. Although they may be overwhelmed initially, eventually their actions should overcome the seemingly chaotic situation and the incident will be under control. First responders should plan to be a part of the solution and not part of the problem and they should never hesitate to seek additional assistance when needed.

6.2   BOMB THREAT MANAGEMENT

During the last two decades, there have been a number of bombing incidents both overseas and at home targeting both military and civilian facilities and events. If this trend continues, organizations must be able to cope with this situation. The time to prepare is before and not after the threat occurs. The best defense is prior planning and prior training to minimize injuries, loss of life, and destruction of property.

Bomb threat contingency planning and bomb threat management dictate the responsibilities and actions to be taken during a bomb incident that may reduce injury or destruction. The bomb threat contingency or management plan will include actions for the following tasks: operations control, evacuations, searches, locating a device, disposal, detonation or damage control, media exposure, and after-action report. A listing of the questions that should be addressed for each task by the bomb contingency plan is given below.

- Control of operations
  - Who will be in charge of the operation?
  - Where will the command center be located?
  - Considerations of the antiterrorism or force protection officer (for military units):
    - Who is the site commander?
    - What mission will the commander have in the event of an incident?
    - What mission will the unit have in the response to an incident?
    - What is the unit's mission in the day-to-day security of the site? The THREATCON level may be a factor.
    - What role does the unit antiterrorism or force protection officer have?
    - What role do other personnel have?
- Evacuation
  - Which type of evacuation?
  - Who will make the determination of the type of evacuation?
  - Considerations of the antiterrorism or force protection officer (for military units):
    - What type of warning system(s), sirens, telephone, and loud speaker announcement exists?
    - Has the unit been thoroughly briefed on the warning system?
    - Does every soldier understand the action to be taken in the event a warning and evacuation is required?

- – In the event of an evacuation, are all personnel accounted for?
- – Are periodic drills being conducted to test the awareness and actions of the unit?
- – Was one conducted shortly after the unit's arrival?
- – What role do other personnel have?
- Searches
  - What type of search will be conducted?
  - Who will be conducting the search?
  - Considerations of the antiterrorism or force protection officer (for military units):
    - – What type of vehicle and parcel searches must be conducted?
    - – Are all soldiers conducting vigilance to identify potential threats?
    - – Are soldiers reporting suspicious activity?
    - – What role do other personnel have?
- Finding a suspected device
  - What action will be taken in the event a bomb or suspected device is found?
  - Consideration of the antiterrorism or force protection officer (for military units):
    - – Do the soldiers know not to advance to or touch a suspected device?
    - – Were the soldiers briefed on their individual actions in the event of discovering a suspected device?
    - – Do the soldiers know who to immediately notify based on their location?
    - – What role do other personnel have?
- Disposal
  - Who and how will the bomb be disposed of?
  - Considerations of the antiterrorism or force protection officer (for military units):
    - – What is the unit's responsibility in augmenting security during disposal operations?
    - – What is the commander's mission after identification and during disposal operations?
    - – What measures are in place in the event the site has to be evacuated?
    - – What role do other personnel have?
- Detonation and/or Damage Control
  - What procedures will be taken if the bomb detonates?
  - What procedures will be taken to protect the crime scene?
  - Considerations of the antiterrorism or force protection officer (for military units):
    - – What is the unit's responsibility in augmenting security?
    - – What is the unit's responsibility in assisting and recovering the injured?
    - – What security measures are in place to deter or counter an attack during recovery operations?

- – What is the force protection officer's role and that of the commanders during recovery operations?
- – What role do other personnel have?
- Control publicity
  - What information will be released to the news media?
  - Who will be responsible for disseminating that information?
- After-action report
  - Complete the report after any bomb incident at a facility or military installation.
  - Describe the incident that occurred.
  - List corrective actions and lessons learned from the incident.

The bomb scene officer has overall responsibility for all actions at the bomb scene. From the emergency operations center, the bomb scene officer controls communications, search teams, and the media. The bomb scene officer evaluates all information and makes the following decision: evacuate and search for suspicious device; search for a suspicious device and evacuate if found; or conduct business as usual, with no action being taken. However, in certain facilities and locations partial evacuation may be warranted such as in a hospital. On a military installation or highly classified area, a special weapons and tactics (SWAT) team may also be called on to the scene.

In an organization the person receiving calls for the organization, such as an administrative clerk, should be briefed and trained in accordance with organization policies and procedures on what to do, if they were to receive a bomb threat call. Most bomb threats are made by telephone to places of employment. When an employee is prepared for such a call, they can respond in a calm manner, ask for specific information about the bomb, and listen for some identifying characteristics of the caller. While on the telephone, they may be able to initiate a trace of the telephone number of the caller, providing vital information about the caller's whereabouts. The following is a guideline that may be used to record the details of a bomb threat made by telephone:

1. Keep the caller on the phone as long as possible.
2. Attempt to write down everything the caller says.
3. Pay particular attention to background noises.
4. Initiate call trace action (if available) while the call is ongoing.
5. Using a prearranged signal, notify your supervisor while the call is still ongoing. Your supervisor should contact the local police service.
6. Listen for indicators.
   a. Gender
   b. Voice (loud, soft, or other)
   c. Accents and speech impediments
   d. Mannerism (vulgar, calm, emotional, or other)
   e. Speech (fast, slow, or other)
   f. Diction (good, nasal, lisp, or other)

7. Use bomb threat card or procedure card to obtain information; the U.S. Federal Bureau of Investigation and the Royal Canadian Mounted Police, Bomb Data Center, maintain procedure cards for recording bomb threat information.

If the decision is to evacuate from the facility, the following actions will be undertaken by personnel to reduce damage, minimizes losses, and protect life:

1. Open windows to reduce the destructiveness of the device.
2. Inspect the immediate area for any suspicious packages or devices.
3. Unlock all drawers and doors.
4. Evacuate with all personal belongings.
5. Move to an assembly area at least 300 ft or 100 m from the building or area.
6. Account for all personnel.

If the decision is to undertake a search for the suspicious device, there are two different types of searches that may be utilized: a supervisor search and a search team search. A supervisor search is the most expedient method for searching for a device since it only involves the supervisor(s) and a few coworkers. This method is also less disruptive to company operations; however, it is only 60% effective in locating a device. The second type of search is a search team search. This method is preferred for larger areas and requires that the facility be evacuated first. The search team consists of those personnel that are most familiar with the work area. In addition, team members should be volunteers and must have had previous training and have conducted drills and exercises. Police officers including military police officers and explosive ordnance disposal personnel are typically not used to conduct searches of suspicious devices. Their skills are better utilized in other areas, such as in the case of police officers who would be involved in scene, crowd, and traffic control. The bomb scene officer will determine what search method will be used.

Publicly accessible areas are likely hiding locations for explosive devices; therefore, any search should start 25–50 ft outside of the facility and move inward. Inside the facility the search should start from downstairs to upstairs. Prior to entering or searching a room the searcher(s) should stop and listen for any suspicious noise(s). The room should be searched by height and each searcher should have an assigned area, overlapping for better coverage. For example, the room should be divided by height for searching: zone 1, floor to hip; zone 2, hip to chin; zone 3, chin to ceiling; zone 4, false ceiling to true ceiling. Also, when searching, search internal public areas first, since they are the most accessible.

Once a device has been located, halt, do not touch the device, and notify the bomb scene officer. The bomb scene officer will have several options available to him or her at this point. He or she may decide to terminate the search, evacuate, and notify explosive ordnance disposal personnel; mark the location and continue the search for a secondary device; or if no device is found authorize personnel to return to the facility.

If a device is found, explosive ordnance disposal personnel take charge of the scene. At this point, they decide on removing the device or destroying it in situ, depending on the situation. After explosive ordnance disposal neutralizes, the device the search should continue to locate any secondary devices. Fire department personnel should be deployed near the scene or involved in damage control, and medical personnel should be on standby for casualty assistance. After the incident, an after-action review should be undertaken. The after-action review should contain the nature of the incident, actions taken, outcome of the situation including lessons learned, and additional information as needed.

6.3   MANAGING A TECHNICAL SURVEILLANCE AND COUNTERMEASURES DETAIL

The electronic age provides access to information and products from around the world. It is more important than ever before for an individual to understand the enormous exposure their organization has to loss of their proprietary communications and information. The following statistics were compiled from data collected by the American Society for Industrial Security and National Security Agency:

- Forty-nine percent of businesses larger than $10 million had known a case of corporate espionage.
- Twenty-two percent had known successes.
- Customer list was the number one (1) loss.
- Pricing was the number two (2) loss.
- Reported incidents have increased over 900% in the past 10 years.
- Thirty percent of the assailants were employees and 28% were former employees.

With the advances made in the information technology and communication industry, much of this technology has been turned to eavesdropping devices. In turn, security is becoming a board-level issue as the number of cyberattacks and corporate espionage incidents are growing significantly each year. Taking advantage of privileged information is illegal. As one can imagine, access to privileged financial and stock information could easily be used for insider trading. The sensitive information and financial data must be controlled in order to comply with the Securities and Exchange Commission disclosure requirements.

What people may not think of are the discussions around information security, which has become a board-level issue. Cyberattacks and corporate espionage are growing significantly year after year. In a training program developed by "Spy-Ops," the company notes that corporate espionage worldwide is now more than a trillion-dollar problem annually and growing. Data breaches, theft of intellectual property, insider trading, and other criminal acts now demand the attention of the board of directors.

How vulnerable is an organization or individual to corporate espionage is greatly dependent upon what the individual or organization is involved in. For instance, is the individual or organization involved in a litigation suit? A dispute involving the corporate officers or senior partners? A buyout, stock purchase, or merger and acquisition? A situation where they have become newsworthy? A situation where they have become a threat to their families and themselves?

The following are examples of cases involving illegal wiretapping, bugging, and/or eavesdropping. Anthony Pellicano, born March 22, 1944, in Chicago, Illinois, is a former high-profile Los Angeles private investigator who served a 30-month federal prison sentence for illegal possession of explosives, firearms, and homemade grenades. Pellicano was known as the "go-to-guy" in Hollywood whose clientele included an A-list cast of celebrities including Garry Shandling, Sylvester Stallone, Warren Beatty, Don Simpson, and Michael Jackson. On February 4, 2006, Pellicano was arrested for wiretapping and racketeering.

On May 15, 2008, after representing himself and nine days of jury deliberation, Pellicano was found guilty on 76 of 77 counts related to racketeering, along with four codefendants. "If the government has no plans to go higher than Pellicano, this is a depressingly pedestrian effort that shows a lack of ambition," commented John C. Coffee, a professor at Columbia Law School and an expert on white-collar crime, as quoted in the *New York Times* story on the verdict (Hall and Abdollah 2008). After a six-week trial in the District Federal Court in Los Angeles, Pellicano was convicted of wiretapping and conspiracy to commit wiretapping. The court denied Pellicano's request for concurrent sentencing on the multiple counts and imposed 15 additional years in prison in December 2008 and ordered Pellicano and his two codefendants to forfeit $2 million.

In a story published in *Newsweek* (Pelisek 2011) on August 7, 2011, Anthony Pellicano, in his first interview since going to prison, revealed new details about his activities. On July 5, 2012, the *Hollywood Reporter* published a story that Pellicano's bail hearing was postponed. Pellicano's lawyer requested the delay because of a personal health issue. The next day it was reported that Tom Cruise had been "accused of wiretap conspiracy with convicted criminal Anthony Pellicano during Nicole Kidman divorce" (Howard 2013).

The vice president and the son-in-law of the chief executive officer who was caught on tape with a girl from Gold Club was blackmailed into letting his company offices be bugged. Faced with the loss of his wife, family, and job, he escorted the bugging team into his offices. His father-in-law requested he have the offices swept so he had the bugs removed. The father-in-law then had cameras installed and caught his son-in-law letting them back into the offices to reinstall bugs.

An antique auction house had their delivery truck hijacked and lost several million dollars' worth of antique art. Further investigation into the matter proved that their competitor had placed a transmitter on their Christmas tree inside Santa's head providing them with shipping information allowing them to steal the art.

American Express Corporation's Vice President David Goldenberg was arrested for allegedly participating in corporate espionage practices against a competing manufacturer's representative firm. David A. Goldenberg, born on May 18, 1962, was arrested on charges of unlawful access to a computer system or network, unlawful access of computer data or theft of data, and conducting an illegal wiretap. Goldenberg pleaded guilty and received three years' probation.

The arrest stemmed from an investigation concerning corporate espionage by the Paramus Police Department. The Paramus Police Department received a complaint from a Paramus-based corporation known as Sapphire Marketing, which specializes in high-end audio/visual systems. Representatives of Sapphire reported that they were being suspiciously and consistently underbid for contracts by a competitor for whom David Goldenberg was working. They expressed suspicion of corporate espionage. Based on anomalies that the complainant noticed within their computer network and more specifically their e-mail system, they suspected that the company's e-mail system had been compromised and that e-mail was being intercepted. The Paramus Police Department, a member of the Computer Crimes Task Force, and the Bergen County Prosecutor's Office Computer Crimes Unit initiated an investigation.

The investigation revealed that Goldenberg had engineered the passwords protecting several of the complainant's e-mail accounts. For a period of time, Goldenberg was intercepting and reading e-mails that related to potential contracts. Goldenberg then established a free e-mail account that he had control over, and created an automatic forward of the victim's e-mail so that they would be sent to him directly. This afforded Goldenberg advanced knowledge of Sapphire's customers and bid prices, thus further affording him an opportunity to underbid Sapphire. Sapphire Marketing estimates the loss in revenue from Goldenberg's actions exceeded $1 million.

However, not all espionage cases are as dramatic as the above cases; some are very conspicuous. For example, a college professor admitted to bugging 56 offices in the Atlanta Financial Center for more than 4 years. The professor stated that a former student had paid him to hardwire transmitters into the offices of the buildings. In one incident, a hotel reported that someone was coming to their business center and putting a key stroke monitor on their computers and coming back the same night and retrieving them. In another case, the company wanting to gain access to a building bought the cleaning company that had the contract on the building. This gave them total access to all the offices and 66 blocks in the utility areas of the building. The 66 blocks are the telephone connecting blocks where individual lines are split off from the mail trunk lines coming into the building or coming up from the lower-floor utility rooms.

Many governments such as those of France, North Korea, and China use their intelligence agencies in situations that impact their economy. France invited 350 U.S. businessmen to fly to France on the Concord to attend the opening of a manufacturing plant. They bugged every seat on the plane and recorded conversations between chief executive officers and chief financial officers. Chinese engineers touring a defense contractor painted their shoes with a tacky glue to pick up metal shavings to be analyzed to see what the metal was made of. In another case, a French hotel provided a shredder to visiting businessmen; the shredder was modified with a scanner to scan everything that the businessmen shredded.

An organization or its people should be aware of some of the indicators that they are being bugged, which may include a break-in with little or nothing stolen, lost contracts by a small margin or to the same company, competition that is one step ahead, and inexplicable loss of key clients.

Within the last two decades, the communications technology market has seen a flood of products that can be used as bugging devices. Some of the hottest new bugging devices are now among the least expensive. GSM or SIM bugs are like cell phones, but without the keypad. Eavesdroppers call and listen from anywhere in the world. At one time these devices sold for $250–$500. The price has plummeted to $35–$55. This is because their sister products, cell phones, PDAs, smart phones, androids, and iPhones are often giveaway items due to economy of scale, with consumer demand fueling mass production. MicroEyes DVR BallCam is the world's first DVR in the shape of a ball. It has real-time recording speed. Unlike digital cameras or camcorders with limited recording time for the battery and memory, these devices go up to 12 hours to 2 days and the unit gets activated on motion detection.

Some devices can be hidden in places no one would notice such as a modified electric plug equipped with a microphone and transmitter. Other devices can be less conspicuous such as a bug that was mounted in plain view on the wall of a boardroom during a hostile takeover meeting, or a device in plain view that reads "radon gas detector" but in actuality is a microphone and transmitter hidden in a plastic case with label on it.

Other methods that have been used to eavesdrop may be considered crude by today's standards, yet they still get the job done. For example, a two-way radio placed under a receptionist's desk by a salesman while she was out of the room making him a copy. The radio had a rubber band holding down the talk button and a two-sided tape connecting it under the front of the desk. In another case, there was a recording device mounted in the utility room of a business with a microphone in the smoke detector that was then mounted in the boardroom. The device could be activated on command remotely using a garage door opener. Similar cases involve a phone system being altered with an FM transmitter that was received with a Sony Walkman player that was placed beneath a central processing unit (CPU) cover and a tape recorder connected to the phone lines in an office and hidden over the drop ceiling.

It is common for eavesdroppers to install locking utility boxes on the walls in the utility room to disguise and conceal recording devices. A major insurance company had three tape recorders installed on the chief executive officer's, the president's, and the senior attorney's phone lines with a label stating "Music on Hold." The devices were in place for months before being located. In another example, a standard wall plug was cut into the wall of an office to hide a microphone. The outlet was installed and the device was active for over two years before the company had a sweep completed and it was located. In addition, outside utility lines, phone, cable, DSL, and electric lines are all vulnerable to attacks. Being outside the building provides the eavesdropper access to the communication and information lines without the risk of breaking into the facility. Furthermore, devices are typically hidden in items that are in plain view but people never seem to notice, such as plants, clock radios, bags and purses, fashion accessories such as lapel pins, information boards, baby monitors, electric reciprocals, smoke detectors, and radon gas detectors.

At one time, government experts stated that a laser was not a threat as it was too expensive to build until a German college student built one for a trade show for less than $500. Lasers can now be purchased in many areas as well as on the Internet. Lasers allow an eavesdropper to record the information on a computer screen without having to enter an office or building. The eavesdropper can set themselves up in an adjacent office or building and direct the laser to the target computer terminal.

By separating themselves from the target, it eliminates the risk of breaking into someone's office.

Many of the techniques and equipment used around the world today were born in the Cold War days. Now that these agencies have downsized, thousands of ex-spies from all intelligence agencies are now working for private corporations, utilizing the techniques they learned while in government service. Russia and the United States both developed hundreds of devices for gathering intelligence from the other side. They used things such as bugs in martini olives, watches, eyeglasses, and cane bugs.

If it is the intent of an organization or individual to hire a technical surveillance and countermeasures (TSCMs) team to conduct a sweep of their facilities or property, there are a number of questions to be asked prior to selecting an appropriate firm. The following is meant to serve as a guideline for selecting a TSCM firm:

- Ask to see their insurance certificate that specifies TSCM.
- See proof of experience for the personnel who will be conducting the sweep.
- See technical data sheets on the equipment they will use to conduct the sweep.
- Ensure that their equipment will pick up microwave transmitters.
- Ensure that they can provide you with a printout and visual and audio records (videotapes) of any activity noted.
- Ensure that they have the capabilities to break out data from a burst transmitter.
- Ensure that they are conducting a complete phone line analysis, not just checking the voltage (the new electronics cannot be picked up by voltage).
- Ensure that they have the ability to conduct a real-time phone line analysis.
- Ensure that they can conduct a nonalerting analysis.
- Determine if they have credible references? Who else have they worked for?
- Request a copy of a sanitized report.
- Request an overview of what inspections will be conducted.
- Determine if they are licensed to perform TSCM sweeps; provinces and states require licenses.
- Determine if they have the personnel available to conduct a proper sweep; a properly completed sweep will take a team several hours to complete.
- Request a curriculum vitae for all team members.
- Check to see where and when they received their initial training and continued professional development training.
- Request the specification sheets for the TSCM equipment that you will be using.
- Request a report that you have completed in the past.
- Ensure that a professional TSCM team has at least $50,000 invested in their equipment; some may have well over $300,000 worth of electronics.

A professional TSCM team will have current up-to-date equipment and training to keep up with the most current industrial espionage threats. They will need to have at a minimum a spectrum analyzer with capability to read a minimum of 25 GHz, a wideband receiver, an electromagnetic field detector, an analyzer capable of receiving microwave transmittances, a telephone analyzer that can analyze break percentages and recognize blue streak taps, and equipment to recognize laser transmittances. Less experienced investigators conduct TSCM sweeps using frequency counters and wideband radio-frequency (RF) detectors. While these are tools, they are not the only type of devices that should be used to conduct the sweep. Although these tools do have some use and do what they were designed to do, they should not be used as the primary TSCM tool. They should always have a spectrum analyzer as well as wideband RF detectors and carrier current analyzers.

When an organization or individual decides to call a TSCM team, they should be sure to call from a payphone or use a cell that does not belong to one of the targets as well as ensure that they are not overheard. Most bugging is completed with the assistance of trusted employees. It is suggested that they not tell anyone that they are having the building swept and that they be there to escort the TSCM team to ensure that the bugs are not removed prior to the sweep. Once they spot a problem, they must decide what their risks are. This will assist them in deciding the level of sweep they will need. For example, if they are dealing with National Security, their risk may be greater than for an individual whose spouse is suing them for divorce. If they are protecting their information from a foreign government, they will need a much better information security program than if an individual was defending against neighbors that is trying to put them on YouTube.

A professional TSCM team will have years of experience and training and will have at least $50,000 thousands of dollars invested in their TSCM equipment. Professional TSCM firms typically charge $5,000–$10,000 for a quality sweep. Most companies charge an average of $600 for an average-size room, $200 per phone line, and $200 per telephone. Many companies charge a $2000 fee for setting up their equipment. A maintenance contract is very important to ensure ongoing information security.

Some TSCM professionals may recommend conducting a sweep in every room in the building. This may not be necessary, since janitor's closets and storage rooms do not keep a lot of important business information or hold important business meetings. It is important to sweep any room where proprietary information is discussed including conference rooms and break rooms. In addition, an organization may also need to sweep their administrative staff and receptionist areas as well as all utility rooms with 66 blocks. The TSCM team should arrive one day early and establish surveillance of the building. The team should also establish real-time monitoring of the RF waves in the area to determine if audio or video is being transmitted from the target area. The team should have the ability to monitor RF signals emitting from the target area providing evidence that someone is wearing a body microphone and transmitting to someone outside the building. Real-time monitoring provides them with the ability to locate an eavesdropper before they transmit the entire meeting data. It is important to ensure that the TSCM team has the capability to record any transmission that is detected so that it can be used as evidence in court. Running real-time analysis allows the team to monitor for "burst transmit ions," a common type of government transmitters that record all day and then send out the days recording in one short burst. Many of the transmitters coming out of Asia right now are transmitting on X band from 7.5 to 12 GHz. Most detection devices only reach 3.6 GHz. Government microwave can reach as high as 18–24 GHz.

In 1986, Congress broadened the wiretapping law to prohibit the interception of electronic communications, to include cell phone use. With the passing of this law, manufacturers of scanners had to place a block in all scanning devices for the cell frequencies. While TSCM manufacturers in the United States are required to block these frequencies, most do not advise purchasers of the fact, and if asked they say that "everyone has to do that." Special permission from a regulating authority is required, such as the U.S. Federal Communication Commission (FCC), to the manufacturer stating that one has the authority to have access to those frequencies and for the manufacturer to leave out the block. Eavesdroppers will find the weakness in a system and exploit that weakness. It is important to ensure that the TSCM team has that issue covered.

Commercial technical countermeasures should include an RF analysis of target areas with a minimum of 10 GHz, telephone analysis, telephone line analysis, carrier current analysis of electric lines, video analysis, and a complete physical search. The frequency analysis should include microwave and laser inspection. The phones and telephone lines should be analyzed by performing complete analyses, not just voltage checks. The following sections contain samples of TSCM sweep checklists; this information is used to complete the TSCM report.

Phones and Telephone Lines

1. DC Resistance Monitor Mode_____ DC Resistance Talk Mode_____
2. AC Impedance Monitor Mode_____ AC Impedance Talk Mode_____
3. DTMF Dial Level_____ DTMF Twist_____
4. DTMF Frequency Deviation_____ Dial Pulse Dialing Speed_____
5. Dial Pulse Break%_____ Ringer Equivalency_____
6. Minimum d/p Receive Inter Digit Time_____ In-Band Tone Level Deviation_____
7. Long Pulse (Hook Flash) Minimum pres_____ Zero Volt Minimum Interval_____
8. Off-Hook Dial Pulse Threshold_____ Decibel Bar Graph Level_____
9. DC Voltage Range_____ AC Voltage Range _____
10. DC Current Range_____ AC Current Range_____

Electronic Analysis

1. RF_____ Electro Magnetic Field_____ Carrier Current_____
2. Spectrum Analysis_____ Laser_____ Micro Wave_____

3. Real   Time_____   Times   Conducted_____   Date
   Conducted_____
4. External Analysis_____   Non Alerting_____   Video
   Analysis_____

Physical Examination Notes

1. Desk_____ Conference Tables _____
2. Outlets # Electric_____   # of Phone_____
   Speakers_____ Vents _____
3. Setting of Screws_____   Phones_____
   Curtains_____ Pictures_____
4. Lights_____ Appliances_____
5. File Cabinets_____ Walls_____
6. Chairs_____ Ceiling_____ Tables_____
   Computers_____Printers_____
7. Cell Phones_____ Copy Machines_____

In this technologically driven age, computers, laptops, fax machines, copy machines, printers, Blackberries, cell phones, MP3 players, and any other electronic devices are potential targets. Therefore, a professional TSCM team will include a computer forensic technician with experience on their team and as part of the TSCM sweep. There are many people who claim that they have expertise in TSCM; however, a professional TSCM company needs to have an electronic engineering background, in addition to knowing how to use their equipment. More eavesdropping devices are located during a thorough physical inspection than with electronics. If a group only knows equipment and if the equipment malfunctions, they will have no idea of any threat that is in the target area. If they claim having worked for a federal agency and do not have documentation to back up their claim, they are not legitimate. Therefore, it is important to always check a company's or team's credentials.

6.4   MANAGING A CLOSE PROTECTION DETAIL

There are three main steps in providing an effective close protection program: a comprehensive threat assessment, a thorough and complete security plan, and a specialized training to neutralize the known threats revealed in the assessment. When undertaking a threat assessment, there are three conditions that need to be satisfied: get the principal where they need to be in a safe manner, be nonobtrusive, and be as inconspicuous as possible. To evaluate the threat, one will need to conduct a security survey of the venues that will be visited by the principal and a route survey of the transportation routes that will be taken by the protection team with the principal.

The security survey will involve undertaking a physical inspection of the event venue, hotel, corporate headquarters, residence, secondary residences, and surrounding areas. In addition, the security survey may involve an inspection and analysis of phone lines and 66 blocks of the venues and facilities being visited and phone lines of the principal's residences. If there is suspicion that the principal is being bugged or under surveillance, a complete TSCMs sweep may be necessary. In addition, a route survey of all transportation routes should also be undertaken. The route survey will identify road changes, detours, road construction, close lanes, scheduling for public transportation, and traffic choke points. Through the survey the team should be able to identify congested areas such as churches, shopping malls, schools, and playgrounds. During the survey, the team should make note of landmark points that may be visited, sporting arenas, and public demonstrations.

The close protection manager or special-agent-in-charge (SAC) is responsible for selecting, assembling, and training the necessary members including specialists that are part of the close protection team. Specialists may include paramedics, TSCMs technicians, high-risk search advisors, intelligence specialists, counterassault team members, designated marksmen, and observers and the escort group. It is the responsibility of the SAC to ensure everyone is properly equipped. Special training for specific job areas or environments may also be necessary. For example, if traveling with the principal to a nonpermissive environment, agents should be familiar with the country including terrain, climate, customs, culture, and language up to an operational level. Agents may also require special training in subjects such as basics of land warfare, survival training including hostage survival, surveillance detection, insurgency tactics and operations, improvised explosive devices and high-risk evasive driving. Most of these subjects are reserved for members of a high-risk protective services detail (PSD) or members of a counterassault team. Agents must also have the ability to travel independently or in teams around the urban centers if necessary. Therefore, the SAC must ensure agents have the necessary coins or call cards needed for phones and necessary change for buses.

Prior to disembarking with the principal, an advance team should be sent to undertake the security survey and route survey as part of the preoperational planning process. The following is a list of tasks that should be undertaken by the advance team as part of the preoperational planning process:

- Conduct liaison preoperational planning with municipal law enforcement officials.
- Obtain the name and direct number for shift commander when the principal will be visiting.
- Liaison with all fire stations that will be encountered in areas the principal will travel.
- Check ambulances with areas that will be covered.
- Check county medical centers.
- Check which hospitals specialize in emergency and trauma medicine.
- Know which hospitals specialize in what type of emergencies. Some hospitals are better in treating trauma cases such as ballistic wounds or burn victims, and some are better with cardiac cases.
- Determine the nearest major trauma and cardio centers to the venues the principal will be visiting and route paths.
- In case of death, determine regulations to ship a body back to the principal's home country.

- Design, develop, and prepare to execute an emergency first-aid training program for the principal's group leaders.
- Conduct liaison and operational planning with transportation agent(s).
- Design, develop, and communicate a specific security briefing for the principal's staff or group leaders.
- Provide personal emergency contact information for each member (visitor or guest) of the principal to be protected.
- Provide biographical information for the principal and family.
- Provide biographical information on the guests of the principal.
- If possible, make a video tape of the principal and each visitor or guest, together with the minimum biographical information.
- Make audio recording of the principal's voice for later comparison in case of kidnapping.
- Develop and brief the principal's management or assigned agent on the operational plan; 45 days prior to the date work must be completed.

A TSCMs sweep should be conducted of the principal's office, home, and vehicles to ensure that the plan is not compromised. Microwave is the major threat today; it is not at all uncommon for the signal to be as high as 1900 GHz. TSCM equipment needs to be able to go high enough to cover the threat. Clock radios in hotel rooms have become popular places to conceal a video transmitter. New technology has made transmitters much smaller than a penny and capable of much better sound quality. Computers are constantly being monitored. The items outlined are a minimum procedure for developing a plan. Each item may be expounded upon in length. Additionally, they are not tactical operational plans. Tactical plans fall under the actual mission.

Everyone has to work within a budget, and optimum security in most cases is not going to be properly covered. This is why it is so important to validate and build the value in what is being done for the principal, regardless of whether the source of funding is from public or private funds. As a result, more person-hours are put into the advance work than the actual protection assignment. The principal only sees the few hours that the close contact protective agent spends with them and not all of the backup personnel, specialists, technicians, follow-up, and advance work that goes on in the background. In order to succeed in an assignment, close protection agents must learn to work with the principal. They must also recognize that the principal may not have the in-depth knowledge of security procedures and practices that they do. Therefore, a close protection officer's duties may include educating the principal on standard security practices, ensuring their safety and simultaneously giving them the freedom to carry on with their daily lives and business. Ultimately, by working with the principal the close protection officer becomes part of the management process.

6.5   MANAGING DISASTERS IN THE DATA CENTER

Information technology disasters whether they are natural or human-induced have to be corrected as soon as possible to limit data and revenue loss. Examples of disasters that can impact information technology systems and compromise data include fires in the building or in the data center; hurricane, including floods and tornados; earthquakes; bombings such as explosions and fires; power outages, both short term and long term; and sabotage in the data center.

Two major concerns for crisis managers are the frequency and the fact that earthquakes may bring with them fires from broken gas lines, power outages, and flooding from failing dams to mention a few. In addition, unlike a hurricane or most other natural disasters, an earthquake does not give emergency or crisis managers days of radar tracking advanced notice. Furthermore, they may find themselves blocked from entering the area due to the structural damage or fear of aftershocks.

Sabotage by a disgruntled, careless, or negligent employee as well as mistakes by employees can also cause power outages. In addition, attacks from external hackers, denial of service attacks, website changes, or attacks on the command and control center to shut down or delay business operations are all too common. To manage the risk of data loss not covered by a traditional disaster recovery plan, businesses must also have in place an administrative plan commonly referred to as a business continuity plan (BCP).

An information technology BCP must detail exactly how an organization will recover from occurrences such as upgrades to both hardware and software. Organizations must understand the differences between the BCP and a disaster management plan in order to develop documents that will help the organization in a crisis event. Traditionally, disaster recovery has been defined as the ability to recover from a catastrophic outage of information technology systems and services. Not knowing the differences and having a crisis could become the negative turning point or an organization and lead that organization to significant losses. Planning a recovery from catastrophic disasters still leaves an enterprise exposed to the risk of lost revenue and lost productivity resulting from occurrences that are far more mundane. In fact, at least 80% of all data loss result from human error.

The primary purpose of the BCP is to recover the business including the recovery of mission-critical, time-sensitive business functions and services. A BCP is more comprehensive than a disaster recovery plan, in that it addresses the risk of lost revenue and productivity from all sources of data loss. As a result, disaster recovery is a subset of business continuity. In fact, an effective BCP includes a catastrophic disaster recovery plan. For instance, it is possible to have a crisis without a disaster. A crisis can exist with no physical damage to facilities or technologies. Similarly, it is possible to also have a disaster without a crisis such as a loss to physical facilities or technologies with the absence of a crisis. However, both crisis and disaster have the potential to escalate. In addition, in crisis management the intent is to limit the intensity of a negative threat or event to an organization's employees, products, services, financial condition and reputation, whereas in business continuity the focus is on recovering mission-critical business services and processes.

A BCP consists of the following:

- To recover mission-critical business services and processes
- Focus on Limited scenarios
- Focus on technology facilities and/or data

A crisis management plan (CMP) consists of the following:

- To limit intensity, manage and control negative results of an event
- Focus on many scenarios
- Focus on people, products, services, and reputation

The plan strategy needs to address recovery from a catastrophic event as part of the total continuity plan. Furthermore, information technology needs to be built into the core of this plan.

Ten basic components of a CMP are as follows:

1. Document introduction
2. Crisis scenarios/situations
3. Crisis considerations
4. Crisis management team
5. Crisis management facility
6. Notification procedures
7. Action procedures
8. Postcrisis analysis
9. Plan exercising
10. Appendix

6.5.1 SECTION 1—DOCUMENT INTRODUCTION

This section includes a cover page, table of contents, a letter from the CEO, and a plan acknowledgment form to be filled out by the recipient. As in all confidential documents, the plans should be carefully controlled and managed.

6.5.2 SECTION 2—CRISIS SCENARIOS/SITUATIONS

This section contains the results of the risk assessment. It also includes detailed scenarios and likely examples that this plan was developed to address.

6.5.3 SECTION 3—CRISIS CONSIDERATIONS

This section includes documentation, proprietary information, financial and legal considerations, and media relations. The documentation section covers the need for careful documentation during the entire crisis event. Because a crisis will usually involve litigation, careful reports, and forms for press notification, team notification and formal notes are developed and managed for present and future use. This section will discuss the process and explain the use of the documentation forms. It will also answer future questions from the press, regulators, neighbors, and shareholders. Formal notes stating who called whom, who stated what, and the date and time of all major events are included. Finally, events will need to be reconstructed for insurance and liability claims.

The proprietary information section contains selected confidential information including state and federal statutes that preclude certain data. This information may be required during a crisis. This section will advise what corporate information may and may not be given to the media during the event. Confidential information includes asset and liabilities of the firm, the net worth of the organization, estimates of damage, and the number of government or private contracts. After the crisis and after the team has reviewed the situation, the next step would be to acknowledge that an incident has occurred and to estimate the extent of damage to the organization.

The financial and legal consideration section covers issues that may be considered by the firm during a crisis such as suspending or acquiring company stock, communications to brokers, vested interest groups, clients, and employees. This section will list the concerns and develop guidelines for management.

The media relations section is a critical area and will include communications with the press, who is the primary contact, media packages, and how to manage and disseminate the information. Noted in this area will be consultants, service providers, and media experts who may be required and what their roles and responsibilities would be in specific scenarios.

6.5.4 Section 4—Crisis Management Team

This section includes basic team information such as who is in charge, local and regional responsibilities, team members, and backup personnel.

6.5.5 Section 5—Crisis Management Facility

This section includes basic site information, purpose, location, onsite inventories, and implementation procedures.

6.5.6 Section 6—Notification Procedures

Notification procedures include primary and secondary notification. The plan should also include criteria for activating the plan or part(s) of the plan and who makes the call.

6.5.7 Section 7—Action Procedures

This section identifies the responsibilities and actions with the assigned functions or personnel. It will assign responsibilities for notifications, media services, legal services, personnel services, advertising, and financial and legal services to the chosen members and service providers.

6.5.8 Section 8—Postcrisis Analysis

This section includes the directions and procedures for the team to evaluate the crisis and the actions taken and not taken. It is implemented after the crisis has ended and will be used to enhance the plan and future training.

6.5.9 SECTION 9—PLAN EXERCISING

Plan exercising includes training and education, exercising methodology, guidelines and procedures, forms and descriptions, and disaster recovery planning or testing. This component could be a regulatory requirement.

6.5.10 SECTION 10—APPENDIX

This is the section of the plan that houses the documents, forms, and information that will need regular updates such as notification numbers, team names and information, and vendors and inventories.

With the threat of terrorism, all plans must be appraised and detailed disaster recovery plans to address all threats and hazards must be developed. It is not sufficient to just simply add business continuity or information technology to an existing disaster plan. There are a number of questions that need to be asked in formulating a plan:

* Has the threat changed since your original plan was written? Was time to recover carefully considered?
* In the event of a catastrophic failure, was a distinction made between getting critical systems back online and full recovery?
* Is protection from the deliberate destruction of data addressed?
* Is there a new technology available that will help?
* Does the enterprise have partners that could help with recovery?

As a result, input needs to be obtained from all areas of an organization when building a BCP. When formulating your BCP, ensure that input is sought from top levels of management. All lines of business should be represented. Calculate the cost of the system or network being down. This will assist in determining which functions are the most valuable and critical. If needed, determine the cost of an outage down to a "per second" basis. Determining the cost of an application outage will help determine the applications that are most important and critical. Determine the order in which an application should be recovered first. Part of this involves determining what applications are dependent on other applications to function properly. This also provides strong guidance for how to apportion the budget for business continuity. This aspect of a BCP is essential for enterprise resource management suites. Some applications may seem less critical than others until one realizes that a critical process may stop functioning without data from other application sources. If those sources are remote, the communication links that deliver data must also be included in the plan.

What is the critical time to recover specific applications when a disaster failure occurs? Information technology staff within affected enterprises realize that there are varying levels of recovery based on time elapsed since the actual event. Some critical applications could be brought back online quickly, but not with full functionality. Backup and restore solutions can play a crucial role in minimizing the time to recovery with full functionality. The time to recovery of commercially available backup and restore solutions should be closely examined.

Data loss can also result from malicious acts. Any BCP should address this possibility. Depending on the type of incident, information technology staff may not be immediately available to recover from an outage. Noninformation technology staff members should also be trained to recover from outages when possible. At least parts of the plan should be documented so that noninformation technology staff can understand it and perform crucial functions.

Developing a plan is not enough; an organization must ensure that the plan functions. An organization's information technology system or network cannot be considered functional until it is tested. The business needs of the enterprise change especially quickly; the technology the enterprise relies on will help to mitigate risk changes as well. Both should be reviewed periodically; this can be done by companies that provide an organization with software to check the system or conduct on-site system audits. Periodic outside audits keep a check and balance on an organization's enterprise system. Develop partnerships with vendors and other information technology organizations. Back up mission-critical applications. Backups should be stored in a safe location, preferably off-site or out of area, in a secure fireproof, climate-controlled environment.

Business models depend heavily on the ability of employees, customers, and suppliers to access a variety of information technology resources such as routers, firewalls, cables, switches, software, people, electricity, specialized equipment, tape drives, and connectors. Although most of these resources are reliable, failures still occur sometimes, making it prudent to have a comprehensive BCP in place. Recovery plans often fail to anticipate the unique stresses and strains that a true catastrophe can place on information technology resources. One of the lessons learned from Hurricane Katrina was that the public records, along with the current disaster management plans, were out of date.

There are many ways to recover data. It is important that the information technology department briefs the response teams on exactly how they are to proceed. Normally they only get one chance to recover data from a hard drive. They should have a security company under contract that can draw personnel from other areas as local law enforcement and security will be unavailable. If the organization has important public records such as deeds or restraining orders, they may want to have a scanned copy in their backup facility to assist in any future legal actions.

During a crisis or emergency situation, employees of an organization are essential. First responders especially will have a hard time staying at work if they are worrying about their families needing help. When failures escalate beyond the point of a system crash or a brief network outage, the impact can be devastating; in fact an extensive data loss can threaten an organization's ability to continue doing business. A comprehensive approach to business continuity should plan for a worst-case disaster. By planning for the worst case, anything less will be easier to handle. In planning for the worst-case scenario, the following questions should be taken into consideration:

- Can the data center be relocated to an alternate site?
- Can some employees work from home?
- Which resources are lost and unrecoverable?

- Which resources are partially functional?
- Which applications and databases are most critical for business continuance?

Organizations should establish contracts with vendors to determine the services that they will be provided with prior to an emergency. Disasters can create high demands on organizational resources; in some cases, the demand can exceed the available resources, further exacerbating the situation. Organizations need to plan ahead to determine what resources they will need in order to deal with a crisis or emergency situation. The plan should be developed with key staff members to enable the organization to keep operating even during a crisis situation. It will be more cost-effective in the long run for an organization to develop a comprehensive plan than waiting until the disaster occurs to act.

6.6   HOSTAGE SURVIVAL AND CRISIS NEGOTIATIONS

This section is subdivided into two parts: hostage survival and crisis negotiations. The first part will discuss hostage survival and the second part will discuss crisis negotiation. The portion on hostage survival will discuss the psychology of hostage taking, personal contingency planning, the actions to be taken by a hostage at the moment of capture, and the effects of the Stockholm syndrome, and describe the techniques for adjusting to captivity, and the actions to be taken by a hostage during rescue or release. Hostage taking represents a unique bargain struck over the value of human life. Whatever the immediate motivation, the basic purpose remains the same. Hostage taking is a way of setting up a bargaining position to achieve an otherwise unattainable objective.

A victim may be chosen because of their value to someone. Government will not negotiate with terrorists. This does not diminish the victim's value to their family, employer, or themselves. There are numerous examples in which ransom demands have been met by private efforts. This may free the hostage but unfortunately further promulgates the act. A victim may also be chosen because he or she is prominent, because of their job position and their social status, or simply because of their ethnic or cultural origins.

A victim may be hated by their captors. The terrorist organization may blame them directly for any setbacks it has suffered or may foresee disaster in the near future due to this individual's work. The U.S. advisors in El Salvador, especially the combat advisors, are not liked by the terrorists of the Farabundo Marti National Liberation Front (FMLN). Because of their desire for publicity, terrorists do not kidnap a victim and then not tell anyone. The higher the status of the victim, the more publicity the event will solicit. The victim may be seen as a source of trouble. In Colombia, the terrorist groups M19 and the Revolutionary Armed Forces of Colombia (FARC), which make more than US$100 million each year from cocaine, could look at the special agents of the U.S. Drug Enforcement Administration from this point of view. Most of the time, the hostage is just an innocent victim of circumstances who happened to be in the wrong place at the wrong time.

Hostage taking, by the very nature of the act, forces the terrorist into stereotyped responses. The hostage becomes a pawn, caught between the terrorist and the authorities. The terrorist becomes violent and strives to control the situation and move the event toward completion of his or her objective. The hostage, by his or her conduct, can enhance or diminish their chances for survival. The more the hostage understands about their capture, the better they will be able to predict the hostage holder's behavior and feel some degree of control, which can assist in diminishing fear.

All classification by category of hostage takers is an artificial matter. We categorize hostage takers merely to provide guidelines. The real identity of the terrorist can only be examined in the context of his or her relationship to the people, the organization, places, ideas, and the historical context with which he or she is associated. The question one must ask when examining any terrorist action is, "does the ideology support the actual beliefs of a group, or is it merely a tool used to justify their actions?" For example, the Symbionese Liberation Army (SLA) leader Donald DeFreeze used Marxism as a cover label for criminal actions.

Terrorist organizations are usually stratified. At the top of a pyramid-like structure, single, urban, bright, and dedicated idealists can be found. Often this layer of terrorists comes from professions such as medicine and law. The lower down the ranks of a terrorist organization one looks, the greater the diversity of personality styles one sees such as the disillusioned, the mentally ill, sociopaths recruited from prisons, and the ideologically motivated. There are also those who are monetarily motivated, the "mercenary"-style terrorists, and those seeking some form of personal revenge. Almost all terrorist organizations form a cross section of the population from which they emerge. The group performing the hostage taking will in turn be a representative subgroup of the organization. This subgroup performs its operations in support of the larger organization's long-range goals.

Two categories of hostage takers that pose a major threat to government(s): political extremists and religious fanatics. Political extremists are hostage takers that often operate within a military-like structure. This control factor may hold violence within planned limits. Still, within the group there may be one or more terrorists whose propensity for cruelty makes them more dangerous. While they may be held in check by their group leadership, it is important to identify them early in the event and spend extra effort to avoid a confrontation.

The average age of most political terrorists ranges between 19 and 35. This youth factor heightens their fanaticism, blinds them to reasonable dialogue, and insulates them to appeals based on morals, decency, or fear for their own safety. While the political extremists are often prepared to die for their cause, they are not necessarily suicidal. When the chances for success dwindle, their primary concern often changes to escape. Political extremists generally only take hostages when they have some control over the general terrain where the event will take place or when a friendly or neutral country may provide shelter. Hostage taking by this group of terrorists can be viewed as a barometer of political extremism in a geographic area.

Religious fanatics are hostage takers that generally fit one of three categories. They are usually a member of a recognized religion or a radical offshoot of that religion, a member of a cult, or a religious loner. No religion has been without its excesses. Rather than become cynical toward religions, it should be acknowledged that unsavory aspects of religious extremists result from human failings, corruption, and inadequacy rather than an imperfection of things divine.

Religious fanatics share a common, unshakable belief in the righteousness of their cause and the appropriateness of their actions. Hostage taking and other terrorist-like actions seem to increase in number when traditional value systems are threatened, change, or appear to collapse (Iran, Egypt). With the exception of the loner, religious fanatics have a charismatic leader and preach a form of exclusivity by which only the select can enter heaven. Therefore, if you are not a member of the sect, you are an enemy. Like political extremists, religious fanatics can admit no fault in their dogma. If they perceive hostility toward their religion by the hostage, they may react with violence. Feeling superior because of their beliefs, they may be inflexible and express a preference for death. The death wish is strong in many religious fanatics.

A second category of religious fanatics is found in the cult. Cults can be defined as religions without political power, such as Jim Jones's church in Guyana. While cults do not present as significant a threat as more conventional and widely known religious extremist groups, they do contain the potential for violent confrontations and hostage takings for a wide variety of reasons.

Some religious extremists who engage in terrorism may seek a violent death at the hands of a nonbeliever. Terrorists of this variety see themselves answerable only to God. Their conduct is often irrational and extremely defensive. They feel threatened by any improper description of their leader, their beliefs, or their activity. They may believe that to die at the hand of a nonbeliever is the holiest achievement possible. Because of that belief, they may be suicidal. They may seek violent resolution to the situation by killing hostages to satisfy that drive.

The lone religious fanatic, while not often encountered, could present the gravest threat. He or she will be even more detached from reality and less rational than the one belonging to a group. This religious fanatic claims to be operating on direct divine orders. Because of that, he or she is probably the most dangerous and the least susceptible to reason. David Berkowitz took orders from "God," who spoke to him through his neighbor's dog Sam.

A criminal takes hostages on impulse to avoid immediate apprehension and to have a bargaining chip for their escape. The criminal hostage taker does not want to die; they have no cause to die for. This hostage taker tends to be impulsive. They will often settle for much less than originally demanded when they recognize their no-win situation, provided they are able to save face, maintain their dignity, and not experience a sudden loss of power. A too sudden loss of power can create agitation, despair, or panic, which can lead to the impulsive killing of a hostage. Time is on the side of a peaceful resolution. Criminal hostage takers are also more likely to resort to political rhetoric during the negotiation process.

The wronged person is the type of hostage taker that seeks to notify society of the defects in the "system" or in the "establishment" because of some disagreeable experience. They may be seeking redress of the experience or publicity of that wrong. This hostage taker tries to take justice into their own hands. Group dynamics outside the hostage situation become complicated, because there are bound to be some groups of people who have suffered the same type of social injustice who will create backing and support for the hostage taker. This situation is different from the political hostage taking in that it is motivated by a wish for personal revenge; therefore, the hostage taker is convinced that they are absolutely right and behave in an ostentatious manner.

These factors are accentuated by media coverage and also by the media coverage of sympathetic outside groups. In the hostage takers' mind, this may justify violence toward the hostages. The hostages may represent the "system" or whatever the hostage taker wishes to avenge, which therefore puts the hostages in a dangerous situation. This type of situation occurs in clusters because of the media and outside sympathetic group dynamics; thus, media restraint is required to avoid a future epidemic of "wronged person" hostage taking. Gentle persuasion is required to convince the hostage taker that what he or she needs to end the situation will be provided.

Many hostage takers are mentally disturbed or under the influence of a narcotic. Hostage taking by a mentally disturbed person may be either spontaneous or planned. It is surprising to see how well planned the situation may be in spite of the hostage takers' obvious psychosis. For many mentally ill persons, there are intermittent periods of lucidity in the psychosis, so it may take an extended period of contact before madness is revealed in the hostage takers' speech. The delusions and the hallucinations probably will not impair their ability to do what they wish with the hostage-holding situation. Rapport may be difficult to achieve and maintain, but efforts to that end should be consistent. In almost all cases, the mentally ill hostage taker will also have a death wish that can only be satisfied by the murder of the hostages, suicide, or both. Hostages should conduct themselves in a very relaxed, laidback, nonaggressive, nonantagonistic manner, and they should avoid prolonged eye contact with the hostage holder and allow the hostage holder to control the situation and conversation.

There are specific strategies, as well as some general rules of behavior, a hostage should consider if they can identify the category of their hostage taker:

1. Determine the area of particular sensitivity, such as politics or religion, and avoid conversation in these subjects. If confronted, become an active listener, and adapt to, rather than adopt, the hostage takers' value system.
2. Because religious hostage takers are very touchy and defensive about their religion, do not improperly describe or make assumptions about the hostage taker's religion.
3. Make every effort to establish rapport with the hostage taker, but never attempt to assume the role of a hostage negotiator.

To be taken as a hostage by anyone is dangerous enough; however, political extremists represent the most danger to military personnel because of their political ideology and value as a logical and symbolic target. The hostage should make every attempt to establish rapport with their abductors. This might be the deciding factor that saves their life when comes the time for the hostage takers to carry out their threat of execution if their demands are not met.

The hostage should make eye contact when possible without being obvious, especially during casual conversation. The U.S. Department of Defense (DOD)

Directive 1300.7 gives guidance on what is considered acceptable conversation. Small talk is better than no talk. If the hostage has photographs of their family, they should show them to their captors. If the hostage takers want to talk about their cause, the hostage should show an interest even if they are not sincere. Conversations pertaining to religion or politics, however, will probably put the hostage on dangerous ground and should be avoided. The hostage should explain that while they might not agree with their captors, they are interested in their point of view. They must not argue with their captors. They must accept the hostage taker for who they are according to their beliefs. This is a common human courtesy. The hostage should accept their captors' culture in a gracious manner, yet not degrade their own. They should disregard the political or religious differences between themselves and their captors and search for the common ground between themselves and the hostage takers.

One of the strongest indicators of accepting an individual's culture is to learn their language. As a start, the hostages should learn short phrases in their captors' language. While increasing their knowledge of the language, the hostages should work on other cultural traits that may please their captors and reinforce the feeling that they accept and understand their culture. Many people are afraid to eat the common dishes of other countries simply because they do not look or smell appetizing. Nothing could be more disrespectful than refusing to eat the very food that the hostage taker may be eating.

A potential hostage can take certain steps to prepare themselves for the rigors of captivity. To lessen the trauma on both themselves and their family, the potential hostage should maintain their family and personal affairs in good order. They should keep their will current, draw up appropriate powers of attorney, and take measures to ensure family financial security. They should also discuss plans and instructions with their family in the event they are abducted. They may have a packet made up containing instructions for their family, money, airline tickets, credit cards, insurance policies, and the name of the person to contact for survivor assistance, with instructions to open the packet in the event they are taken hostage.

The potential hostage should not carry classified documents or other sensitive or potentially embarrassing items in their briefcase or on their person. If taken hostage, they should be prepared to explain phone numbers, addresses, names, and other items carried at the time of capture. During captivity, they should try to convince their captors that they have kidnapped the wrong person. The kidnappers will not be convinced, but they should not give up. They cannot use this approach if the kidnappers intended to kidnap the defense attaché and the hostage is carrying documents on their person or in their briefcase that proves they are the defense attaché. A compromising document found on this person could destroy their argument that they are innocent or that the kidnappers took the wrong person.

The potential hostage should carry a week's supply of any essential medication when traveling or when stationed in a high-risk area. If taken hostage, they should explain the importance of this medication to their captors. If necessary, they should request additional medication; after all, they are more valuable to their captors alive than dead. They should not be reluctant to accept what is provided. When in a foreign country, they should know the generic name of their medication.

At the moment of capture, the victim must make an instantaneous decision whether to resist or surrender. It is dangerous to resist, but for a number of reasons there are circumstances in which it is more risky to be captured. If they decide in advance to escape, they should plan and practice their escape. For example, they can observe possible points of interception and judge their best course of action. They can also mentally visualize plans to resist at their home or at their office.

The initial moment in a capture is the most dangerous time because the captors are tense and their adrenaline is flowing. In these circumstances, the assailants may commit unintentional violence with the slightest provocation. Innocent acts, such as the victim reaching into a coat or purse to produce identification or raising a hand to scratch his scalp, may be wrongly interpreted and precipitate a deadly response. The victim may reassure their captors that they are not trying to escape. It is unrealistic to believe that a target can escape at this initial point in the capture when faced with determined and well-armed abductors. Many have tried to evade their captors and have been killed or injured. For example, on August 28, 1968, the U.S. Ambassador to Guatemala, John Gordon Mein, was killed as he tried to escape. He took only a few steps before he was shot by a burst of submachine gun fire. Other individuals who resisted were fortunate to receive only cuts, bruises, or smashed spectacles, an unfavorable way to begin an indeterminate period of incarceration.

Faced with such overwhelming odds, it is vital that the victim not panic or try any sudden movement that might unnerve an already anxious gunman. The hostage should keep in mind that the assailant has meticulously planned and executed the hostage operation. The initiative, the time, the location, and the circumstances of the incident all favor the kidnappers. The manpower and firepower brought to bear on the incident leave little opportunity for escape.

Even though the kidnappers may use blindfolds, gags, and drugs at the time of abduction, the victim should keep in mind the fact that the kidnappers want them alive. They should therefore not be alarmed or resist unduly. The kidnappers may use drugs on the victim to physically control them, put them to sleep, or keep them pacified. The "truth serum" drugs, if used on the hostage, have an effect similar to that of alcohol. Since the hostage has no choice in drug application, they should not physically resist.

The kidnappers may use blindfolds or hoods on the victim to keep them from knowing where they are being taken or to prevent them from identifying them. The victim should not remove the blindfold, even if an opportunity should arise, as this could leave the kidnappers no alternative but to kill them. With a hood over their head, there is a pretty good chance that the victim will not make it out alive. Unable to make eye contact, the victim is no longer seen as a human.

The victim should always stay alert. Even if they are blindfolded and gagged during transport, they have been deprived of only one or two senses. They should use their other senses to obtain information and help control fear and panic. They should occupy their mind by noting sounds, direction of movement, passage of time, and their abductors' conversations. Any information they can acquire could be useful for later reference. Another consideration is to find surfaces where they can leave full fingerprints to assist law enforcement authorities in their recovery.

Stress is defined as the reaction of the body to any demand made upon it. Stress can be further defined as a body condition that occurs when a person faces a threatening or unfamiliar situation. Stress causes a person's energy and strength to increase temporarily. Not all stress is bad. A certain level of stress is necessary for optimal performance. Stress causes bodily changes that may help an individual overcome challenges and danger. A person's physical condition affects their ability to handle stress. An individual's response to stress also depends on whether they feel in control of the situation. A difficulty may cause little stress if a person can predict, overcome, or at least understand it.

There are several ways of exerting control to influence one's physiological reaction and limit harmful stress:

1. Maintain confidence in one's ability to survive with honor, maintain confidence in one's country's ability and desire to end one's captivity, and maintain faith.
2. Maintain your physical condition. Install a program of exercise on the first day of captivity. Eat what is given, no matter how strange or unusual it may seem.
3. Take care of any family matters that would cause one to worry if they were captured. Accept what one cannot change because of captivity.
4. Maintain sleep discipline and avoid sleeping too much.
5. Practice recreation by being inventive about reading, writing, radio or TV usage; interaction with other hostages in a positive sense can provide opportunities for periods of recreation.
6. Maintain a daily schedule. Keep yourself self-occupied and have a purpose each day. Plan a schedule for each day and attempt to stick to it.
7. Keep your sense of humor. Give nicknames to one's captors; joke about one's situation.
8. Maintain a positive mental attitude. The most deadly aspect of long-term captivity is the mental lassitude, exhaustion, lethargy, and depression that set in. One must constantly and vigorously oppose it.
9. Verbally talk yourself down; talk yourself through a situation.
10. Do not let depressing thoughts take over. Maintain a silent "NO" to stop depressing thoughts and thoughts of catastrophe. Accent the positive.
11. Maintain religious values. Religious values will provide a source of strength.
12. Learn or devise methods to communicate.
13. Keep faith with fellow captives. They will be a major source of strength. Nurture this faith constantly. Remember that kidnappers may attempt to divide hostages.

The stress induced when one is taken hostage has a significant psychological impact and may lead the hostage to alter their behavior. An unexpected behavior displayed by a hostage is that of aligning themselves with the hostage taker. This unusual phenomenon has been termed the "Stockholm syndrome."

The Stockholm syndrome first came to public attention as a result of a bank robbery that turned into a hostage-barricade situation in Sweden in 1973. At 10:15 a.m.

on Thursday, August 23, 1973, the quiet early routine of the Sveriges Kredit Bank in Stockholm was destroyed by the chatter of a submachine gun. As clouds of plaster and glass settled around the 60 stunned occupants, a heavily armed lone gunman called out in English, "The party has just begun." The party was to continue for 131 hours, permanently affecting the lives of four young hostages and giving birth to a psychological phenomenon, subsequently called the Stockholm syndrome. During the 131 hours from 10:15 a.m., August 23, 1973, until 9:00 p.m., August 28, 1973, four employees were held hostage: three females, ranging in ages from 21 to 31, and one 25-year-old male. They were held by a 32-year-old thief, burglar, and prison escapee named Jan-Erik Olsson. Their jail was an 11- by 47-foot carpeted bank vault, which they came to share with another criminal and former cell mate of Olsson's, Clark Olofsson, age 26, who joined the group after Olsson demanded and got his release from prison. During their captivity, a startling discovery was made. Contrary to what had been expected, it was found that the victims feared the police more than they feared the robbers. Media attention was attracted when hostage statements, such as "The robbers are protecting us from the police," were released. Scientific investigation after the event as to why the hostages felt emotionally indebted to the bank robbers led to the discovery of the phenomenon.

This phenomenon, which can affect both the hostage and the hostage taker, seems to be born in the high-stress environment of a siege room. An emotional bond forms and leads to the development of a philosophy: "It's us against them." The Stockholm syndrome seems to be an automatic, probably unconscious, emotional response to the trauma of becoming a victim. The physical and psychological stress induced by the hostage situation causes the hostage to react in a manner totally against their normal beliefs, values, and ethics.

When the hostage experiences a great deal of stress, their mind seeks a means of survival. One avenue it may take is the use of defense mechanisms. These are defined as essential unconscious psychological adjustments made in the presence of danger. Three of the most common defense mechanisms are denial, regression, and identification.

Denial occurs when the mind is overlooked by a traumatic experience. The mind responds as if the incident were not happening. It may be verbally expressed by phrases like "Oh no! Not me!" "This must be a dream!" or "This is not happening!" One person may deal with stress by believing they are dreaming and will soon wake up and it will all be over. Another person deals with stress by sleeping. An example of denial can be compared with the unexpected death of a loved one. Frequently, hostages gradually accept their situation, but find a safety valve in the thought that their fate is not fixed. They view the situation as temporary, convinced that the police will come to their rescue. This gradual change from denial to delusions of reprieve reflects a growing acceptance of the facts.

Regression results in behavior adjustments in which the mind unconsciously selects a behavior that has been used successfully in the past when confronted with total dependency. For example, children are totally dependent on a primary care giver for psychological and physiological needs. In a hostage situation, the hostage taker assumes the role of primary care giver.

Identification with the hostage taker occurs on the unconscious level. The mind seeks to avoid wrath or punishment by emulating behaviors and adapting to the hostage takers' values. For example, a young undergraduate engineering student identifies with his or her professor. Identification initially is to seek a passing grade in the course, but eventually changes to the adoption of the professor's standards and values.

Victims of the syndrome share common experiences, including positive contact, sensing, and identifying with the human qualities of their captor, and a willingness to tolerate situations far beyond what they considered their logical limits.

Positive contact is generated by a lack of negative experiences. It appears that positive contact is reinforced to a greater degree if there has been a negative experience such as beatings, rapes, and murder early in the event, followed by positive contact.

The second common experience is sensing and identifying with the human qualities of the captors. The hostages may relate to their captors. Dr. Fredrick Hacker called this the "poor devil" syndrome. The terrorists may talk about their own mental and physical suffering and their perceptions of being victims of circumstance rather than aggressors. The hostages may then transfer anger from the hostage takers to the society or the situation that created the dilemma in which they are now victims. The hostages may feel the terrorists are entitled to their protection and care, possibly even their help and support.

The third common experience of victims of the Stockholm syndrome deals with preconceived levels of tolerance and expectations of actions. Before a situation develops, most individuals anticipate a limit to the extent they will allow themselves to be pushed or abused. That extent is the logical limit. Because the need to survive is strong, hostages rationalize an extension of their preconceived limits far beyond what they expected.

When all the elements of a hostage situation and the possible emergence of the Stockholm syndrome are considered, a common observation is that the hostages view the event in the same perspective as the hostage taker. The Stockholm syndrome may have been at work during the ordeal of TWA Flight 847. Flight 847 passengers spoke of being angry and frustrated. Several spoke sympathetically of their captors' demands. Passengers reported of being angry, scared, and unsure of their future. The immediate effects of the event include anxiety. The hostages were traumatized and their body chemistry was prepared to fight or flee, yet they could do neither. That frustration for many may have served as a precursor to a bond. Flight 847's first 3 days were marked by extreme violence and death, in sharp contrast to the remainder of the ordeal, a reinforcement to the later, positive contact.

Allyn Conwell, a self-appointed spokesperson for the passengers, endorsed his captor's demand that Israel release the Lebanese prisoners and spoke of his profound sympathy for the Amal's cause. Hostage Thomas Murray (Smith 2001) said in a broadcast: "I really feel that my being taken hostage was a way the people here had ... to point out that there are hostages in Israel."

Passenger Jimmy Dell Palmer said of the Amal that the most frightening moment was when he and other passengers were herded off the plane in the dead of night and had to grope their way into a dark cellar. "When they turned the lights on, there were half a dozen machine guns looking at us, and a lot of us thought ... this was it." Later, Palmer said of the Amal (Smith 2001), "They went out of their way to be nice." This same occurrence took place in the case of Reverend Lawrence Jenco. Jenco, released after being held for 18 months by the same group responsible for the 1983 suicide truck bombing of the Marine barracks in Beirut and the Flight 847 skyjacking, said, "They were basically gentle people, very religious, very prayerful" (Smith 2001).

James McLoughlin told CBS News about the hostages "getting impatient with the lack of initiative" and Washington's concern for special interests taking precedence over the hostage situation. Flight 847 pilot John Testrake said he was not a victim of the Stockholm syndrome (Smith 2001): "I think it's more a sense of merely seeing the other fellow's viewpoint."

The Stockholm syndrome produces a variety of responses. At the minimal level of response, the victim sees the event through the perspective of their captor. At a higher level, the victim respects and recognizes the terrorist for their gallant efforts. Responses have ranged from hostage apathy to actual participation by the hostage in terrorist activity. Other responses of the Stockholm syndrome include losing touch with reality, impeding efforts of the rescue forces and negotiation teams, and suffering long-term emotional instability. However, it should be noted, that the hostage is not the only one susceptible to the Stockholm syndrome. The hostage holder can also be affected, which could contribute to the enhanced survivability of the hostage.

During the seizure of an office or a residence, hostages may find themselves in familiar, comfortable surroundings in which they have worked or lived. Kidnap victims, on the other hand, are frequently forced to live in makeshift "people's prisons" in attics or basements or in remote hideouts. The cells of these prisons are usually quite small and in some cases prevent the hostage from standing up or moving around. Sleeping and toilet facilities may be scarce, consisting of a cot or mattress and a bucket or tin can for body waste. In some instances, toilet facilities may not be provided, forcing the hostage to soil their living space and themselves.

Such an experience may be further compounded by a total lack of privacy and a feeling of utter helplessness and dependency on the kidnappers for every necessity of life, which is what the kidnappers want. It will be very difficult for the hostages to maintain their dignity and self-respect under such conditions, but it is very important that they strive to do so. By maintaining their dignity and self-respect, the hostages may cause their captors to empathize with them, which in turn could lessen their aggression toward them. Most people are unable to inflict pain unless their victims remain dehumanized.

It is essential to maintain a good appearance as much as possible under the existing conditions. If the hostage is unable to wash his or her clothing over extended periods of time, such measures as brushing off dust, straightening their clothes, tucking in their shirttails, and smoothing out wrinkles are necessary to help their overall appearance.

The hostage should bathe whenever possible. However, if this is not possible, they should keep their hands and face clean, not only for appearance but also for health. One of the most prevalent health-care problems that occurs during extended periods of captivity is that of oral hygiene. If a toothbrush is not available, the hostage may fabricate one by wrapping a rag around their index finger or shredding the end of a stick. Emphasis should be placed on massaging the gums. Other measures that will improve the hostage's appearance are keeping their hair, mustache, and/or beard well groomed and keeping their fingernails clean.

One of the most important aspects of sustaining dignity is to maintain a positive attitude. The hostage should mingle and converse with other captives, if allowed, while going about their daily routines. They should assist others who need help and assume the role of a leader. They should maintain an optimistic attitude that will allow them to confront the problems and stress of captivity and deal with them accordingly. Smiles are contagious; the hostage should wear one. In contrast to this, a hostage who has given up and has no interaction with other hostages and is only concerned with their own well-being will certainly stand out in a negative way. The pessimist or introvert will find these times much more difficult. Their ability to maintain their dignity and self-respect will be weakened, even though self-respect and dignity may be the keys to retaining status as a human being in the eyes of the kidnappers.

Two excellent examples of individuals who were able to maintain dignity and self-respect were Sir Geoffrey Jackson and Dr. Claude Fly. Both were held by the Tupamaros in Uruguay. Sir Jackson, the British Ambassador, was abducted and held for 244 days. He remained, in thought and action, the Ambassador, the Queen's representative. Dr. Fly was held for 208 days. During that time he wrote a 50-page Christian checklist, which he used to analyze the New Testament. Like Jackson, Fly was able to create his own world and insulate himself against the hostile pressure around him. Both developed an inner peace and acceptance of their surroundings while not giving up their desired goal of freedom.

The passage of time without rescue or release can be depressing, but it does work to the hostage's advantage. The longer a hostage is held in captivity, the greater their chances of release or rescue. Time is a factor in the development of the Stockholm syndrome and in rapport-building efforts. A hostage should avoid setting anticipated release dates or allowing their captor to establish these types of milestones for them. Thinking that they will be home by the holidays is setting themselves up for an emotional fall if expectations of release are not met.

Boredom is the one companion a hostage will have in captivity. The hostage must aggressively face the challenge of captivity by engaging in creative mental and physical activities. The hostage should develop and maintain a daily physical fitness program. Exercise may be difficult due to cramped space or physical restraints on the arms and legs. If possible, however, the hostage should start and maintain a regular program of running in place, push-ups, and sit-ups. Isometric exercises may be substituted to overcome space or physical restraints. Staying physically fit might be the deciding factor should an escape opportunity present itself. However, the hostage should ensure that their caloric expenditure does not exceed their caloric intake.

The hostage should engage in creative mental activity, such as reading, writing, or daydreaming. They should ask for reading and writing material and request permission to listen to a radio or a phonograph or to watch television. Such requests have been granted. This may enhance the hostage's rapport-building activities. Other ways that they may keep active are to use deliberate and slow methods when brushing their teeth, take an hour to make the bed, or perhaps study the activities of a file of ants parading in and out of the cell. One hostage imagined he had different members of his family join him for the day. If it is a day of worship, the hostage may mentally walk through the various parts of the service.

A side effect of captivity for some hostages is weight loss. Although this loss may be considerable, it generally does not cause health problems. Weight loss may occur even with an adequate food supply, since captives often lose their appetite. In some cases, hostages may suffer gastrointestinal upset, including nausea, vomiting, diarrhea, and/or constipation. Although these symptoms can be debilitating, they usually are not life-threatening. Since the kidnappers are primarily concerned with keeping their hostages alive and well, the hostage should not hesitate to complain and ask for medication. Kidnappers want their hostages alive and are not likely to take chances by providing the wrong medicine. In a number of cases, terrorists have provided medical care for hostages who were suffering from illness and/or injury.

Social isolation may be better understood by the term "solitary confinement." The captive is completely denied interpersonal contact. Perceptual isolation is the denial or overuse of one of the senses to deny the hostage outside stimuli, thus creating social isolation. Headphones, blindfolds, and tents or small enclosures may be used to accomplish this goal. Here, the captive must rely completely on mental activities to adjust to and overcome each type of isolation. In isolation the hostage may have less opportunity to build rapport with their captor, so they must concentrate their efforts.

Having adjusted to his or her captivity, the hostage is now faced with a new possibility: rescue or release. Hostages are taken generally to establish a bargaining position. Eventually that position will lead to rescue or release. Each eventuality has special considerations. Often the first hostage execution will serve as a green light for rescue forces. Hostages must be mentally prepared for rescue attempts, as they represent a significant danger. Most hostages who die are killed during rescue attempts. It is therefore crucial that the hostage be especially alert, cautious, and obedient to instructions should he or she suspect such an attempt is imminent or is occurring. When the rescue begins, the safest response is for the hostage to drop immediately to the floor and avoid any sudden movement, especially with his or her hands.

During a rescue operation at Entebbe, a female hostage threw her hands up to praise the Lord as the commandos came bursting in, and she was shot by the commandos. Sudden movement also caused the deaths of two hostages in the South Moluccan train incident when Dutch commandos assaulted the train. The hostage should not attempt to run or pick up a gun and attempt to assist the rescue forces, as they may be mistaken for a terrorist. After order has been restored by the rescue forces, the hostage may be handled roughly. This is a common procedure for the rescuers who must separate the hostages from the terrorists. Some of the terrorists may have thrown away their weapons and masks in an attempt to disguise themselves as hostages. This occurred during the May 1980 rescue operation at the Iranian Embassy in London. British SAS discovered terrorists hiding among the hostages after the initial assault.

The moment of release, like the moment of capture, is dangerous. The terrorists are losing their bargaining chip. The hostages may feel threatened and even panic from the disruption of normal activities. The hostages might know something is happening, but may not know it is a release. The hostages of TWA Flight 847 were herded into a school yard at night, but did not know they were being released until they were turned over to Syrian representatives. The rules the hostages should adhere to are simple: pay close attention to the instructions the terrorists are giving when the release is taking place, and do not panic or attempt to run.

Media may be the first challenge encountered after the release or rescue. The hostage should ask for an official media spokesperson, such as a public affairs officer. They will provide guidance and act as a media buffer. If the hostage is confronted by the media and chooses to respond without a public affairs officer, they should say nothing that might be harmful to their fellow hostages who are still in captivity. They should keep their comments as short as possible and limit them to statements such as "I am thankful to be home" or "... alive" or "... out of captivity." They should say nothing that is sympathetic to the terrorists' cause, which might endanger hostages still in captivity.

The hostage should be mentally prepared to be debriefed by government and military personnel. Once released from captivity, they should write down everything they can remember about the incident to aid in the debriefing process: the location of guards, the description and placement of weapons and explosives, and any other information that might be of value to the authorities. They should keep in mind that although the information they provide the authorities might seem insignificant, it could aid in the release or rescue of other hostages who remain in captivity or assist in bringing the captors to trial. Hostages may emerge from these ordeals with hostile feelings toward their government. They may feel that their government should have been more active in their release. Many victims, after having undergone a hostage ordeal, experience feelings of guilt for the way they conducted themselves during captivity. A hostage may suffer defeats in captivity, and once released, may need professional help to help them sort through their feelings and emotions. If this assistance is available, the hostage should ask for and use it.

Barricaded subjects, hostage takers, or persons threatening suicide represent especially trying and stressful moments for law enforcement personnel who respond to them. Officers responding to the scene must

- Assess the totality of the situation.
- Secure the area.
- Gauge the threat to hostages/bystanders.
- Request additional units as appropriate.

Crisis negotiations must

- Establish contact with subjects.
- Identify their demands.
- Work to resolve tense and often volatile standoffs without loss of life.

SWAT teams must prepare to neutralize subjects through swift tactical means. Field commanders assume ultimate responsibility for every aspect of the police response. Supervisors who understand the purpose behind the actions taken by negotiators will avoid delays at the scene that occur when negotiators must stop and explain or justify their intended courses of action. Society requires that law enforcement exhausts all means available prior to launching a tactical resolution to an incident. If these means prove unsuccessful, then the transition from negotiation to tactical assault must be a smooth one.

To enhance cooperation, negotiators and personnel from tactical teams should train together on a regular basis. These training sessions include four fully enacted crisis scenarios. Members of the department's command staff are encouraged to participate, and through this training, will learn how the two teams work together. Law enforcement agencies generally place a premium on the training provided to tactical teams. Administrators should place no less emphasis on the training provided to their negotiations teams. At the very minimum, negotiators should complete a Law Enforcement Hostage/Crisis Negotiations course at the federal, state, or provincial level. Because the department's training qualifications may become subject to critical review in the courts should negotiations fail, negotiators should further their training through advanced courses, seminars, basic psychology classes, and detailed critical analysis of past incidents.

Field commanders should remember that the peaceful resolution of a barricade situation is as important to negotiators as the resolution of an incident involving a person threatening to jump from a bridge or a hostage taking with extensive media coverage. Each incident takes on a personality of its own. Field commanders can be sure of only one thing: Their decisions will be scrutinized by every official. Therefore, they should base their decisions on an understanding of the negotiations process and the many factors that affect it.

A successful negotiations process requires a good foundation. Often, circumstances force the first responding officers to initiate some type of negotiation with the subject's. Once line officers or first-line supervisors realize that an incident appears to be heading for something other than a prompt resolution, they should immediately terminate negotiations and call in trained negotiators. Too many tragedies in communities demonstrate how negotiations should not be initiated. A bad start by well-intentioned, but untrained, personnel can have negative effects throughout the process. Simply put, personnel who are not trained negotiators should not negotiate.

A negotiator's most important ally in all situations is time. Field commanders should not rush anything unless the loss of life appears imminent. It may seem as if nothing is happening because a suspect is not negotiating; this is not so. During these quiet times, many things occur that will eventually lead to a peaceful resolution. Negotiators refer to these quiet intervals as "dynamic inactivity." As long as time passes without any harm to persons involved, they are making progress. The passing of time works for the police in many ways and only means that a resolution is closer at hand. Field commanders should keep in mind that patience is a virtue.

Generally, the negotiations team consists of at least three main negotiators. Each team member plays a vital role in the successful resolution of critical incidents. The primary negotiator actually communicates with the subject. The secondary or backup negotiator assists the primary negotiator by offering advice, monitoring the negotiations, keeping notes, and ensuring that the primary negotiator sees and hears everything in the proper perspective. The intelligence negotiator interviews persons associated with the suspect to compile a criminal history and a history of mental illness and gather other relevant information. Often, an additional negotiator will act as the chief negotiator. Their primary responsibility is to act as a buffer between command personnel and the negotiations team.

Invariably, field commanders want to offer their advice to the negotiations team. Whenever possible, suggestions should be routed to the negotiations team via the chief negotiator. Typically, the negotiations team sets up away from the rest of the activity and maintains communications with the command post via a liaison. A member of the Emergency Response Team generally monitors the negotiations and provides tactical intelligence to the arrest, entry, and perimeter teams. Often, well-meaning civilians offer to negotiate with subjects. Sometimes, these civilians insist that they be allowed to negotiate. A wide range of individuals from parents, spouses, and lovers to friends, members of the clergy, attorneys, counselors, and mental health professionals might offer to speak.

As a general rule, direct civilian participation in negotiations is entirely unacceptable. The tactical negotiations process is a police operation. When faced with these offers, field commanders should keep in mind that the individual now willing to help might have been directly or indirectly responsible for the subject's behavior. While these individuals might be a useful source of information, only in very rare circumstances should they be allowed to speak directly with subjects. Instead, they should be escorted to the intelligence negotiator and kept well clear of the actual negotiations process.

Police procedure dictates that any crisis incident be contained using both inner and outer perimeters established and maintained. Critical incidents such as hostage takings, barricade situations, or suicide attempts must be contained prior to the start of negotiations. Mobile negotiations should not be attempted. While the need for a secure inner perimeter is obvious, crisis incidents also require an emphasis on a well-controlled outer perimeter.

When arriving at the scene of a hostage taking, barricade situation, or suicide intervention, negotiators often encounter a large crowd made up of bystanders, the press, and the subject's family members. It is important that the subject not be given an audience to "play to." Negotiations cannot succeed if negotiators must compete with outside influences for the subject's attention. Individuals with potentially helpful information about a subject should be secured in an area where they can provide details to the intelligence negotiator. The press should be provided a designated gathering area away from the perimeter and be briefed regularly regarding the status of the negotiations process. Field commanders should remember that reporters have a job to do. They will do that job, with or without the help of the police. It is far more preferable to provide them with the accurate information they need than to force them to gather it for themselves.

During protracted incidents, supervisors should request the assistance of the department's media relations personnel to help deal with the press. The highly unstable nature of these incidents also makes it imperative that an arrest team be prepared to take the subjects into custody at a moment's notice. In fact, the surrender phase represents the most critical stage in any negotiated incident. In some cases, surrender can occur very rapidly. Depending on the severity of the incident, the arrest team can be made up of patrol officers or members of specialized teams. Once the SWAT team sets up at a scene, it should assume the arrest duty.

During a crisis situation, it is essential that the police control the telephone lines. One of the first actions negotiators take when arriving at an incident is to arrange with the telephone company to deny origination to telephones at the subject's disposal. Once origination is denied, the subject's telephones will no longer get a dial tone. At the negotiators' request, the telephone company then establishes a new number that serves as a direct line between negotiators and the subject. Restricting telephone access in this way prohibits the subject from talking to family, friends, attorneys, and, most important, the press. It also prevents the suspect from gathering intelligence about police maneuvers from associates.

When there is no telephone accessible to the subject, or the telephone has been disabled as a tactical move by SWAT, the police must reestablish a means of communication. Because of the potential danger posed to negotiators, face-to-face negotiations do not represent an acceptable option. In these situations, the SWAT team often tactically delivers a "throw phone"—a standard telephone linked to a hard-line system connected to the hostage phone system. Because telephone delivery places members of the SWAT team in dangerous situations, it should be practiced regularly during joint negotiator–SWAT training exercises.

In many instances, the negotiations team might determine a need to control the electricity and water, as well. Only the on-scene commander can make the final decision to interrupt these services. Some of the most common reasons include taking away a subject's ability to monitor the incident on television; darkening the environment to provide a tactical advantage for SWAT; and eliminating comforts, such as toilet facilities. Tactical teams might also call for disconnection of plumbing services to deny subjects the ability to neutralize chemical agents. In turn, the denial or resumption of utilities provides negotiators with an effective bargaining tool. Different perspectives exist concerning the appropriate time to deny subjects utility services. Some experts believe that utilities should be disconnected before negotiations begin. Others believe negotiators should save such steps for use as bargaining tools later. While this is a matter of individual agency policy, administrators should ensure that the department adopts well-established policy guidelines in this pivotal area.

It is preferable for field commanders to resist the tendency to monitor the negotiations process personally. Supervisors who monitor negotiations or hear demands, deadlines, and death threats related during briefings should not become overly concerned. They should remember that the negotiating team is trained to deal with such scenarios. When a subject demands "$1 million," the negotiators actually hear "a 6-pack of soda." Likewise, if the on-scene commander hears a subject say, "If I don't get the car by 2:00, I'll kill a hostage." Negotiators actually hear, "Good, now we are really negotiating." Remarkably few hostages have ever been harmed as a result of missed deadlines. Of course, negotiators take deadlines and demands very seriously; however, skilled negotiators generally can work around them and even make them work to law enforcement's advantage.

A member of the negotiations team keeps the field commander informed of the negotiations. Commanders who find it absolutely necessary to monitor the negotiations need to inform the negotiations team, which should have the capability to wire a speaker to the command post to enable supervisors to listen to exchanges with the subject. Decisions should be based on the law, departmental policy, and the need for preservation of life and property. They should not make decisions based on exchanges they overhear between subjects and negotiators. The decision-making ability of commanders who personally monitor the negotiations process may be affected by any number of factors that have little actual bearing on the situation.

Much of the insight into the minds of troubled subjects comes from the specialized psychological training that crisis negotiators receive. As part of their training, negotiators learn a great deal about personality types, personality disorders, and the psychological motivations of hostage takers, suicidal persons, and subjects who barricade themselves. This training enables negotiators to manipulate a subject through their understanding of that person's state of mind. Accordingly, negotiators rely primarily on mental rather than physical tactics to resolve conflicts.

Each field commander should carry a pocket-sized checklist of actions that must be performed during a negotiated crisis. The checklist assists on-scene commanders to accomplish in an orderly fashion the various tasks required during a crisis. During times of extreme pressure, even the most prepared and composed professionals might not always remember to do everything at the right time. A checklist can prove invaluable in assisting supervisors to keep tense situations under control.

Agencies should conduct debriefings after the resolution of any crisis incident. Whenever possible, these debriefings should take place immediately following an incident, when details are still fresh in the participants' minds. Debriefing should focus on how the various units handled their roles during the incident. Each component must be represented, and officers should feel free to offer criticism—both positive and negative. Debriefings of this type should not be confused with or conducted in place of critical incident stress debriefings. Both serve valuable but distinct purposes.

Despite moves toward proactive policing methodologies, law enforcement remains an inherently reactive profession. When violent or troubled subjects create a crisis, they force the police to react to a situation in which the offenders already hold many of the cards. The press and the public judge the police by how well they respond to such situations. Concerns for hostage and officer safety, in addition to the well-being of often mentally disturbed subjects, dictate that the police respond at the lowest force level possible. Therefore, on-scene commanders should be prepared to supervise a negotiated settlement. The negotiations process can be tedious, complex, and, at times, confusing. The better field commanders understand the many factors that affect it, the more likely negotiators will get the support necessary to resolve critical incidents peacefully.

6.7 MANAGING VIOLENT BEHAVIOR IN THE WORKPLACE

Homicide is the leading cause of death for women in the workplace. Two-thirds of all injuries result from assaults, 25% by people they know and 16% stem from domestic violence. In the United States, homicide is the second leading cause of death overall. On average, one of four workers per year suffers some form of abuse in the workplace; 80% are males, with 61% in private companies and 30% in government agencies. Workplace violence costs businesses an average of $36 billion a year. The perception "that it will never happen to me" is a myth. Statistically, workplace homicide is the fastest-growing category of murder in the United States.

Violence is defined as any verbal, physical, or psychological threat or assault on an individual that has the intention or results in physical and/or psychological damage. The different types of violence include the following:

- Threats or obscene phone calls
- Intimidation
- Harassment of any nature
- Being followed or verbally assaulted
- All forms of physical contact with an intent to cause harm
- Psychological trauma
- Communicating a threat (to include nonverbal, vague, or covert)
- Disorderly conduct, shouting, throwing, pushing or punching inanimate objects, doors, and walls
- Insubordination

There are four different categories of attackers: Type I, criminal intent; Type II, customer and/or client; Type III, worker-on-worker; and Type IV, personal relationship.

6.7.1 Type I Attacker—Criminal Intent

With this type of attack, the attacker has no legitimate relationship to the business or employees in the business. This could be an office prowler, thief, rapist, or perpetrator looking for a crime of opportunity. However, the victim may be closely associated to the work environment and occupation, for example, taxi driver, police officer, store clerk, and bank teller. A deadly weapon is often involved, increasing the risk of fatal injury to the victim. Primary motive is usually theft, but may also involve a fleeing criminal. This type of attacker poses the greatest risk to jobs that involve exchange of cash, late hours, and/or individuals who work alone or in isolation. For example, on May 2000, two men entered a Wendy's in Flushing, New York, intending to rob the fast-food restaurant. They left with $2400 in cash after shooting seven employees, killing five and injuring two.

6.7.2 Type II Attacker—Customer and/or Client

They have a legitimate relationship with the business and have a reason to be in the building or on the campus. Problems stem from the fact that the person feels that they had lost money due to contracts being canceled and returned to kill the buyer. In this scenario, the act generally occurs in conjunction with the worker's normal duties. At greatest risk are police officers, mental health workers, and social services–related occupations. Violence in this category may be constant and even routine. They are often associated with occupations that project or possess a level of authority to withhold, deny, or approve service and/or care.

6.7.3   Type III Attacker—Worker-on-Worker

Personality problems in the workplace are common. Some people just can't seem to control the level of the disagreement. The perpetrator is an employee or past employee who attacks or threatens another employee(s) or past employee in the workplace. No specific occupation or industry is more prone to this type of attack than others. Motivating factor is often one or a series of interpersonal or work-related disputes. Managers and supervisors may be at greatest risk of victimization.

6.7.4   Type IV Attacker—Personal Relationship

The most common and dangerous is the personal relationship. Be it from a spouse coming into the workplace or tension from a relationship between coworkers, these situations can end very badly. Perpetrator usually does not have a legitimate relationship with the organization, but has or had a personal relationship with the intended victim. This may involve a current or former spouse, lover, relative, friend, or acquaintance. The perpetrator is motivated by perceived difficulties in the relationship or by psychosocial factors that are specific to the perpetrator. A violent outburst can be better characterized as the result of a "slow burn" or an accumulation of unresolved personal problems that can or have gone on for years. This may include a failing relationship, economic hardship, feelings of personal failure, or actual or perceived injustice in the workplace.

The spectrum of violence for an individual can range from low to high and is defined as follows:

Low: Verbal abuse, harassment, and intimidation (corrected by first-line supervisor)
Medium: Threats and assaults (intervention by crisis team and/or security)
High: Criminal assault and homicide (criminal prosecution)

Table 6.1 lists some defining characteristics of the two more commonly found workplace violence scenarios: Type III and Type IV employees. Table 6.2 lists the personality, signs, warning signs, and indicators for a Type III employee and Table 6.3 for a Type IV employee.

Type IV employees are one of the most difficult to foresee and prevent when it comes to violence in the workplace scenarios. In this situation, it is absolutely

---

TABLE 6.1
Personality Traits of Type III and Type IV Employees

Type III Employee (Worker-on-Worker)	Type IV (Personal Relationship)
• History of interpersonal conflict	• Distraught employee
• Argumentative or uncooperative	• Evidence or claims of harassment
• Tends to blame others for problems	• Suspicious person on property
• Significant change in behavior, performance, and appearance	• Domestic spillover
• Substance abuse problem	• Significant change in behavior, performance, and appearance

TABLE 6.2

Characteristics of a Type III Employee

Personality Traits	What to Look For	Contributing Factors	Warning Signs and Indicators	Key Events and Benchmark
• Low self-esteem	• Obsessive behavior	• High workplace stress	• Fired, laid off or suspended; passed over	Bonus periods, holidays, counseling, contract closures, promotions Calendar events Significant dates
• Low productivity	• Increased absenteeism	• Unusual stress outside of work	• Disciplinary action, criticism, poor performance review	
• Low impulse control	• Chemical dependency	• Lack of empathy or consideration	• Financial and/or legal action	
• Lacks empathy	• Verbal threats or threatening actions	• Personality conflicts	• Failed or spurned relationship	
• Social withdrawal	• History of discipline problems	• Lack of policy	• Personal crisis (divorce or death in family)	
• Feelings of rejection	• Depression and isolation	• Lack of training		
• Resists change	• Defensiveness	• Hostile/ threatening work environment		
• Feelings of being picked on	• Emotional outbursts	• Failure to respond		
• Easily frustrated	• Interests in weapons			
• Challenges authority	• Self-destructive behavior			
	• Fascination with weaponry			
	• Unwarranted anger			
	• Lashing out			

critical that an employee be referred to proper support channels. A policy should be established encouraging and require reporting of domestic abuse and the workforce should be trained to recognize signs of potential domestic abuse.

A list of action items from the workplace perspective in dealing with violence in the workplace is as follows:

- Management commitment to developing effective policy
- Prehiring checks
- Employee involvement
- Zero-tolerance policy
- Risk assessment and crisis team
- Training
- Documentation
- Emergency action plans

TABLE 6.3
Type IV Employee Characteristics

Signs and Indicators	Personal Relationship Difficulty and Dangers
• Partner has ongoing/increasing drug use, alcohol, gambling	• Complex psychological trauma
• Demonstrated or threatened physical violence	• Survival instinct and parental bonds
• Threatened to harm family or children	• Discovery and embarrassment
• Threatened or tried suicide	• Isolation and desertion
• Violently and constantly jealous or accusatory	• Desperation, fear of confinement, loss of control
• Does the partner follow, spy, or restrict your liberty?	• No options, anger, and/or suicide
• Does any history of violence exist?	
• Partner-forced relations	
• Volatile, spontaneous, and/or violent outburst of anger	
• Does partner routinely assign blame to you?	
• Has/does partner abused/abuse family pet/animals?	
• Severe changes in status—fired, unemployed, losses	

- Employee assistance plan
- Reporting requirements
- Controlling access
- Communication
- Postincident response

A list of precautionary measures from the employee and supervisor perspective in dealing with violence in the workplace is as follows:

- Report and communicate developments to human resources, security, and management.
- Avoid trouble spots such as stairwells, corridors, elevators, and parking lots.
- Avoid isolation and nonoffice hours.

The following is a list of action items that can be taken when being threatened or harassed by a customer or coworker, from someone who is being verbally abusive, and from someone who is threatening with a weapon.

For any angry hostile customer or coworker:

- Stay calm.
- Maintain eye contact.
- Listen attentively.
- Be courteous and patient.
- Keep the situation under control.

For a person shouting, swearing, and threatening:

- Signal a coworker or supervisor that help is needed; utilize a prearranged call sign.
- Do not make any calls in the presence of the perpetrator.
- Have someone call security, police, and the supervisor.

For someone threatening with a weapon:

- Stay calm and signal duress at available opportunity.
- Maintain eye contact, and stall for time.
- Keep talking, and follow instructions.
- Do not grab the weapon, challenge the person, move suddenly, or try to escape.

Avoid:

- Speaking in ways that shows
  - Apathy
  - Brushing off
  - Condescension
  - Giving the run-a-around
- Rejecting all demands
- Challenging or daring
- Posing or posturing in challenging stances
- Interrupting

Resolution:

- Attempt to bargain.
- Try to devalue seriousness of situation.
- Make false statements.
- Take sides or agree with distortions.
- Avoid redefining or restating the grievance.
- Do not interrupt.
- Allow the aggrieved party to suggest a solution.
- Speak of consequences and not of threats.

Develop and implement effective policy within the organization for dealing with violence in the workplace. Management, employees, and supervisors should actively address violence in the workplace issues. Assess the risk of violence in the workplace for the organization. Involve employees in the planning process and keep them abreast of developments. Consistently apply standards to work and maintain a happy and safe work environment. Document all incidents and take threats against employees, supervisors, and managers seriously. Create a zero-tolerance policy for abusive and harassing behavior in the workplace. Train all employees to a standard so that they can recognize abusive and harassing behavior. In order for the policy, standards, and practices to be effective, everyone in the organization must play a role.

6.8   EVENT AND CROWD MANAGEMENT

Crowd management must take into account all the elements of an event especially the type of event such as a circus, sporting, theatrical, concert, rally, and/or parade. Characteristics of the facility, size and demeanor of the crowd, methods of entrance, communications, crowd control, and queuing are all taken into consideration. As in all management, it must include planning, organizing, staffing, directing, and evaluating. Particularly critical to crowd management is defining the roles of parties involved in an event, the quality of the advance intelligence, and the effectiveness of the planning process.

To have an effective plan, facility management must be aware of the characteristics of the audience attracted by a particular event. Once the facility operator, police commander, and event promoter know their crowd they must plan accordingly. Sociologist Dr. Irving Goldaber has pointed out that the way patrons perceive the environment and the various "sociological signals" they receive at an event whether consciously or unconsciously can escalate or deescalate patron emotion and influence their behavior. For example, the general attitude of the facility staff and of the interior and exterior security and law enforcement personnel as well as the promulgation and enforcement of patron house rules combine to produce additional "signals" to influence patron behavior. Other signals include reliable door opening policy and truthfulness in communicating about alterations in event programming. When people are informed of changes and delays and the reasons for them, they can more readily accept those delays. While patrons are waiting, the provision of necessary comforts becomes crucial and can diminish discomfort and impatience.

Hundreds of thousands of events are held nationally and few, if any, have problems. But unquestionably, new and unexpected difficulties have been arising. In major cities, for example, some police officers have informally estimated that at any one time anywhere from 1.5% to 2% of the spectators at sporting events carry handguns. Dr. Goldaber speaks of four types of conditions that can create crowd management problems: problems created by a crowd from within, problems created for a crowd from outside, environmental catastrophes, and rumors. These threats must be considered by those responsible for managing crowds.

Schools and governmental and social service agencies have prepared the public to confront many situations that pose serious threats to personal safety. Fire drills teach effective escape procedures, driver education courses encourage safe driving, and first aid encourages saving lives. Yet, there is little to guide the public to anticipate and respond to danger signals in crowds. Education about crowd dynamics and the role of individuals in crowds is sorely needed on a national and international basis. The consequences of the various modes of individual and group behavior should be afforded equal importance with other safety programs by governmental, educational, and public services agencies. It is time to include this safety concern with others taught to the public.

The media can also play a significant role in public education by promoting special features, programs, and public service announcements relating to crowd safety and personal and group responsibilities. They can help discourage present safety hazards at large events such as the use of open flames and firecrackers. They can also monitor the crowd management techniques of facilities at indoor and outdoor events for their audiences. Facilities, too, can educate the public by publicizing and enforcing their house rules and by setting a courteous, professional level of conduct for their staff.

Drug and alcohol abuse is a serious crisis. Recognition alone does not diminish the problem at events where patrons use illegal drugs or abuse alcohol. The difficult task of enforcing drug and alcohol laws at major events without violating individual's rights has facilities and law enforcement agencies directing their attention to drug sellers rather than to users. This, in turn, has created a belief among patrons that the illegal use of drugs or alcohol is possible if not acceptable at major events. New and equitable methods of enforcing relevant laws are needed. This is an area where facility operators and law enforcement agencies must cooperate and patrons, regardless of age or social standing, must assume the consequences of breaking the law.

The sale of alcoholic beverages at events where rowdy audiences are expected or where a high percentage of the audience will be under the legal age for consuming alcohol can have adverse effects. When these conditions exist, rowdiness, a high level of excitability, and the potential for and detrimental effects of alcohol abuse become very real. Even though a prohibition on alcohol sales may reduce concession profits, many facility operators by such action reflect their concern for the safety of their patrons.

The role and responsibility of those parties involved in an event should be specified in writing and known to all prior to an event. There must be a clear understanding by all involved of the chain of command and the duties that each person is to perform. An important aid in this endeavor is an event management plan produced by the facility or promoter with the cooperation of public agencies that specifies names, duties, and location of the people at the event; the lines of communication; contingency plans; door opening; method of plan implementation; a checklist of personnel, equipment, and procedures; the expected crowd size and characteristics; and normal and emergency egress/ingress procedures. Those with a role in planning, organizing, and controlling events cooperatively must find ways to anticipate potential sources of danger in public gatherings, take steps to prevent trouble when and where possible, and be prepared to respond to trouble quickly and effectively when, and if, necessary.

Through laws and their enforcement, local government influences the character of event management by establishing building and safety codes and by determining facility capacity, seating configurations, and other related items. Government also influences an event by the manner by which it provides such services as police, waste collection, and traffic control.

In 1972, an American Bar Association report, *The Urban Police Function*, noted that police responsibilities are frequently the result of "design and default." Because it is often assumed that police can and will take on all manner of broad responsibilities, they sometimes carry out duties and functions for which there are no written policy directives. While the need for law enforcement remains the paramount duty of the police, there is an ever-increasing demand in the other areas of policing.

This is especially true where crowd management is required. Generally speaking, the role of police at events is to enforce laws and to manage crowds on or adjoining public property in cooperation and with the necessary support of the facility operator and/or event promoter.

The Fire Division is responsible for making unscheduled and routine inspections of facilities to enforce local fire and building codes. It also has the responsibility of citing a facility operator or patron for violation of safety laws. Their authority to require safe exiting conditions, as well as to enforce capacity and safety regulations, and their relationship to other personnel should be clearly defined in advance. Fire personnel, like other appropriate city personnel, should be involved in the advance planning of an event to ensure an acceptable level of compliance with fire and life safety codes.

Next to local government, facility management has the most influence on crowd safety and on the activities of promoters, speakers, entertainers, and performers. Regardless of how a contract between a facility and promoter is written, local facility management must acknowledge and accept its obligation for the safety of the community that it serves. Facility management has primary responsibility for ensuring safe conditions in compliance with applicable statutes and reasonable standards. That responsibility also requires cooperative efforts with law enforcement and other event managers. But that cooperation should not relieve facility management of its accountability for providing resources for safe and successful events. Of course law enforcement officials can take over direction and control in emergencies, but that should not dilute management responsibility for taking all reasonable steps to ensure that emergencies do not happen.

The establishment of house rules and the strict enforcement of those rules and local laws determine how the patrons, promoters, and the performers will behave. Many facilities train their crowd management personnel and provide orientation manuals for staff and security. These manuals describe audience characteristics, problem areas, staff functions, house rules, and emergency plans and facility layouts. They deal with the types and levels of security and familiarize personnel with management objectives. The use of such manuals underscores the notion that the best crowd management results are obtained when there is active cooperation between facility management and personnel, promoters, and public agencies.

The promoter is the broker between the performer(s) and the facility, and plays a critical role in the preparation of contracts. The promoter obtains the use of the desired facility, prepares appropriate contracts between the facility and the performer(s), arranges for event promotion and ticket sales, and pays for security requirements. The promoter is also likely to pay the taxes on the performers' profits and may even arrange to provide the performers' meals and snacks. Promoters are paid by the performers to organize the event and most often work independently of facilities. The promoter's responsibilities are to coordinate all aspects of an event with facility and government officials and to ensure that an event complies with local safety laws. Promoters often prepare their own event management plan for an event, listing personnel responsibilities and an event timetable, and usually share this material with the other parties in an event.

Entertainers, performers, and speakers have varying degrees of influence over the promotion and execution of their performances. The most popular can often demand a certain type of seating, determine the audience size, within the legal capacity of a facility, set ticket prices and promotional arrangements, and stipulate when the doors will be open prior to their show. Most realize the influence they maintain over their audiences and do not exploit it. With their support, a facility is better able to discourage open flames, blocking aisles, use of fireworks, and drug and alcohol abuse. There are, however, those who intentionally and irresponsibly incite their audiences to a level of behavior where fighting, vandalism, or rowdyism may occur. If this happens, they must be held fully accountable for their actions.

In addition to seating patrons, an usher's duties include enforcement of the house rules, maintaining order, reporting security problems to security, police, or others, keeping people out of the aisles, and enforcing open flame and smoking regulations. Ushers should remain at their posts until the event is completed.

Peer security are hired by promoters to protect the stage area, screen patrons for contraband, and to do other special assignments; private security personnel are people of similar age and background to the patrons and, therefore, presumably have good rapport with them. Peer security can also serve as an effective buffer or mediator between uniformed security and patrons in tense situations. They are usually recognizable by the specially designed T-shirts or outfits that they wear.

Though a careful and elaborate crowd management plan may be implemented, it cannot be fully effective without patron cooperation. In addition, it cannot protect individuals from self-inflicted harm. In a crowd, patrons should always be aware of the possible effect of their actions on the safety of the whole group. Pushing, fighting, spreading rumors, and the use of firecrackers or projectiles all can cause severe repercussions that the instigator may never have considered. An audience's tolerance of abusive actions further jeopardizes its own safety. Responsible patrons will acquaint themselves with local laws and facility house rules and should not hesitate to report situations that threaten their safety to the facility management, promoter, and/or the media. In many instances, the pressure of public opinion is the best regulator of private industry.

The elimination of festival seating and restrictions on general admission seating may have unexpected repercussions at ticket outlets, especially for "superstar" performances or major sporting events such as the Superbowl. While reserved seating largely removes the factors that cause early and overwhelming crowds to gather hours before an event, reserved seating can instead result in the early gathering of large crowds at ticket outlets who have come to purchase tickets for the limited prime seating areas. These factors can cause problems and difficulties for ticket outlets. To help relieve this problem, two options are recommended: first, the actual date, time, and location that the tickets are to go on sale should not be announced prior to the time that tickets are released for sale; second, when the demand for tickets is expected to exceed the available seating capacity, a mail order system of ticket sales should be implemented.

Most tickets are similar in color and overall appearance. As a result, it may be difficult for ticket takers and others to screen patrons with bogus tickets, especially when the rate of patron flow is high. A variation in ticket color or format would aid those facility and security officials attempting to prevent patrons with invalid tickets from gaining access to an event at which they do not belong. A ticket should also state the specified entrance the ticket holder is to enter.

In determining the number of ticket takers to be employed, most facility operators use a ratio of one ticket taker for about every 1000 ticket holders. The actual ratio may vary and depends on the actual crowd size, location of contraband searches, the type of entertainment, and the architectural design of the building. The efficient movement of ticket holders is critical in preventing crowds from gathering outside a facility. Limiting entrances and using fewer doors or opening and closing doors to control crowd movement are very dangerous practices. They only serve to increase anxiety in a crowd and make it more difficult to manage. It is much more effective to separate people in a crowd by using many entrances, by queuing, and by providing for the proper ratio of ticket takers and doors to patrons. Dispersing entering crowds through multientrances is particularly effective in processing people efficiently into a facility.

Whenever large crowds gather for the purpose of peaceably entering an area it is vital that the processing of those people be organized, orderly, and disciplined, and, if ticket taking is going to take place, that it be coordinated with the queuing of patrons. There are two major types of queues, linear and bulk, as described by pedestrian planner Dr. John Fruin in his book, *Pedestrian Planning and Design* (1971). In linear queuing, people line up in single file to gain entrance to the facility, whereas in bulk queuing there are no defined lines, but simply a large amorphous mass trying to enter the facility.

Many facilities in cooperation with law enforcement agencies queue their patrons in zigzag lines, around buildings, and on sidewalks. Often queues are further organized by metering, when sections of a queue enter a facility in a measured and regulated manner. This way, patrons can claim a particular space, feel less anxious about their ability to enter in an orderly fashion, and can judge better the length of time it will take them to enter, as they progress in a line. Using a queue means having control over a large crowd. It also prevents the potential hazard of a mob craze—the sense of urgency causing a rush toward an entry point. This sense of urgency or anxiety is the crucial factor that must be removed.

Searching patrons for contraband has become increasingly prevalent. Preadmission screening is a reasonable preventive measure to prohibit or reduce such items as weapons, dangerous objects, alcohol, drugs, and other undesirable objects and substances from being introduced on to the premises. Municipal councils should specify by ordinance contraband materials not allowable at major events and also require that contraband prohibition be posted at the event and on tickets. Legal considerations suggest that the screening of patrons for contraband is best performed by private security and not public law enforcement officers.

Safety aspects at facilities should be routinely inspected by the Fire Division and the Municipal Building or Engineering Department to ensure their compliance with municipal regulations. The adherence to numerous municipal regulations is pivotal to providing a safe environment for the public. What is needed beyond that is a method for assessing a facility management's or an event promoter's preparedness to accommodate its patrons safety. Having a formal crowd management plan is equally as important as compliance with safety regulations. The municipalities should require crowd management plans of all facilities and/or event promoters contemplating hosting or sponsoring events attracting 2000 or more people. These plans should be prepared in writing and presented to the municipality for public filing. Plans could be written for categories of events and, when necessary, for specific events. The format and requirements of a plan should be determined by the municipality, facility operators, onsite security, promoters, and other stakeholders. A copy of a facility's crowd management plan should be on file with the municipality and accessible to the public so they may understand what kind of crowd management to expect. The required filing of a plan will make it difficult for complacency to return to the issue of crowd safety.

7 Contingency Plans

7.1 DEVELOPING A FORCE PROTECTION PLAN

Force protection is preventive measures taken to mitigate hostile actions in specific areas or against a specific population, usually military personnel, resources, facilities, and critical information. In the Canadian Forces and U.S. military, those covered by force protection include family members and chaplains. The purpose of this section is to provide a guideline/outline that assists organizations to develop a force protection plan that details protective measures to mitigate hostile actions. In a military context or when defending a high-risk facility, a force protection plan consists of the following elements: threat assessment, vulnerability assessment, protective measures, conduct of routine security or base operations, and conduct of contingency operations. A general outline of a force protection plan would be as follows:

- Estimate the threat.
- Assess vulnerabilities.
- Develop protective measures that
  - Safeguard personnel.
  - Safeguard information.
  - Safeguard facilities and equipment.
- Conduct routine security/base defense operations.
- Conduct contingency operations.

7.1.1 ESTIMATE THE THREAT

- Collect, evaluate, and process all intelligence including law enforcement information from all sources, consistent with oversight restrictions.
- Develop threat analysis to include
  - Threat existence, capabilities, intentions, history, and targeting area of responsibility.
  - A review in the context of the security environment.
  - Threat models.
- Classify threats according to the organization threat level classification system.
- Disseminate threat information.
- Publish threat statement.

7.1.2   Assess Vulnerabilities

- Conduct physical security surveys of installations to determine overall strengths and weaknesses.
- Conduct physical security inspections of mission-essential/vulnerable areas (MEVA).
- Identify political/soft targets of opportunity that represent a potential for mass casualties.
- Conduct personal security vulnerability assessments for designated personnel identified as being at greater risk than the general population.
- Conduct information systems vulnerability analysis.

7.1.3   Develop Protective Measures

Based on threat estimate and assessment of vulnerabilities,

- Develop courses of action.
- Develop plans for normal conditions.
- Develop plans for contingency situations.
- Identify resources required to support courses of action.
- Exercise and test plans.
- Document findings.

7.1.4   Conduct Routine Security Operations

- Conduct law enforcement operations.
- Execute a physical security program.
- Implement an antiterrorism program
  - Antiterrorism training and awareness
  - Terrorist force protection condition system
  - Random antiterrorism measures program (RAMP)
- Conduct information operations.
- Protect high-risk personnel (HRP).
- Conduct intelligence operations in support of above security programs.

7.1.5   Conduct Contingency Operations

- Initiate notifications/reports.
- Monitor/establish terrorist force protection condition levels.
- Operate crisis action cell.
- Activate response forces.
- Implement terrorist force protection condition measures/other responses.

7.2   DEVELOPING A SECURITY PLAN

The purpose of this section is to provide a guideline/outline that assists organizations to develop a security plan that details decisions to managing security risks and outlines strategies, goals, objectives, priorities, and timelines for improving organizational security. It provides an outline to developing the security plan that is based upon a process of security risk management and to ensure that the decisions for managing security risks are substantiated through thorough analyses and supported by processes that are rigorous, repeatable, and documented. Developing a security plan and documenting the analysis and findings of the security risk management process helps to ensure that the results are reproducible and provides evidence of due diligence so that anyone, including managers and auditors, can understand the thinking that led to action being taken, and trace those actions to management decisions, plans, and policies.

Both the security risk assessment and security risk treatment processes should be documented to capture the analysis, findings, and resulting actions, and they should also provide a basis for review, priority setting, decision making, and performance measurement. The security plan should demonstrate the relationship between the selected security controls in the plan and the results of the security risk assessment and security risk treatment processes. While the extent and format of a security plan will differ from one organization to another given each one's size, complexity, operations, and internal management practices, each organization should maintain evidence of undertaking a security risk assessment and ensure that the more common elements of a security plan are adhered to. Each plan must contain at a minimum the following elements: approvals, executive summary, communications and consultations, context, security risk assessment, security risk treatment, and implementation.

7.2.1   Approvals

A statement from the organization head (or equivalent) to endorse decisions regarding risk treatment and support the implementation of controls and performance indicators

7.2.2   Executive Summary

Nontechnical synopsis of the security plan that highlights the main points, issues, and conclusions and contains enough information to familiarize the reader with what is discussed in the full plan

7.2.3   Communications and Consultations

- Brief description of the governance structure and process associated with the development of the security plan, for example, who was engaged and how and roles, responsibilities, and authorities of key stakeholders

- Identification of individuals and organizations that were consulted and/or who participated in the development of the security plan
- List of reference documents/authoritative sources that were consulted in the development of the security plan

7.2.4 CONTEXT

- Business context
    - A brief description of the department's mandate, priorities, strategic outcomes, program activities, and program subactivities
    - An overview of departmental resources and services associated with these program activities and subactivities
- Organizational context
    - A brief description of the departmental, organizational, geographical, and governance structure
    - A brief description of internal policies, processes, and operations in place to help the department achieve its business objectives
- Security context
    - A general description of the role that security plays in enabling the department to achieve its mission and support government priorities
    - A brief description of security constraints and requirements derived from legal, regulatory, policy, contractual, or other obligations
    - An overview of security threats that are specific to the organization's business and organizational context
    - A description of other factors or constraints that may impact security risk decisions
    - A description of the relationship between the departmental security and other internal management practices and program activities
- Approach to developing the security plan
    - Description of the scope of the security plan, for example, organization-wide or pertaining to only a subset of the organization or programs
    - Description of the structure and organization of plans and security risk assessment that collectively provide a consolidated view of security risks to the organizations, for example, for large organizations where the security plan may actually consist of a set of plans
- Approach to security risk management
    - Brief description of the department's approach to security risk management including description of alignment between the security risk management process and the organization's program(s) and activities
    - Criteria for evaluating significance of security risks
- Description of how the results of the risk assessments and the risk treatment are considered

7.2.5 SECURITY RISK ASSESSMENT

- Risk identification
  - List of the key security risks to organizational resources and services, based on risk assessments conducted
  - Identification of risk owners and stakeholders
- Risk analysis
  - Identification of the likelihood and impact of security risks
- Risk evaluation
  - Identification of risks that have been deemed unacceptable along with rationale
- Identification of risks that have been deemed acceptable and approach that will be used for monitoring and periodic review.

7.2.6 SECURITY RISK TREATMENT PROCESS

- Security risk treatment decision
  - Identification of the selected risk treatment option for each risk
- Security control objectives
  - Description of the control objectives for risks for which the selected treatment option is to apply controls
- Security controls
  - Description of security controls for achieving the control objectives
  - Controls that may be categorized as administrative, technical, or physical or as common to all activities and program(s) of an organization or program-specific
- Performance indicators
  - Description of performance indicators for monitoring the effectiveness of security controls
  - Identification of the sources of data, frequency, and targets for each performance indicator
- Gap analysis
  - Identification of security control objectives that remain unmet and for which controls need to be implemented, improved, or otherwise adjusted, or for which performance indicators have not yet been established
  - Identification of controls that are deemed excessive and will be eliminated
- Priorities
- A list of recommended priorities for implementing additional controls and establishing performance indicators

7.2.7 IMPLEMENTATION

- Implementation strategy
  - A summary of short- and long-term goals for implementing security controls and performance indicators
  - Activities and roles and responsibilities

- Timelines and milestones
- Resources
- Considerations for transition period
- Monitoring and reporting
  - Description of approach for monitoring and reporting on
    - Progress at implementing controls and performance indicators
    - Effectiveness of security controls at achieving control objectives using the performance indicators
    - Changes in the business, organizational, and security environment
    - Effectiveness of security risk management processes
- Description of how residual risks will be monitored
- Update
- Description of process and timelines for updating the security plan

7.3   DEVELOPING A BUSINESS CONTINUITY PLAN

Business continuity planning (BCP) identifies an organization's exposure to internal and external threats and integrates assets to provide effective prevention and recovery for the organization, while maintaining operations. A BCP is a roadmap for continuing operations under adverse conditions such as a storm or a criminal act. Any event that could impact operations is included, such as supply chain interruption and loss of or damage to critical infrastructure such as major facilities, machinery, and computing or network resources. As such, risk management must be incorporated as part of the BCP. A BCP is made up of the following phases: the analysis phase, the solution design phase, the implementation phase, the testing phase, and the maintenance phase.

7.3.1   Analysis Phase

The analysis phase consists of an impact analysis, threat and risk analysis, and impact scenarios. A business impact analysis (BIA) differentiates critical and noncritical organization functions or activities. Critical functions are those whose disruption is regarded as unacceptable. Acceptability is affected by the cost of recovery solutions. A function may also be considered critical if dictated by law. For each critical function, two values are then assigned:

1. Recovery point objective (RPO): The acceptable level of unused data that will not be recovered
2. Recovery time objective (RTO): The acceptable amount of time to restore the function.

The RPO must ensure that the maximum tolerable data loss for each activity is not exceeded. The RTO must ensure that the maximum tolerable period of disruption (MTPD) for each activity is not exceeded.

Next, the impact analysis results in the recovery requirements for each critical function. Recovery requirements consist of the following information:

- The business requirements for recovery of the critical function
- The technical requirements for recovery of the critical function

After defining recovery requirements, each potential threat may require unique recovery steps. Common threats include the following:

- Epidemic
- Earthquake
- Fire
- Flood
- Cyberattack
- Sabotage: both internal and external threats
- Hurricane or other major storms
- Utility outage
- Terrorism/piracy
- War or civil disorder
- Thefts, both insider and external threats, vital information, or material
- Random failure of mission-critical systems

The impact of an epidemic can be regarded as purely human and may be alleviated with technical and business solutions. However, if people behind these plans are affected by the disease, then the process can stumble. For instance, during the 2002–2003 SARS outbreaks, some organizations grouped staff into separate teams and rotated the teams between primary and secondary work sites, with a rotation frequency equal to the incubation period of the disease. The organizations also banned face-to-face intergroup contact during business and nonbusiness hours. The split increased resiliency against the threat of quarantine measures if one person in a team was exposed to the disease.

After defining threats, impact scenarios form the basis of the business recovery plan. In general, planning for the most wide-reaching impact is preferable. A typical impact scenario such as "facility loss" encompasses most critical business functions. A BCP may document scenarios for each building. More localized impact scenarios, for example, loss of a specific floor in a facility may also be documented.

After the analysis phase, business and technical recovery requirements precede the solutions phase. Asset inventories allow for quick identification of deployable resources. For an office-based, information technology-intensive business, the plan requirements may cover desks, human resources, applications, data, manual workarounds, computers, and peripherals. Other business environments, such as production, distribution, and warehousing, will need to cover these elements, but likely have additional issues.

7.3.2 SOLUTION DESIGN PHASE

The solution design phase identifies the most cost-effective disaster recovery solution that meets two main requirements from the impact analysis stage. For information technology purposes, this is commonly expressed as the minimum application and data requirements and the time in which the minimum application and application data must be available.

Outside the IT domain, preservation of hard copy information, such as contracts, skilled staff, or restoration of embedded technology in a process plant must be considered. This phase overlaps with the disaster recovery planning methodology. The solution phase determines the following:

- Crisis management command structure
- Secondary work sites
- Telecommunication architecture between primary and secondary work sites
- Data replication methodology between primary and secondary work sites
- Applications and data required at the secondary work site
- Physical data requirements at the secondary work site

7.3.3 IMPLEMENTATION PHASE

The implementation phase involves policy changes, material acquisitions, staffing, and testing.

7.3.4 TESTING PHASE

The purpose of testing is to achieve organizational acceptance that the solution satisfies the recovery requirements. Plans may fail to meet expectations due to insufficient or inaccurate recovery requirements, solution design flaws, or solution implementation errors. Testing may include the following:

- Crisis command team call-out testing
- Technical swing test from primary to secondary work locations
- Technical swing test from secondary to primary work locations
- Application test
- Business process test

Tabletop exercises typically involve a small number of people and concentrate on a specific aspect of a BCP. They can easily accommodate complete teams from a specific area of an organization. Another form involves a single representative from each of several teams. Typically, participants work through simple scenarios and then discuss specific aspects of the plan. For example, a fire is discovered out of working hours. The exercise consumes only a few hours and is often split into two or three sessions, each concentrating on a different theme.

A medium exercise is conducted within a "Virtual World" and brings together several departments, teams, or disciplines. It typically concentrates on multiple BCP aspects, prompting interaction between teams. The scope of a medium exercise can range from a few teams from one organization colocated in one building to multiple teams operating across dispersed locations. The environment needs to be as realistic as possible, and team sizes should reflect a realistic situation. Realism may extend to simulated news broadcasts and websites. A medium exercise typically lasts a few hours, though they can extend over several days. They typically involve a scenario that adds prescribed actions throughout the exercise.

A complex exercise aims to have as few boundaries as possible. It incorporates all the aspects of a medium exercise. The exercise remains within a virtual world, but maximum realism is essential. This might include no-notice activation, actual evacuation, and actual invocation of a disaster recovery site. While start and stop times are preagreed, the actual duration might be unknown if events are allowed to run their course.

7.3.5 Maintenance Phase

Biannual or annual maintenance cycle maintenance of a BCP manual is broken down into three periodic activities:

1. Confirmation of information in the manual, rollout to staff for awareness, and specific training for critical individuals
2. Testing and verification of technical solutions established for recovery operations
3. Testing and verification of organization recovery procedures

Issues found during the testing phase must often be reintroduced to the analysis phase.

The BCP needs to be organic, and it must evolve with the organization. Activating the contact and communication list verifies the notification plan's efficiency as well as contact data accuracy. Types of changes that should be identified and updated in the manual include the following:

- Staffing
- Important clients
- Vendors or suppliers
- Organization structure changes
- Investment portfolio and mission statement
- Communication and transportation infrastructure such as roads and bridges

Specialized technical resources must be maintained. Checks include the following:

- Virus definition distribution
- Application security and service patch distribution
- Hardware operability
- Application operability
- Data verification
- Data application

As work processes change, previous recovery procedures may no longer be suitable. Checks include the following:

- Are all work processes for critical functions documented?
- Have the systems used for critical functions changed?
- Are the documented work checklists meaningful and accurate?
- Do the documented work process recovery tasks and supporting disaster recovery infrastructure allow staff to recover within the predetermined RTO?

8 Response and Recovery Operations

8.1 RESPONDING TO NATURAL AND HUMAN-INDUCED DISASTERS

Disasters can be naturally occurring or human-induced. Human-induced disasters could be intentional such as an act of sabotage or terrorism or unintentional, that is, accidental, such as a hazardous spill or a dam break. Disasters may encompass more than weather. They may involve cyberthreats or take on other human-induced manifestations such as subversion, vandalism, or theft.

A natural disaster is a major adverse event resulting from the earth's natural hazards. Examples of natural disasters include floods, tsunamis, tornadoes, hurricanes, cyclones, volcanic eruptions, earthquakes, heat waves, and landslides. Other types of disasters include epidemics or infestations and the more cosmic scenario of an asteroid hitting the earth.

Human-induced disasters are the consequence of technological or human hazards. Examples of human-induced disasters include stampedes, urban fires, industrial accidents, oil spills, nuclear explosions or nuclear radiation, and acts of war. Other types of human-induced disasters include the more intense scenarios of catastrophic global warming, nuclear war, and bioterrorism.

The following list categorizes different types of disasters, both natural and human-induced, and notes first response initiatives for each disaster scenario. It should be noted that in some cases the sources of a disaster may be natural, for example, heavy rain, or human-induced such as a dam break; however, the results for both scenarios is similar—flooding.

8.1.1 NATURAL DISASTERS

Avalanche: The sudden, drastic movement of snow down a slope, occurring when either natural triggers, such as loading from new snow or rain, or artificial triggers, such as explosives or backcountry skiers, overload the snowpack. Shut off utilities, evacuate building if necessary, and determine the impact on the equipment and facilities and any disruption.

Blizzard: A severe snowstorm characterized by very strong winds and low temperatures. Power off all equipment; listen to blizzard advisories; evacuate area, if unsafe; and assess damage.

Earthquake: The shaking of the earth's crust caused by underground volcanic forces of breaking and shifting rock beneath the earth's surface. Shut off utilities, evacuate building if necessary, and determine impact on the equipment and facilities and any disruption.

Wild fires: Fires that originate in uninhabited areas and that pose the risk of spreading to inhabited areas. Attempt to suppress fire in early stages; evacuate personnel on alarm, as necessary; notify the fire department; shut off utilities; and listen to weather advisories.

Flash floods: Small creeks, gullies, dry streambeds, ravines, culverts, or even low-lying areas flood quickly. Listen to flood advisories, determine the flood potential to facilities, prestage emergency power generating equipment, and assess damage.

Freezing rain: Rain occurring when outside surface temperature is below freezing. Listen to weather advisories, notify employees of business closure, go home, and arrange for snow and ice removal.

Heat wave: A prolonged period of excessively hot weather relative to the usual weather pattern of an area and relative to normal temperatures for the season. Listen to weather advisories, power off all servers after a graceful shutdown if there is imminent potential of power failure, and shut down main electric circuit usually located in the basement or the first floor.

Hurricanes: An extreme weather event triggered by heavy rains and high winds. Power off all equipment; listen to hurricane advisories; evacuate area, if flooding is likely; check gas, water, and electrical lines for damage; do not use telephones, in the event of severe lightning; and assess damage.

Landslides: Geological phenomenon that includes a range of ground movement, such as rock falls, deep failure of slopes, and shallow debris flows. Shut off utilities, evacuate building if necessary, and determine impact on the equipment and facilities and any disruption.

Lighting strike: An electrical discharge caused by lightning, typically during thunderstorms. Power off all equipment; listen to hurricane advisories; evacuate area, if flooding is likely; check gas, water, and electrical lines for damage; do not use telephones, in the event of severe lightning; and assess damage.

Limnic eruption: The sudden eruption of carbon dioxide from deep lake water. Shut off utilities, evacuate building if necessary, and determine impact on the equipment and facilities and any disruption.

Tornado: Violent rotating columns of air that descend from severe thunderstorm cloud systems. Listen to tornado advisories, power off equipment, shut off utilities (power and gas), and assess damage once storm passes.

Tsunami: A series of water waves caused by the displacement of a large body of water, typically an ocean or a large lake, usually caused by earthquakes, volcanic eruptions, underwater explosions, landslides, glacier calvings, meteorite impacts, and other disturbances above or below water. Power off all equipment; listen to tsunami advisories; evacuate area, if flooding is likely; check gas, water, and electrical lines for damage; and assess damage.

Volcanic eruption: The release of hot magma, volcanic ash, and/or gases from a volcano. Shut off utilities, evacuate building if necessary, and determine impact on the equipment and facilities and any disruption.

8.1.2 Human-Induced Disasters

Bioterrorism: The intentional release or dissemination of biological agents as a means of coercion. Get information immediately from your public health officials via the news media as to the right course of action; if you think you have been exposed, quickly remove your clothing and wash off your skin; also put on a high-efficiency particulate air filter (HEPA) to help prevent inhalation of the agent.

Civil unrest: A disturbance caused by a group of people that may include sit-ins and other forms of obstructions, riots, sabotage, and other forms of crime, and which is intended to be a demonstration to the public and the government, but can escalate into general chaos. Contact local police or law enforcement.

Urban fires: Even with strict building fire codes, people still perish needlessly in fires. Attempt to suppress fire in early stages; evacuate personnel on alarm, as necessary; notify the fire department; shut off utilities; and listen to weather advisories.

Hazardous materials spill: The escape of solids, liquids, or gases that can harm people, other living organisms, property or the environment, from their intended controlled environment such as a container. Leave the area and call the local fire department for help. If anyone was affected by the spill, call the local Emergency Medical Services (EMS) line.

Nuclear and radiation accidents: An event involving significant release of radioactivity to the environment or a reactor core meltdown, which leads to major undesirable consequences for people, the environment, or the facility. Recognize that a chemical, biological, radiological, and nuclear (CBRN) incident has or may occur. Gather, assess, and disseminate all available information to first responders. Establish an overview of the affected area. Provide and obtain regular updates to and from first responders.

Power failure: Caused by summer or winter storms, lightning, or construction equipment digging in the wrong location. Wait 5–10 minutes; power off all servers after a graceful shutdown; do not use telephones, in the event of severe lightning; and shut down main electric circuit usually located in the basement or the first floor.

In the field of information technology, disasters may also be the result of a computer security or cybersecurity exploits. Some of these events include computer viruses, cyberattacks, denial-of-service attacks, hacking, and malware exploits. These are ordinarily attended to by information security experts.

8.2 RESPONDING TO CHEMICAL, BIOLOGICAL, RADIOLOGICAL, NUCLEAR, EXPLOSIVE, AND INCENDIARY EVENTS

The following sections provide a list of recommended action items that can be undertaken by first responders when responding to chemical, biological, radiological, nuclear, and explosive (CBRNE) events including biological agents, nuclear and radiological agents, incendiary devices, chemical agents, and explosives.

8.2.1 BIOLOGICAL AGENTS

Biological agents may produce delayed reactions. Unlike exposure to chemical agents, exposure to biological agents does not require immediate removal of victims' clothing or gross decontamination in the street. Inhalation is the primary route of entry. Self-contained breathing apparatus (SCBA) and structural firefighting clothing provides adequate protection for first responders against biological agents. Department of Transportation Emergency Response Guide No. 158 (DOT-ERG #158) provides additional information on what to do when exposed to a biological agent. Position uphill and upwind and away from building exhaust systems. Isolate and secure the area. DOT-ERG #158 recommends an initial isolation distance of 80 ft. Do not allow unprotected individuals to enter the area. Be alert for small explosive devices designed to disseminate the agent.

The first responder should gather as much information as possible about the agent including the type and form of agent, that is, whether it is a liquid, powder, or aerosol; the method of delivery; and the location in structure. The following is a list of recommended action items when an agent is being delivered from a point source: a heating, ventilation, and air conditioning (HVAC) system with no physical evidence and a confirmed agent placed in an HVAC system.

8.2.1.1 Wet or Dry Agent from a Point Source

Personnel entering the area must take care of the following:

- Wear full personal protective equipment (PPE) including SCBA.
- Avoid contact with puddles or wet surfaces or areas.
- Keep all potentially exposed individuals in close proximity, but out of the high-hazard areas.
- Isolate that area of the building and restrict movement and traffic.
- Shut down the HVAC system that services the area that may be contaminated.
- If victims have visible agent on them, wash the exposed skin with soap and water.
- If the area is highly contaminated and the facility is equipped with showers, make the victims take a shower and change clothes as a precaution.
- Ask the HazMat team to conduct bioassay field tests (limited to the number of agents).
- If possible, collect a sample of the material for testing. If test results are positive, decontaminate in the shower facility with warm water or soap.
- Provide emergency covering or clothing and bag personal effects.
- Refer to medical community for treatment.

8.2.1.2 Threat of Dry Agent Placed in HVAC or Package with No Physical Evidence

Personnel entering the area must take care of the following:

- Initiate a search of the building.
- Wear full PPE, including SCBA.

- Avoid contact with puddles and wet surfaces.
- If any evidence of an agent is found in or near the HVAC system, remove occupants from the building and isolate them in a secure and comfortable location.
- If a suspicious package is found, handle as a point source location.
- Make the contaminated victims shower and change. No decontamination should take place in unprotected areas and in the open. Tents or other sites should be used.
- Make the exposed victims shower and change at their discretion.
- Investigate all HVAC intakes and returns for evidence of agent or dispersal equipment.
- Refer exposed individuals to the medical community for treatment.

8.2.1.3 Confirmed Agent Placed in HVAC System (Visible Fogger, Sprayer, or Aerosol Device)

Personnel entering the area must take care of the following:

- Wear full PPE and SCBA.
- Avoid contact with puddles, wet surfaces, and so on.
- Shut down HVAC systems.
- Remove occupants from the building or area, and isolate them in a secure and comfortable location.
- Ensure that the HazMat teams conduct a bioassay field test (limited number of agents) on the scene.
- If possible, collect a sample of the material for testing.
- If test results are positive, make the contaminated victims shower and change. No decontamination should take place in an unprotected area or out in the open. Tents or other sites should be used.
- Gather all decontaminated victims in a specific holding area for medical evaluation.

8.2.2 Nuclear or Radiological Agents

Radiological agents may produce delayed reactions. Unlike exposure to chemical agents, exposure to radiological agents does not require immediate removal of victims' clothing or gross decontamination in the street. Inhalation is the primary route to entry for particulate radiation. In most cases, SCBA and structural firefighting clothing provides adequate protection for first responders. One major concern or threat for exposure to nuclear and radiological agents is the sabotage of an existing nuclear power facility. This can be accomplished with minimal expertise. Other threats include natural disasters that can cause direct damage to the facility and human-induced unintentional threats, such as a nuclear meltdown.

Alternately, gamma sources require minimizing exposure time and maintaining appropriate distance as the only protection. Exposed or contaminated victims may not demonstrate obvious injuries. DOT-ERG #163 and #164 provide additional information with regard to exposure of nuclear and radiological agents. In case of exposure, an individual should position themselves upwind of any suspected event. Isolate and secure the area; DOT-ERG #163 recommends a minimum distance of 80–160 ft. Be alert for small explosive devices designed to disseminate radioactive agents. Use time, distance, and shielding as protective measures. Use full PPE including SCBA in any nuclear response situation. Try to avoid contact with the agent and stay out of any visible smoke or fumes. Establish background levels outside of suspected area and continuously monitor radiation levels.

Remove victims from the high-hazard area to a safe holding area and be alert to their reactions. Triage, treat, and decontaminate trauma victims as appropriate. Detain or isolate uninjured persons or equipment. Delay decontamination until instructed by radiation authorities. Use radiation detection devices, if possible, to determine if patients are contaminated with radiological material.

8.2.3 INCENDIARY DEVICES

Incendiary devices and fires may present intense conditions such as rapid spread, high heat, multiple fires, and the presence of chemical accelerants. In dealing with incendiary devices, first responders must be aware that terrorists may sabotage fire protection devices. They must be alert for booby traps. They must be aware of the possibility of multiple devices. Be aware of an explosive device known as "Armstrong's mixture," a mix of red phosphorous and potassium chlorate. Solid rocket fuel has been used in several arson fires in the recent past. It burns very hot and causes extensive damage in a short time. This is a device commonly used by the Irish Republican Army (IRA) to destroy retail establishments in England. Trash cans are also common targets for vandalism and terrorists. Therefore, municipalities have installed bomb-resistant and fire-proof containers.

8.2.4 CHEMICAL AGENTS

If persons were exposed to a chemical agent, two questions that need to be asked to identify the signs and symptoms of hazardous substance are the following: Are there unconscious victims with minimal or no trauma? Are there victims exhibiting SLUDGEM signs or seizures? SLUDGEM is an acronym that stands for salivation, lacrimation, urination, defecation, gastric distress, emesis, and miosis. Other signs and symptoms include blistering, reddening of the skin, discoloration or skin irritation, and difficulty breathing. Physical indicators and other outward warning signs include medical mass casualty or fatality with minimal or no trauma; first responder casualties; dead animals and vegetation; and unusual odors, color of smoke, and vapor clouds.

Different types of delivery systems can be used to disperse chemical agents including irritants and toxins into the atmosphere such as air handling systems, heating and air conditioning systems, and misting or aerosol devices, either individually or on a large scale. In the Tokyo Subway system, a simple plastic milk carton and a paper bag were used to expose over 5000 people to sarin (GB aerosol) gas. Sprayers, or for larger areas a crop dusting plane or a modified small aircraft, may be used to disperse chemical agents. Another type of dispersion device is a gas cylinder. Intelligence reports have indicated that there have been incidents where thousands of smoke detectors were purchased and the small amount of radioactive material used in these devices were recovered to manufacture a "dirty bomb." Additional information on various agents and toxins can be found in DOT-ERG for the following agents: nerve agent (Guide #153), blistering agent (Guide #153), blood agents (Guide #117, #119, and #125), choking agents (Guide #124 and #125), and irritant agents (Guide #153 and #159).

The following is a list of recommended action items when responding to the scene of a chemical agent attack or chemical spill:

- Approach from uphill and upwind.
- Ensure that victims exposed to chemical agents remove clothes immediately; gross decontamination and definitive medical care must be undertaken too.
- Upon arrival, stage at a safe distance away from site.
- Secure and isolate the area and deny entry.
- Complete a hazard and risk assessment to determine if it is acceptable to commit responders to the site.
- Be aware of larger secondary chemical devices.
- Do not enter areas of high concentration, unventilated areas, or below-grade areas for any reason.
- Ensure that personnel in structural PPE or SCBA enter the hot zone near the perimeter (outside of areas of high concentration) to perform life-saving functions.
- Move ambulatory patients away from the area of highest concentration or source.
- Confine all contaminated and exposed victims to a restricted or isolated area at the outer edge of the hot zone. Keep in mind that these people may not want to be confined.
- Segregate symptomatic patients into one area and asymptomatic patients in another area.
- Law enforcement should establish an outer perimeter to completely secure the scene.
- If a particular agent is known or suspected, forward this information to EMS personnel and hospitals so that sufficient quantities of antidotes can be obtained.
- Notify hospitals immediately that contaminated victims of the attack may arrive or self-present at the hospital.
- Begin emergency gross decontamination procedures starting with the most severe symptomatic patients. Use soap and water decontaminate.
- Check whether decontamination capabilities are provided at the hospital to assist with emergency gross decontamination prior to victims entering the facility.
- If available, seek the help of HazMat personnel in chemical PPE for rescue, reconnaissance, and agent identification.
- Decontaminate asymptomatic patients in a private area (tent or shelter) and then forward them to EMS for evaluation.

8.2.5 Explosives

Explosive devices may be designed to disseminate chemical, biological, or radiological agents. Explosives may produce secondary hazards, such as unstable structures, damaged utilities, hanging debris, void spaces, and other physical hazards. Devices may contain antipersonnel features such as nails, shrapnel, and fragmentation design. First responders should always be alert for potential secondary devices. Outward warning signs of an explosive device scene include oral or written threats; a container or vehicle that appears out of place; devices attached to compressed gas cylinders, flammable liquid containers, bulk storage containers, pipelines, and other chemical containers such as a dirty bomb; and oversized packages with oily stains, chemical odors, excessive postage, protruding wires, excessive binding, and no return address. Additional information on explosives can be obtained from DOT-ERG #112 and #114.

A list of recommended action items when responding to the scene of a suspected explosive device is given in Sections 8.2.5.1 through 8.2.5.7.

8.2.5.1   Unexploded Device and Preblast Operations

Personnel entering the area must take care of the following:

- Ensure that the command post is located away from the area where improvised secondary devices may be placed, such as mailboxes and trash cans.
- Stage incoming units
  - Away from the line of sight of target area.
  - Away from buildings with large amounts of glass.
  - In such a way as to utilize distant structural and/or natural barriers to assist with protection
- Isolate and deny entry.
- Secure perimeter based on the size of the device.

All activities should be coordinated with law enforcement, and first responders should be prepared to respond if the device were to be activated.

8.2.5.2   Explosive Device Preblast

Personnel entering the area must take care of the following:

- Device characteristics
  - Type of threat
  - Location
  - Time
  - Package
  - Device
  - Associated history
- Standoff distance should be commensurate with the size of the device:
  - Car bomb = 500 m
  - Package bomb (1–10 kg) = 300 m
  - Pipe bomb = 150 m

- Use extreme caution if the caller identifies a time for the detonation. It is very possible that the device will be activated prior to the announced time.
- Discontinue use of all radios, mobile data terminals (MDTs), and cell phones in accordance with local protocols.
- Evaluate scene conditions:
  - Potential number of affected people
  - Exposure problems
  - Potential hazards: Utilities, structures, fires, chemicals, and so on
  - Water supply
- Evaluate available resources (EMS, HazMat, and technical rescue).
  - Review preplans for affected buildings.
  - Make appropriate notifications.
  - Develop an action plan that identifies incident priorities, potential tactical assignments, and key positions in the ICS/unified command.

8.2.5.3 Explosive Device Postblast

Personnel entering the area must take care of the following:

- Command post should be located away from areas where improvised secondary devices may be placed, such as in mailboxes and trash cans.
- Initial arriving units
  - Stage a safe distance from reported incident (or where debris was first encountered).
  - Away from line of sight of target area.
  - Away from buildings with large amounts of glass.
  - Utilize distant structural and/or natural barriers to assist with protection.
  - Stage incoming units at a greater distance. Consider using multiple staging sites.
- Debris field may contain unexploded bomb material.
- The use of all radios, MDTs, and cell phones should be discontinued in accordance with local protocols.
- All citizens and ambulatory victims should be removed from the affected area.
- On-scene conditions should be determined and resource requirements evaluated.
  - Explosion
  - Fire
  - Structural collapse or unstable buildings.
  - Search and rescue nonambulatory or trapped victims.
  - Exposure to other agents that may have been brought to or in the area
  - Utilities, power, and gas
- The number of patients and extent of injuries need to be determined as soon as possible to assist in recovery logistics.
- Other hazards, such as if there is a threat of armed dissidents present, must be ascertained.
- Make notifications to
  - Law enforcement.

- Hospitals.
- Emergency management.
- Municipal, provincial, state, and federal agencies as appropriate.
- Hazard and risk assessment should be completed.
- Personnel should only be allowed to enter the blast area for life safety purposes.
- Viable patients should be removed to a safe refuge area as soon as possible.
- Ambulatory patients should be handled with care.
- No command post should be set up near any buildings that may become collateral damage to the main target.
- The number of personnel should be limited and the exposure time minimized. Personnel entering the blast area should
  - Wear full protective clothing, including SCBA.
  - Monitor the atmosphere for flammability, toxicity, radiation, and pH.
- Emergency gross decontamination should be established.
- Area should be evacuated of all emergency responders if there is any indication of a secondary device.
- Patients should be removed from the initial blast site to a safe refuge area.
- Triage and treatment area should be established at the casualty collection point (CCP), if the latter has been established.
- Hospitals should be notified.
- A mass casualty plan should be implemented.
- Rescuers should not be allowed to enter unsafe buildings or high-hazard areas.
- Utilities should be controlled and exposures from a defensive position must be protected.
- Evidence must be preserved and maintained.
- There must be an awareness of the need to become tactical at any time.

8.2.5.4   Agency-Related Actions, Fire Department

Personnel entering the area must take care of the following:

- Isolate and secure the scene, deny entry, and establish control zones.
- Establish command.
- Evaluate scene for safety and security.
- Stage incoming units.
- Gather information regarding the incident and number of patients.
- Assign ICS positions as needed.
- Initiate notifications to hospitals, law enforcement, municipal, provincial, state, and/or federal agencies.
- Request additional resources.
- Use appropriate self-protective measures:
  - Proper PPE
  - Time
  - Distance
  - Shielding

- Minimize the number of personnel exposed to danger.
- Initiate public safety measures:
  - Rescue
  - Evacuate
  - Protect in place
- Establish water supply:
  - Suppression activities
  - Decontamination
- Control and isolate patients (away from the hazard, at the edge of the hot/warm zone).
- Coordinate activities with law enforcement.
- Begin and/or assist with triage, administering antidotes, and treatment.
- Begin gross mass decontamination.
- As the incident progresses, prepare to initiate unified command system.
- Establish unified command post including representatives from the following organizations:
  - Law enforcement
  - Hospitals and public health
  - EMSs
  - Emergency management
  - Public works
- Establish and maintain a chain of custody for any evidence that is found.
- Protect and preserve evidence found at the scene.
- Preserve HazMat, chain of evidence.

8.2.5.5 Emergency Medical Services

Personnel entering the area must take care of the following:

- If EMS is first on the scene
  - Isolate and secure the scene; establish control zones.
  - Establish command.
  - Evaluate scene safety and security.
  - Stage incoming units.
- If command has been established report to and/or communicate with command post.
- Gather information regarding
  - Type of event.
  - Number of patients.
  - Severity of injuries.
  - Signs and symptoms.
- Assign medical incident command positions as needed.
- Notify hospitals.
- Request additional resources as appropriate:
  - Basic life support (BLS) and advanced life support (ALS)
  - Medivac helicopter (trauma/burn only)

- Medical equipment and supply caches
- Metropolitan medical response system
- National medical response team
- Disaster medical assistance team
- Disaster mortuary response team
- Use appropriate self-protective measures:
  - Proper PPE
  - Time, distance, and shielding
- Minimize the number of personnel exposed to the danger.
- Initiate mass casualty procedure.
- Evaluate the need for CCP and patient staging area (PSA).
- Control and isolate patients (away from the hazard, at the edge of the hot or warm zone).
- Ensure that patients are decontaminated prior to being forwarded to the cold zone.
- Triage, administer antidotes, and treat and transport victims.
- Evidence preservation/collection.
- Recognize potential evidence.
- Report findings to appropriate authority.
- Consider embedded objects as possible evidence.
- Secure evidence found in ambulance or at hospital.
- Types of evidence
  - Debris from the bomb.
  - Damage that is out of the ordinary such as melted iron pipes
  - Tracks entering or leaving the area
  - Tire tracks
  - Fingerprints
  - Tool marks
- Establish and maintain chain of custody for evidence preservation.
- Ensure participation in unified command system when implemented.

8.2.5.6 Law Enforcement

Personnel entering the area must take care of the following:

- If law enforcement is first on scene
  - Isolate and secure the scene; establish control zones.
  - Establish command.
  - Stage incoming units.
- If command has been established, report to command post.
- Evaluate scene for safety and security:
  - Ongoing criminal activity
  - Victims considered to be possible terrorists
  - Secondary devices
  - Additional threats
- Gather witness statements and observations and document.
- Initiate law enforcement notification:

- • Federal, provincial, or state police
- • Federal bomb squad or national CBRNE response team
- • Explosive ordnance disposal (EOD)
- • Private security forces
- Request additional resources.
- Secure outer perimeter.
- Traffic control considerations:
  - • Staging areas
  - • Entry areas
  - • Egress areas
- Use appropriate self-protective measures:
  - • Proper PPE
  - • Time, distance, and shielding
- Minimize the number of personnel exposed to the danger.
- Initiate public safety measures:
  - • Evacuate.
  - • Protect in place.
- Assist with control and isolation of patients.
- Coordinate activities with other response agencies.
- Evidence preservation:
  - • Diagram the area.
  - • Photograph the area.
  - • Photograph the crowd.
  - • Prepare a narrative description.
  - • Maintain an evidence log (if using a microrecorder to take notes, be sure to keep some written notes also in case of mechanical failure).
- Participate in a unified command system with
  - • Law enforcement.
  - • Hospitals and public health.
  - • EMSs.
  - • Emergency management.
  - • Public works.

8.2.5.7   HazMat Group

Personnel entering the area must take care of the following:

- Establish the HazMat group.
- Provide technical information assistance to
  - • Command.
  - • EMS providers.
  - • Hospitals.
  - • Law enforcement.
  - • Emergency management.
- Detect and monitor to identify the agent, determine concentrations, and ensure proper control of the area.
- Continually reassess control zones.

- Enter the hot zone (chemical PPE) to perform rescue, product confirmation, and reconnaissance.
- Product control and mitigation may be implemented in conjunction with expert technical guidance.
- Improve hazardous environments:
  - Ventilation
  - Control HVAC
  - Control utilities
  - Implement a technical decontamination corridor for hazardous materials response team personnel.
- Coordinate and assist with mass decontamination.
- Provide specialized equipment as necessary, such as tents for operations and shelters.
- Assist law enforcement personnel with evidence preservations or collection and decontamination.

8.3 RESPONDING TO A TERRORIST EVENT

When responding to a terrorist event, initial security assessment should be undertaken before approaching the scene. Some of the questions a first responder should be asking themselves prior to approaching the scene are as follows:

- Is there only one indicator or are there multiple indicators?
- Is the response to a target hazard or target event?
- Has there been a threat?
- Are there multiple (nontrauma-related) victims?
- Are responders victims?
- Has there been an explosion?
- Has there been a secondary attack or explosion?

If there is only one indicator, respond with a heightened level of awareness. However, if there are multiple indicators, it may be a scene for a terrorist incident. If it is a scene for a terrorist incident, the following action items need to be undertaken:

- Initiate response operations with extreme caution.
- Be alert for actions against responders.
- Evaluate and implement personal protective measures.
- Consider the need for maximum respiratory protection.
- Make immediate contact with law enforcement for coordination.

Establish a safety route for first responders, by taking the following actions into consideration:

- Approach cautiously, from uphill/upwind if possible.
- Consider law enforcement escort.

- Avoid choke points, for example, regrouping areas different from staging areas for responders.
- Identify safe staging location(s) for incoming units.

During a terrorist incident, command and control is vital; therefore, the following need to be considered for command:

- Establish command.
- Isolate area/deny entry.
- Ensure scene security.
- Initiate on-scene evaluation and hazard risk assessment.
- Provide, identify, and designate safe staging location for incoming units.
- Ensure the use of personal protective measures and shielding.
  - Assess emergency egress routes.
  - Position vehicles for rapid evacuation.
- After egress, have designated rally point(s).
- Ensure personnel accountability.
- Designate incident safety officer.
- Assess command post security.
- Consider assignment of liaison and public information positions.
- Assess decontamination requirements.
  - Gross decontamination.
  - Mass decontamination.

Part of responding to a terrorist incident is determining the need for additional specialized resources including the following:

- Fire
- Emergency medical services
- Hazardous materials
- Bomb squad
- Public works
- Public health
- Environmental emergency management
- Law enforcement explosive ordnance disposal
- Other emergency units

Everything at the scene should be considered as evidence. First, ensure coordination with law enforcement when collecting evidence at the site. First responders should also make the appropriate notifications with dispatch center to include updates and situation reports, hospitals, utilities, and law enforcement. In addition, when communicating with any of the aforementioned entities it is important to state point of contact as appropriate.

If the site is going to become a potential crime scene, then responders need to prepare for transition to a unified command and to ensure coordination of communications and identify needs. Prior to approaching the scene, first responders must undertake an on-scene evaluation including a review of dispatch information and observation of physical indicators and other outward warning signs including biological, nuclear, incendiary, chemical, explosive events, and armed assaults. In the situation where a hazardous substance was released, an on-scene evaluation would be necessary, in which case the responder(s) would need to take note of the following:

- Debris field
- Mass casualty
- Fatality with minimal or no trauma
- Responder casualties
- Severe structural damage without an obvious cause
- Dead animals present
- Dead vegetation
- Systems disruptions
- Utilities
- Transportation
- Unusual odors
- Color of smoke
- Vapor clouds

In addition to the above, first responders will also need to make note of signs and symptoms of victims exposed to a hazardous substance including the following:

- Are there unconscious victims with minimal or no trauma?
- Are there victims exhibiting SLUDGEM signs/seizures?
- Is there blistering, reddening of skin, discoloration, or skin irritation?
- Are victims having difficulty breathing?

First responders need to be able to identify the apparent signs and victim commonality. In addition, victims and witnesses must be interviewed after the event. The reason for this is to collect as much information (intelligence) as possible regarding the event(s) and, therefore, to be able to piece together as to what happened and why? Typical questions asked by the interviewer(s) are as follows:

- Is everyone accounted for?
- What happened (information on delivery)?
- When did it happen?
- Where did it happen?
- Who was involved?
- Did they smell, see, taste, hear, or feel anything out of the ordinary?

By interviewing witnesses and victims and carrying out on-scene evaluation, first responders can identify the type of event(s) or attack(s) including biological, chemical, radiological, incendiary, explosive, nuclear, armed assault or electromagnetic pulse weapon(s), or cyberattacks. In the case of a hazardous substance release or exposure, it is important to monitor weather reports and forecasts to determine downwind exposures. In addition, it is also important to determine life safety threats including to the self, responders, victims, and public. It is also important to determine the mechanism of injury such as thermal, radiological, asphyxia, chemical, etiological, mechanical, or psychological; this will determine the number of EMS personnel needed on the scene. Furthermore, it is important to estimate the number of victims that are in need of ambulatory assistance and nonambulatory assistance. Those that do not require ambulatory assistance will be able to reach a health-care facility or trauma center on their own. However, it is still necessary to maintain a record of all victims since it will directly affect the number of resources needed to respond to the event(s), specifically the number of emergency medical personnel and equipment including ambulances.

In addition to identifying the event(s), the number of victims, and the extent of their injuries, it is also important to identify the affected surroundings to the incident scene including structural damage and exposure, downwind exposure, environmental exposure, below-grade occupancies, below-grade utilities, and aviation or air space hazards. First responders should also consider the possibility of a secondary attack such as an explosive device, chemical device, or booby trap. They will need to determine available and additional resources if needed. Furthermore, incident site management is paramount including the establishment of security and safety zones.

Reassessment of the initial isolation and standoff distances including the establishment of inner and outer perimeter zones should be undertaken. Public protection areas also need to be established including the removal of endangered victims from high-hazard areas, establishing safe refuge areas for both contaminated and uncontaminated zones, evacuating all victims out of contaminated areas, and protecting those in place, such as first responders.

The on-site incident commander needs to identify appropriate PPE options prior to committing personnel. Dedicate EMSs needed for responders. Prepare for gross decontamination operations for responders. Coordinate with law enforcement to provide security and control the perimeters. Designate an emergency evacuation signal. At the scene, there are certain tactical (life safety) options that need to be taken into consideration including the following:

- Isolate and secure area and deny entry.
- Public protection (evacuate/protect in place).
- Implement self-protection measures.
- Commit only essential personnel and minimize exposure.
- Confine and contain all contaminated and exposed victims.
- Establish gross decontamination capabilities.

Prior to taking on any rescue operations the following questions need to be addressed:

- Is the scene safe for operations?
- Can I make it safe to operate?
- Are victims viable?
- Are they ambulatory?
- Can they self-evacuate?

- Are they contaminated?
- Do they require extrication (bombings)?
- Is a search safe and possible?
- Is specialized PPE required?

In addition, in order to stabilize the incident the following defensive operations also need to be taken into consideration:

- Water supply
- Exposure protection
- Utility control
- Fire suppression
- HazMat control

If mass decontamination were to be undertaken, position the area upwind and uphill. First responders wearing full structural gear and SCBA may approach the victims to provide direction and guidance. Avoid contact with any liquids on the ground, victims' clothing, or other surfaces. Remove contaminated or exposed victims from the high-hazard area. They should be isolated and secured in a holding area at the outer periphery of the hot zone. Evaluate their signs and symptoms to determine the type of agent involved. Signs or symptoms of exposure, depending on the agent, may include difficulty in breathing, reddening, burning, itching of the eyes and/or skin; irritation of the nose and throat; runny nose or salivation; coughing; pinpoint pupils; pain in the eyes or head; seizure-like activity or convulsions; and vomiting.

During mass decontamination operations victims should be separated into groups: symptomatic and asymptomatic and ambulatory and nonambulatory. Medical providers may access the patients in the holding area to initiate triage, administer antidotes, and provide basic care in accordance with local protocols. The type of decontamination system is dependent on the number of patients, the severity of their injuries, and the resources available. Large numbers of patients may require engine companies to use the "side by side" system as well as numerous showers to move multiple lines of patients through the process. Several patients may be handled with a single hose line, while numerous patients will require the use of a mass decontamination corridor.

Begin emergency gross decontamination immediately on victims who are symptomatic and have visible (liquid) product on their clothing and were in close proximity to the discharge. In a mass casualty setting, life safety takes precedence over containing runoff. Set up decontamination in an area such that the decontamination water will flow away from your operation and into the grass or soil, if possible. Provide privacy only if it will not delay the decontamination process. Remove all of the victims' clothing. Thoroughly wash and rinse the victims. For limited number of patients use soap, soft brushes, and water from small hose lines at low pressure usually at around 30 psi. For multiple patients use engines parked side-to-side dispersing water at low pressures from discharges or use multiple showers. Patients should remain in the water for several minutes and receive a thorough flushing with their arms up, and spin around. Personnel should be positioned at the exit side of the corridor to manage the patients and ensure they stay in the water for an adequate period of time. Separate lines may be required to process nonambulatory patients. As resources become available, separate decontamination lines may be established for male and female patients, as well as families. Provide emergency covering such as blankets. Upon decontamination, transfer patients to EMSs for triage.

For asymptomatic patients that have been contaminated or exposed, process patients through the gross decontamination showers with their clothes on and have them proceed to separate holding areas by gender. Separate systems should be established for male and female patients. Set up tents and shelters and provide showers or an improvised wash system. Patients should be numbered, and bags should be used to store their personal effects. Provide emergency covering and clothing. Upon decontamination, transfer patients to a holding area for medical evaluation.

Stand-alone decontamination systems may have to be established outside of hospital emergency rooms for patients who self-present at the location. Units with decontamination capabilities should be dispatched to establish a system. Triage the patients and separate them into symptomatic and asymptomatic groups. Patients who are symptomatic or have visible product on their clothes will be a priority. Remove clothes and flush thoroughly. Liaison with the hospital staff to determine where patients will be sent after decontamination.

It is important to recognize potential evidence including unexploded devices, portions of devices, clothing of the victims, containers, and dissemination devices. Note location of potential evidence. Report the findings of evidence to appropriate authorities. Move potential evidence only for life safety and incident stabilization. Establish and maintain chain of custody for evidence preservation.

8.4   DISASTER RECOVERY OPERATIONS

The recovery phase starts after the immediate threat to human life and property has receded. The immediate goal of the recovery phase is to bring the affected population back to some degree of normalcy or at minimum predisaster conditions. Recovery remains the least understood aspect of emergency management. At the operational level there is a lack of knowledge and understanding of what to do in the aftermath of a disaster during recovery operations. Because recovery is not fully understood by practitioners at all levels of government, countries remain largely unprepared to address the challenges associated with recovery operations.

Hurricane Katrina is an example of the perils of failing to understand the recovery needs of a community, a region, a province, and/or a state. Although Katrina struck more than eight years ago, many communities still struggle with the challenges of rebuilding homes, businesses, and infrastructure, as well as a sense of community. Unfortunately, the recovery challenges and problems of communities affected by Katrina are but the most visible evidence of the recovery needs of countless communities that have been in the line of hurricanes, tornadoes, floods, and other disasters, large and small.

Most emergency management practitioners will agree that recovery is a nonlinear and complex process. Following a disaster, stakeholders from all sectors of society are involved in the recovery process. Often these stakeholders operate with little knowledge of what others are doing or understanding of the recovery goals that should guide recovery strategies and priorities. A vast number of decisions are made in a relatively short time, and the pressure to "return to normalcy" sometimes undermines the ability of a community to take advantage of what it has learned. A long-term approach is the only responsible way to make the needed adjustments that are always required.

The recovery process includes five component variables: the physical, the economic, the institutional, the social, and the environmental. As a result, recovery is a complex process. After a major disaster, a community rarely, if ever, "returns to normal." Although the people involved in recovery often look to the past, in reality, recovery from a major event is about inventing the future. The past is gone, and what emerges results from a new (socially and regulatory) set of relationships between those who remain and who join in to influence the system. Inventing the future is a positive term in that it implies that interventions can be useful. For example, in New Zealand, recovery has come to mean "opportunities to reduce risk for the future." Long-term recovery is an opportunity for transformative advancement, emphasizing the importance of culturally sensitive and locally relevant system adjustments. The Chinese call this "better than before."

Complexity requires a holistic approach that accounts for multiple sectors. Elements of this holistic approach are appearing at the national level in New Zealand and in other countries, including in China's response to the 2008 Wenchuan earthquake. A few cities in the United States have also engaged in a holistic approach. Recovery needs management, it is not only a managed process by which a community is assured a better future, but also it is a process of restoring, rebuilding, and reshaping the physical, social, economic, institutional, and natural environment. The process is by definition complex. Recovery requires establishing or reestablishing important relationships within the system and between the system and its environment and requires the community to reassert itself as it picks up the pieces.

A comprehensive and robust theory of recovery should address the people, institutions, structures, and artifacts within a geographic area. Other factors that can have an effect and influence the recovery effort include the scale and magnitude of the recovery effort and economic viability of the impacted geographic area. Adding both of these factors to the Alesch framework would move it in the direction of a holistic approach to disaster recovery. More importantly, in order for the holistic approach to be effective, a community will require a number of coping mechanisms at least equal to the array of environmental challenges it faces. This, however, will be influenced by the relationship between geographic areas and the external environment.

In a broader sense recovery refers to those programs that go beyond the provision of immediate relief to assist those who have suffered the full impact of a disaster to rebuild their homes, lives, and services and to strengthen their capacity to cope with future disasters. Following a disaster, life-saving assistance is the most urgent need. The rapid provision of food, water, shelter, and medical care is vital to prevent further loss of life and alleviate suffering.

During the recovery process the most extreme home confinement scenarios include war, famine, and severe epidemics and may last a year or more. Then recovery efforts take place inside the home. Planners for these events usually buy bulk foods and appropriate storage and preparation equipment and eat the food as part of normal life. A simple balanced diet can be constructed from vitamin pills, wholemeal wheat, beans, dried milk, corn, and cooking oil. One should add vegetables, fruits, spices, and meats, both prepared and fresh-gardened, when possible.

However, even at this stage, relief must be conducted with a thought to the affected community's longer-term benefit. As people begin to recover from a disaster and rebuild their lives, aid agencies need to help them strengthen their resilience toward future hazards. Restoring the predisaster status quo may inadvertently perpetuate vulnerability. Similarly, development programs need to take into account existing risks and susceptibility to hazards and to incorporate elements to reduce them. The two approaches are interdependent, complementary, and mutually supportive. This supports the definition of disaster management as a social issue more than simply a technological one. Although disasters themselves may be naturally occurring, their impacts on society and the surrounding environment are completely influenced by human intervention. Therefore, when undertaking recovery operations there are two simultaneous processes: (1) the immediate relief effort to provide the basic human needs to the impacted population and (2) the long-term plan to strengthen a community's coping capacity by targeting the root causes of a disaster.

8.5 SPECIAL RESPONSE TEAMS

Special response teams are teams of select highly trained personnel with specialized skill sets that are able to deploy rapidly and respond to emergencies on a moment's notice. The function and mission of the special response teams will vary between each team. Special response teams exist at the municipal (local level), provincial, state, and/or federal levels. Many teams, not all, are set up to respond to terrorist-type threats including bombings, hostage or barricaded person incidents, and the deliberate release of weapons of mass destruction (WMD) in a noncombatant role. Other functions or missions of special response teams include guarding of nuclear power plants; fugitive recovery; responding to natural disasters, such as the Canadian Forces Disaster Assistance Response Team (DART); search and rescue; urban heavy search and rescue; hazardous materials incidents; and responding to WMDs.

Special response teams could be either a joint effort consisting of volunteers from a variety of departments and agencies or a stand-alone unit within an organization. For example, the Canadian National CBRNE response team comprises personnel from Health Canada, the Royal Canadian Mounted Police, the Center for Forensic Sciences, and the Canadian Forces. The team is called in to evaluate scenes potentially linked to terrorist or suspected terrorist activity. Such events typically involve the use of devices that can cause damage or injury over a large area, for example, poisonous gases or explosives. The team looks for trace evidence and collects samples to determine if a chemical, biological, radiological, or nuclear WMD was used. The CBRNE team can also be deployed to major events to provide an immediate, on-site analysis of suspicious packages, attacks, or unexpected incidents. When requested, the CBRNE team provides assistance to police agencies following a CBRNE event and is available to provide police agencies with CBRNE training.

Other special response teams are stand-alone units with a specific mission or mandate, such as nuclear response forces. These are stand-alone special weapons and tactics (SWAT) teams used for guarding nuclear power plants against sabotage or terrorist acts. Each nuclear power plant in the United States and Canada has its own nuclear response force. They do not rely on outside assistance during an incident, but instead are completely self-contained and capable of thwarting any attack against a nuclear facility.

Besides various police, security, and public agencies, there are special response teams at the federal or national level whose role is to directly engage terrorists and prevent terrorist attacks. These units are referred to as special mission units. Such units are engaged in preventive action, in hostage rescue, and in responding to ongoing attacks. Countries of all sizes can have highly trained counterterrorist teams. Tactics, techniques, and procedures for combating terrorist activities are under constant development. Most of these measures deal with terrorist attacks that affect an area, or threaten to do so. It is far harder to deal with assassination, or even reprisals on individuals, due to the short (if any) warning time and the quick exfiltration of the assassins.

Special response teams and special mission units are specially trained in tactics, techniques, and procedures for a variety of functions and missions. They comprise personnel from various backgrounds and specialized skill sets. For example, special mission units are very well equipped for close quarter battle (CQB) with emphasis on stealth and performing the mission with minimal casualties. The units include takeover force (assault teams), snipers, EOD experts, dog handlers, and intelligence officers. Other teams may include additional experts such as physicians, scientists, hazardous materials technicians, search and rescue technicians, and forensic experts.

The majority of special response team operations at the tactical level are conducted by provincial, state, federal, and national law enforcement agencies or intelligence agencies. In some countries, the military may be called in as a last resort. Obviously, for countries whose military are legally permitted to conduct police operations, this is a nonissue, and such counterterrorism operations are conducted by their military.

A special response team consists of a headquarters cell or command element, a perimeter containment team, a sniper team (hostage situations), a support team, and a show of force element or assault team. The perimeter containment team is used to cordon off an area of an incident and to provide scene control. The role of the perimeter containment team may be undertaken by conventional police troops. The function of the team is to provide additional tactical support to the assault force and also to prevent suspects and assailants from escaping with hostages. A second function is to cordon off the area to keep the public and media at bay.

The headquarters cell or command element is responsible for making the key decision with respect to command and control of the operations. As with any operation, the commander of the headquarters cell is supported by a variety of specialists including communications specialists to maintain communications equipment, negotiators to extract concessions from the terrorists, psychologists to develop a profile of the subject(s), and intelligence specialists to oversee the placement of covert surveillance devices and to build up a picture of the number of terrorists and their relative position to the hostages.

During a hostage scenario or siege, snipers may be employed to provide on-site intelligence to the commander and also to cover the assault element as they engage the terrorists using precision rifle fire. Each sniper is given a specific quadrant of the building and reports back to the command center the movement of the terrorists in and out of their rifle sights. The snipers are in constant communication with the command center. This is done verbally using radio-linked throat microphones or by depressing a switch on the rifle stock, which changes a light on the console in front of the sniper commander. In the rare event of all known terrorists covered simultaneously, a siege can end safely by precision rifle fire alone.

The support team provides direct support to the assault element, by securing ropes, ladders, abseiling equipment and breaching doors, walls, and windows. The support element may also use distractionary devices, including CS gas (tear gas) and percussion grenades, prior to the assault element entering the building, maintaining the element of surprise. The assault force is made up of the point man, the team leader, and the defensive group, usually made up of two to four men. The assault force is responsible for eliminating the terrorists and rescuing the hostages. However, in police operations, it is preferred that the situation be deescalated and that the assailants be subdued without the use of deadly force. For example, Germany's national counter-terrorist team GSG-9 places an equal emphasis on hand-to-hand combat skills and shooting skills, with the preference being to arrest and subdue the assailants.

Each assault team enters the building from a different position and is given an area responsibility with the building. The teams sweep through the rooms and stairwells, eliminating the assailants using discriminatory fire. An additional challenge to the team is that the terrorists may begin to eliminate hostages upon an eminent assault; therefore, the use of distraction devices deters this by causing confusion and disorientation among the terrorists and maintaining an element of surprise for the team.

Once the hostages are secure and the assailants are either killed or incapacitated the evacuation of all personnel including those gunmen who were wounded during the assault begins. At this point hostages are treated as hostile since the assault team has not yet identified them as friend or foe. The assault team members typically line themselves up in a linear fashion along the corridors and stairwells passing hostages and injured personnel from one man to another and outside of the building. Once outside, intelligence officers and other support personnel identify and interview the hostages, detain prisoners, and provide medical care to the injured. Furthermore, in the case of a CBRNE incident, mass decontamination of the hostages, injured assailants, assault element, and anyone else who has come into contact with a hazardous substance is also necessary.

8.6 STRESS MANAGEMENT AFTER A DISASTER

Disasters are often unexpected, sudden, and overwhelming. In some cases, there are no visible signs of physical injury, but there is a serious emotional toll. It is common for people who have experienced traumatic situations to have very strong emotional reactions. Understanding normal responses to these abnormal events can aid a person in coping with their feelings, thoughts, and behaviors, and help them along the path to recovery.

Shock and denial are typical responses to traumatic events and disasters, especially shortly after the event. Both shock and denial are normal protective reactions. Shock is a sudden and intense disturbance of one's emotional state that may leave them feeling "stunned" or "dazed." Denial involves not acknowledging that something very stressful has happened or not experiencing fully the intensity of the event, at which point an individual may feel numb or disconnected from life. As the initial shock subsides, reactions can vary from one person to another. The following are normal responses to a traumatic event:

- *Feelings become intense and are sometimes unpredictable.* The individual becomes more irritable than usual, and their mood may change back and forth dramatically. They might be especially anxious or nervous or even become depressed.
- *Thoughts and behavior patterns are affected by the trauma.* The individual might have repeated and vivid memories of the event. These flashbacks may occur for no apparent reason and may lead to physical reactions such as rapid heartbeat or sweating. They may find it difficult to concentrate or make decisions or become more easily confused. Sleep and eating patterns also may be disrupted.
- *Recurring emotional reactions are common.* Anniversaries of the event, such as at one month or one year, can trigger upsetting memories of the traumatic experience. These "triggers" may be accompanied by fears that the stressful event will be repeated.
- *Interpersonal relationships often become strained.* Greater conflict, such as more frequent arguments with family members and coworkers, is common. On the other hand, the person might become withdrawn and isolated and avoid the usual activities.
- *Physical symptoms may accompany the extreme stress.* Symptoms may include headaches and nausea, and chest pain may result, requiring medical attention. Preexisting medical conditions may worsen due to stress.

There is no one "standard" pattern of reaction to the extreme stress of traumatic experiences. Some people respond immediately, while others have delayed reactions, sometimes months or even years later. Some have adverse effects for a long period of time, while others recover rather quickly. Reactions can change over time. Some who have suffered from trauma are energized initially by the event to help them with the challenge of coping, only to later become discouraged or depressed. A number of factors tend to affect the length of time required for recovery, including the following:

- *The degree of intensity and loss.* Events that last longer and pose a greater threat, and where loss of life or substantial loss of property is involved, often take longer to resolve.
- *A person's general ability to cope with emotionally challenging situations.* Individuals who have handled other difficult, stressful circumstances well may find it easier to cope with the trauma.

- *Other stressful events preceding the traumatic experience.* Individuals faced with other emotionally challenging situations, such as serious health problems or family-related difficulties, may have more intense reactions to the new stressful event and need more time to recover.

There are a number of steps that can be taken to help restore emotional well-being and a sense of control following a disaster or other traumatic experience, including the following:

- *Time to adjust.* The individual should anticipate that this will be a difficult time in their life, mourn the losses they have experienced, and try to be patient with changes in their emotional state.
- *Support from people who care and who will listen and empathize with their situation.* However, an individual's typical support system may be weakened if those who are close to them have also experienced or witnessed the trauma.
- *Communication of their experience.* Talking with family or close friends or keeping a diary.
- *Local support groups that are often available.* Such as for those who have suffered from natural disasters or other traumatic events. These can be especially helpful for people with limited personal support systems.
- *Groups led by appropriately trained and experienced professionals.* Group discussion can help people realize that other individuals in the same circumstances often have similar reactions and emotions.
- *Engagement in healthy behaviors to enhance their ability to cope with excessive stress.* Well-balanced meals should be consumed and plenty of rest taken. If experiencing ongoing difficulties with sleep, they may be able to find some relief through relaxation techniques. Alcohol and drugs should be avoided.
- *Establishment or reestablishment of routines.* Eating meals at regular times and following an exercise program. Take some time off from the demands of daily life by pursuing hobbies or other enjoyable activities.
- *Avoid major life decisions such as switching careers or jobs if possible.* These activities tend to be highly stressful.

Some people are able to cope effectively with the emotional and physical demands brought about by traumatic events by using their own support systems. A wide range of emotional reactions are common after a disaster or traumatic event, including anxiety, numbness, confusion, guilt, and despair. In and of themselves, these emotions are not the cause for undue alarm. Most will start to fade within a relatively short time. It is not unusual, however, to find that serious problems persist and continue to interfere with daily living. For example, some may feel overwhelming nervousness or sadness that adversely affects their daily functioning including job performance and interpersonal relationships.

Individuals with prolonged reactions that disrupt their daily functioning should consult with a trained and experienced mental health professional. Psychologists and other appropriate mental health providers help educate people about normal responses to extreme stress. These professionals work with individuals affected by trauma to help them find constructive ways of dealing with the emotional impact.

With children, continual and aggressive emotional outbursts, serious problems at school, preoccupation with the traumatic event, continued and extreme withdrawal, and other signs of intense anxiety or emotional difficulties all point to the need for professional assistance. A qualified mental health professional can help such children and their parents understand and deal with thoughts, feelings, and behaviors that result from trauma.

Appendix A
Select Emergency Management Organizations

Organization	Description
International	
The Air Force Emergency Management Association	Affiliated by membership with the IAEM, the Air Force Emergency Management Association provides emergency management information and networking for the U.S. Air Force Emergency Managers.
The European Union (EU)	Since 2001, the EU adopted Community Mechanism for Civil Protection, which started to play a significant role on the global scene. The mechanism's main role is to facilitate cooperation in civil protection assistance interventions in the event of major emergencies that may require urgent response actions. This applies also to situations where there may be an imminent threat of such major emergencies. The heart of the mechanism is the Monitoring and Information Center (MIC). It is part of the Directorate-General for Humanitarian Aid & Civil Protection of the European Commission and accessible 24 hours a day. It gives countries access to a platform, to a one-stop shop of civil protection means available among all the participating states. Any country inside or outside the Union affected by a major disaster can make an appeal for assistance through the MIC. It acts as a communication hub at the headquarters level between participating states, the affected country, and the dispatched field experts. It also provides useful and updated information on the actual status of an ongoing emergency.
The International Association of Emergency Managers (IAEM)	The IAEM is a nonprofit educational organization dedicated to promoting the goals of saving lives and protecting property during emergencies and disasters. The mission of IAEM is to serve its members by providing information, networking, and professional opportunities, and to advance the emergency management profession.
The International Recovery Platform (IRP)	The IRP was conceived at the World Conference on Disaster Reduction (WCDR) in Kobe, Hyogo, Japan, in January 2005. As a thematic platform of the International Strategy for Disaster Reduction (ISDR) system, the IRP is a key pillar for the implementation of the Hyogo Framework for Action (HFA) 2005–2015: Building the Resilience of Nations and Communities to Disasters, a global plan for disaster risk reduction for the decade adopted by 168 governments at the WCDR.

(*Continued*)

Organization	Description
	International
	The key role of IRP is to identify gaps and constraints experienced in postdisaster recovery and to serve as a catalyst for the development of tools, resources, and capacity for resilient recovery. The IRP aims to be an international source of knowledge on good recovery practice.
The Red Cross/Red Crescent	The National Red Cross/Red Crescent societies often have pivotal roles in responding to emergencies. Additionally, the International Federation of Red Cross and Red Crescent Societies (IFRC, or "The Federation") may deploy assessment teams, for example, Field Assessment and Coordination Team (FACT), to the affected country if requested by the national Red Cross or Red Crescent Society. After having assessed the needs, Emergency Response Units (ERUs) may be deployed to the affected country or region. They are specialized in the response component of the emergency management framework.
The United Nations (UN)	Within the UN system, responsibility for emergency response rests with the Resident Coordinator within the affected country. However, in practice, international response is coordinated, if requested by the affected country's government, by the UN Office for the Coordination of Humanitarian Affairs (UN-OCHA), by deploying a UN Disaster Assessment and Coordination (UNDAC) team.
The World Bank	Since 1980, the World Bank has approved more than 500 operations related to disaster management, amounting to more than $40 billion. These include postdisaster reconstruction projects, as well as projects with components aimed at preventing and mitigating disaster impacts, in countries such as Argentina, Bangladesh, Colombia, Haiti, India, Mexico, Turkey, and Vietnam to name only a few. Common areas of focus for prevention and mitigation projects include forest fire prevention measures, such as early warning measures and education campaigns to discourage farmers from slash and burn agriculture that ignites forest fires; early warning systems for hurricanes; flood prevention mechanisms, ranging from shore protection and terracing in rural areas to adaptation of production; and earthquake-prone construction. In a joint venture with Columbia University under the umbrella of the ProVention Consortium, the World Bank has established a Global Risk Analysis of Natural Disaster Hotspots. In June 2006, the World Bank established the Global Facility for Disaster Reduction and Recovery (GFDRR), a longer-term partnership with other aid donors to reduce disaster losses by mainstreaming disaster risk reduction in development, in support of the Hyogo Framework of Action. The facility helps developing countries fund development projects and programs that enhance local capacities for disaster prevention and emergency preparedness.

(Continued)

Country	Description
	International
Australia	Natural disasters are part of life in Australia. Drought occurs on average every 3 out of 10 years and associated heat waves have killed more Australians than any other type of natural disaster in the twentieth century. Australia's emergency management processes embrace the concept of the prepared community. The principal government agency in achieving this is Emergency Management Australia.
Canada	Public Safety Canada (PS) is Canada's national emergency management agency. Each province is required to have legislation in place for dealing with emergencies, as well as establish their own emergency management agencies, typically called an "Emergency Measures Organization" (EMO), which functions as the primary liaison at the municipal and federal level. PS coordinates and supports the efforts of federal organizations ensuring national security and the safety of Canadians. They also work with other levels of government, first responders, community groups, the private sector (operators of critical infrastructure), and other nations. PS's work is based on a wide range of policies and legislation through the Public Safety and Emergency Preparedness Act, which defines the powers, duties, and functions of PS as outlined. Other acts are specific to fields such as corrections, emergency management, law enforcement, and national security.
Germany	In Germany, the federal government controls the German *Katastrophenschutz* (disaster relief) and *Zivilschutz* (civil protection) programs. The local units of the German fire department and the Technisches Hilfswerk (Federal Agency for Technical Relief, THW) are part of these programs. The German Armed Forces (*Bundeswehr*), the German Federal Police, and the 16 state police forces (*Länderpolizei*) have all been deployed for disaster relief operations. Besides the German Red Cross, humanitarian help is dispensed by the Johanniter-Unfallhilfe, the German equivalent of the St. John Ambulance, the Malteser-Hilfsdienst, the Arbeiter-Samariter-Bund, and other private organizations, to cite the largest relief organization that are equipped for large-scale emergencies. As of 2006, there is a joint course at the University of Bonn leading to the degree "Master in Disaster Prevention and Risk Governance."
India	The responsibility for emergency management in India falls to National Disaster Management Authority of India, a government agency subordinate to the Ministry of Home Affairs. In recent years, there has been a shift in emphasis from response and recovery to strategic risk management and reduction and from a government-centered approach to decentralized community participation. The Ministry of Science and Technology, headed by Dr. Karan Rawat, supports an internal agency that facilitates research by bringing the academic knowledge and expertise of earth scientists to emergency management. A group representing a public/private partnership has recently been formed by the Government of India. It is funded primarily by a large India-based computer company and is aimed at improving the general response of communities to emergencies,

(Continued)

Country	Description
	International
	in addition to those incidents that might be described as disasters. Some of the groups' early efforts involve the provision of emergency management training for first responders (a first in India), the creation of a single emergency telephone number, and the establishment of standards for emergency management system staff, equipment, and training. It operates in three states, though efforts are being made in making this a nation-wide effective group.
The Netherlands	In the Netherlands, the Ministry of the Interior and Kingdom Relations is responsible for emergency preparedness and emergency management on the national level and operates a National Crisis Centre (NCC). The country is divided into 25 safety regions (*veiligheidsregio*). Each safety region is covered by three emergency services: police, fire, and ambulance. All regions operate according to the Coordinated Regional Incident Management system. Other services, such as the Ministry of Defence, water board(s), and Rijkswaterstaat, can have an active role in the emergency management process.
New Zealand	In New Zealand, the responsibility for emergency management moves from local to national depending on the nature of the emergency or risk reduction program. A severe storm may be manageable within a particular area, whereas a national public education campaign will be directed by the central government. Within each region, local governments are unified into 16 Civil Defence Emergency Management Groups (CDEMGs). Every CDEMG is responsible for ensuring that local emergency management is as robust as possible. As local arrangements are overwhelmed by an emergency, preexisting mutual-support arrangements are activated. As warranted, the central government has the authority to coordinate the response through the National Crisis Management Centre (NCMC), operated by the Ministry of Civil Defence & Emergency Management (MCDEM). These structures are defined by regulation and best explained in *The Guide to the National Civil Defence Emergency Management Plan 2006*, roughly equivalent to the U.S. Federal Emergency Management Agency's (FEMA) National Response Framework.
Pakistan	Disaster management (in Pakistan) basically revolves around flood disasters with a primary focus on rescue and relief. After each disaster episode, the government incurs considerable expenditure directed at rescue, relief, and rehabilitation. Within disaster management bodies in Pakistan, there is a dearth of knowledge and information about hazard identification, risk assessment and management, and linkages between livelihoods and disaster preparedness. Disaster management policy responses are not generally influenced by methods and tools for cost-effective and sustainable interventions. There are no long-term, inclusive, and coherent institutional arrangements to address disaster issues with a long-term vision. Disasters are viewed in isolation from the processes of mainstream development and poverty alleviation planning.

(Continued)

Country	Description
	International
	For example, disaster management, development planning, and environmental management institutions operate in isolation and integrated planning between these sectors is almost lacking. Absence of a central authority for integrated disaster management and lack of coordination within and between disaster-related organizations is responsible for effective and efficient disaster management in the country. State-level disaster preparedness and mitigation measures are heavily tilted toward structural aspects and undermine nonstructural elements such as the knowledge and capacities of local people, and the related livelihood protection issues.
Russia	In Russia, the Emergency Control Ministry (EMERCOM) is engaged in firefighting, civil defense, search and rescue, including rescue services after natural and human-made disasters.
Somalia	In Somalia, the federal government announced in May 2013 that the cabinet had approved draft legislation on a new Somali Disaster Management Agency (SDMA), which had originally been proposed by the Ministry of Interior and National Security. According to the Prime Minister's Media Office, the SDMA will lead and coordinate the government's response to various natural disasters. It is part of a broader effort by the federal authorities to reestablish national institutions. The Federal Parliament is now expected to deliberate on the proposed bill for endorsement after any amendments.
The United Kingdom	The United Kingdom adjusted its focus on emergency management following the 2000 UK fuel protests, severe flooding in the same year, and the 2001 United Kingdom foot-and-mouth crisis. This resulted in the creation of the Civil Contingencies Act 2004 (CCA), which defined some organizations as Category 1 and 2 Responders. These responders have responsibilities under the legislation regarding emergency preparedness and response. The CCA is managed by the Civil Contingencies Secretariat through Regional Resilience Forums and at the local authority level. Disaster Management training is generally conducted at the local level by the organizations involved in any response. This is consolidated through professional courses that can be undertaken at the Emergency Planning College. Furthermore, diplomas and undergraduate and postgraduate qualifications can be gained throughout the country—the first course of this type was carried out by Coventry University in 1994. The Institute of Emergency Management is a charity, established in 1996, providing consulting services for the government, media, and commercial sectors. The Professional Society for Emergency Planners is the Emergency Planning Society. One of the largest emergency exercises in the United Kingdom was carried out on May 20, 2007, near Belfast, Northern Ireland, and involved the scenario of a plane crash landing at Belfast International Airport. Staff from 5 hospitals and 3 airports participated in the drill, and almost 150 international observers assessed its effectiveness.

(Continued)

Country	Description
	International
The United States	Disaster and catastrophe planning in the United States has utilized the functional all-hazards approach for over 20 years, in which emergency managers develop processes (such as communication and warning or sheltering) rather than developing single-hazard/threat-focused plans such as a tornado plan. Processes then are mapped to the hazards/threats, with the emergency manager looking for gaps, overlaps, and conflicts between processes.
	This has the advantage of creating a plan more resilient to novel events (because all common processes are defined); it encourages planning done by the process owners who are the subject matter experts such as the traffic management plan written by the public works director, rather than the emergency manager, and focuses on processes that are real, can be measured, ranked in importance, and are under our control. This key planning distinction often comes in conflict with nonemergency management regulatory bodies that require development of hazard/threat-specific plans, such as the development of specific H1N1 flu plans and terrorism-specific plans.
	In the United States, all disastrous events are initially considered as local, with the local authority usually a law enforcement agency (LEA) having charge. LEAs typically have situational responsibility as disasters may lead to the normal tenants for lawful instruction being destroyed or in need of extraneous enforcement. Most disasters do not exceed the capacity of the local jurisdiction or the capacity that they have put in place to compensate such as memorandum of understandings with adjacent localities. However, if the event becomes overwhelming or the local government, state emergency management (the primary government structure of the United States) becomes the controlling emergency management agency. Under the Department of Homeland Security (DHS), the FEMA is the lead federal agency for emergency management and supports, but does not override, state authority. The United States and its territories are covered by 1 of 10 regions for FEMA's emergency management purposes.
	If during mitigation it is determined that a disaster or emergency is terror-related or if declared an "Incident of National Significance," the Secretary of Homeland Security will initiate the National Response Framework (NRF). Under this plan the involvement of federal resources will be made possible, integrating it with the local, county, state, or tribal entities. Management will continue to be handled at the lowest possible level utilizing the National Incident Management System (NIMS).

(Continued)

Country	Description

International

The Citizen Corps is an organization of volunteer service programs, administered locally and coordinated nationally by the DHS, which seeks to mitigate disaster and prepare the population for emergency response through public education, training, and outreach. Community Emergency Response Teams are a Citizen Corps program focused on disaster preparedness and on teaching basic disaster response skills. These volunteer teams are utilized to provide emergency support when disaster overwhelms the conventional emergency services.

The U.S. Congress established the Center for Excellence (COE) in Disaster Management and Humanitarian Assistance as the principal agency to promote disaster preparedness and societal resiliency in the Asia Pacific region. As part of its mandate, the COE facilitates education and training in disaster preparedness, consequence management, and health security to develop domestic, foreign, and international capability and capacity.

Most secondary or long-term disaster response is carried out by volunteer organizations. In the United States, the Red Cross is chartered by Congress to coordinate disaster response services, including typically being the lead or largest agency handling sheltering and feeding of evacuees. For large events, religious organizations are able to mount volunteers quickly. The largest partners are the Salvation Army and Southern Baptists. The Salvation Army is usually primary chaplaincy and rebuild services; the Baptists' 82,000+ volunteers do bulk food preparation (90% of the meals in a major disaster) for Salvation Army distribution and homeowner services such as debris and downed limb removal, mold abatement, hot showers and laundry, child care and chaplaincy. Similar services are also provided by Methodist Relief Services, the Lutherans, and Samaritan's Purse.

Unaffiliated volunteers can be counted on to show up at most large disasters. To prevent abuse by criminals and for the safety of the volunteers, procedures have been implemented within most response agencies to manage and effectively use these "Spontaneous Unaffiliated Volunteers" (SUVs).

Appendix B
Top 10 Global Disasters Since 1900

Rank	Death Toll	Event	Location	Date
1	1,000,000–4,000,000	1931 China floods	China	July 1931
2	650,000–779,000	1976 Tangshan earthquake	China	July 1976
3	500,000	1970 Bhola cyclone	East Pakistan (now Bangladesh)	November 1970
4	234,117	1920 Haiyuan earthquake	China	December 1920
5	230,000	2004 Indian Ocean tsunami	Indian Ocean	December 26, 2004
6	229,000	Typhoon Nina—contributed to Banqiao Dam failure	China	August 7, 1975
7	159,000	2010 Haiti earthquake	Haiti	January 12, 2010
8	145,000	1935 Yangtze River flood	China	August 1935
9	142,000	1923 Great Kanto earthquake	Japan	September 1923
10	139,000	1991 Bangladesh cyclone	Bangladesh	April 29, 1991

Appendix C
Select Global Special Operations Teams

Unit	Country/Role Notable Operations
	Argentina
Army Commando Coys	Trained to conduct mountain operations and long-range reconnaissance patrols (LRRPs). Deployed during the Falklands conflict. A forward patrol of army commandos were defeated by the British Royal Marines Mountain and Arctic Warfare Cadre at Top Malo House on May 31, 1982.
Buzo Tactico	Naval Special Forces. Saw action against the British Royal Marines during the invasion of the Falkland Islands (April 1982)
Halcon (Falcon 8)	Permanent cadre of Army Special Forces (40–45 men). Counterterrorism and counterinsurgency operations
	Australia
Army Commando Coys	Party of Citizens Military Force (Reserve). Roles include reconnaissance, raids, special assault tasks, and training indigenous forces. Some NCOs and officers were attached to the American Civilian Irregular Defense Program during the Vietnam War.
Regional Surveillance Units	NORFORCE, the Pilbara Regiment, and the 51 Far North Queensland Regiment are Army Reserve units trained to conduct surveillance and reconnaissance operations and fight in a stay behind capacity.
Special Air Service	Strategic and operational intelligence gathering, harassing the enemy in depth with raids and ambushes, combat–rescue operations, and counterterrorism. The regiment has seen combat in Borneo and Vietnam.
	Belgium
Belgian Federal Police Counter-Terrorism Unit (CGSU)	National Counter Terrorism Unit of the Belgian Federal Police. Deployed in cases of terrorism, kidnappings, hostage taking, and other forms of serious crime. In major terrorist operations outside the country, the Special Units would be replaced by the Belgian Army Special Forces Group.
Special Forces Group	In April 2000, the Special Forces Company became a part of the 3 Régiment de Lanciers Parachutists based in Flawinne. When the regiment was dissolved in February 2003, the unit was renamed the Special Forces Group. Acts as a rapid reinforcement for NATO while maintaining specialist reconnaissance, maritime, and mountain warfare units. The unit also provides support for counterterrorist operations.

(Continued)

Unit	Country/Role Notable Operations

Canada

| Canadian Special Operations Forces Command (CANSOFCOM) | Composed of Joint Task Force 2 (JTF-2), Canadian Special Operations Regiment (CSOR), 427 Special Operations Aviation Squadron (SOAS), and Canadian Joint Incident Response Unit (CJIRU). Provides the Canadian Forces (CF) with a capacity to prevent and react to terrorism in all environments, to provide the CF with a capability to perform other missions as directed by the Government of Canada, such as direct action (DA), special reconnaissance (SR), defense diplomacy and military assistance (DDMA), as well as special humanitarian assistance (such as the evacuation of noncombatants) |
| Emergency Response Teams | Royal Canadian Mounted Police national intervention teams |

France

| Groupement d'Intervention de la Gendarmerie Nationale (GIGN) | GIGN, the French national intervention team, is a special operations unit of the French Armed Forces. It is part of the National Gendarmerie and is trained to perform counterterrorist and hostage rescue missions in France or anywhere else in the world. Notable operations include the release of 30 school children from a school bus in Djibouti in 1976 and the rescue of 229 passengers and crew from Air France Flight 8969 in Marseille. The plane, hijacked by four Armed Islamic Group (GIA) terrorists that wished to destroy the Eiffel Tower, had been completely mined, and three passengers had been executed during the negotiations with the Algerian government (1994). The GIGN was selected by the International Civil Aviation Organisation (ICAO) to teach the special forces of the other member states in hostage-rescue exercises aboard planes. |

Germany

| Grenzschutzgruppe 9 (GSG-9) | GSG-9 is the German national intervention team; members are drawn from the Federal German Border Police. GSG-9 is deployed in cases of hostage taking, kidnapping, terrorism, and extortion. The group may also be used to secure locations, neutralize targets, track down fugitives, and sometimes conduct sniper operations. Furthermore, the group is very active in developing and testing methods and tactics for these missions. Finally, the group may provide advice to the different *Länder*, ministries, and international allies. The group assists the *Bundespolizei* and other federal and local agencies on request. At the time of the 1977 Mogadishu mission, the Commander of the Israeli Border Police Tzvi War described GSG-9 as "The best anti-terrorist group in the world." From 1972 to 2003 they reportedly completed over 1500 missions, discharging their weapons on only five occasions. |

Israel

Mistaaravim	IDF and Border Guard undercover units for foiling terrorism
Shin Bet	Counterterrorism, intelligence, and security agency
Yamam	Elite Israeli police antiterrorist unit used for hostage rescue, counterterrorism, and terrorism foiling operations

(Continued)

Unit	Country/Role Notable Operations

The United Kingdom

United Kingdom Special Forces (UKSF)	UKSF is a Ministry of Defence directorate that provides a joint special operations task force headquarters. UKSF is commanded by Director Special Forces. The Special Air Service, Special Boat Service, Special Reconnaissance Regiment, Special Forces Support Group, 18 Signal Regiment, and the Joint Special Forces Aviation Wing together form the UKSF. UKSF assets undertake a number of roles, with a degree of interaction and interoperability: counterterrorism, direct action, covert and special reconnaissance, unconventional warfare, combat search and rescue, close protection, counterrevolutionary warfare, human intelligence collection, and training of other nation's armed forces.

The United States

Joint Special Operations Command	All units perform highly classified activities and operations, fall under special mission units (SMUs).
	• First Special Forces Operational Detachment-Delta (counterterrorism)
	• U.S. Army Flight Concepts Division (FCD)
	• Intelligence Support Activity (ISA)
	• U.S. Army Skills Evaluation Detachment (USASED)
	• U.S. Naval Special Warfare Development Group (counterterrorism)
	• 66th Air Operations Squadron
	• 427th Special Operations Squadron
	• 724th Special Tactics Group
	• Aviation Tactics Evaluation Group (AVTEG)
	• Joint Communications Unit (JCU)
	• Joint Communications Integration Element (JCIE)
	• Joint Intelligence Brigade (JIB)
	• Joint Medical Augmentation Unit (JMAU)
U.S. Army Special Operations Command	• U.S. Army Special Forces (unconventional warfare)
	• Army Compartmented Element (ACE) U.S. Army military intelligence unit supporting Army special operations forces
	• 75th Ranger Regiment
	• U.S. Army Special Operations Aviation Command
	• Military Information Support Operations Command
Naval Special Warfare Command	• Naval Special Warfare Group 1: SEAL Teams 1, 3, 5, 7
	• Naval Special Warfare Group 2: SEAL Teams 2, 4, 8, 10
	• Naval Special Warfare Group 3: SEAL Delivery Vehicle Team 1
	• Naval Special Warfare Group 4: Special Boat Teams 12, 20, 22
	• Naval Special Warfare Group 10
	• Naval Special Warfare Group 11: SEAL Teams 17, 18 (Reserves)
	• Naval Special Warfare Center
	• U.S. Naval Special Warfare Development Group (DEVGRU/SEAL Team 6)

(Continued)

Unit	Country/Role Notable Operations
Air Force Special Operations Command	• 23rd Air Force • 1st Special Operations Wing • 24th Special Operations Wing • 27th Special Operations Wing • 193rd Special Operations Wing (ANG) • 919th Special Operations Wing (AFR) • 352nd Special Operations Group (Europe) • 353rd Special Operations Group (Pacific) • USAF Special Operations Training Center
CIA Special Activities Division	• Special Operations Group
FBI Hostage Rescue Team	• National intervention team for domestic operations
U.S. Naval Special Warfare Development Group Coast Guard	• Coast Guard Deployable Operations Group
U.S. Naval Special Warfare Development Group Marine Corps Forces Special Operations Command	• Marine Special Operations Regiment • Marine Special Operations Support Group • Marine Special Operations School

Appendix D
Select Global Terrorist Organizations

Organization	Description
Abu Sayyaf Group	One of several militant Islamist separatist groups based in and around the southern Philippines, in Bangsamoro (Jolo and Basilan), where for almost 30 years Muslim groups have been engaged in an insurgency for an independent province in the country. Since its inception in the early 1990s, the group has carried out bombings, kidnappings, assassinations, and extortion in what they describe as their fight for an independent Islamic province in the Philippines. Abu Sayyaf seeks the establishment of an Islamic theocracy in the southern Philippines.
Al-Gama'a al-Islamiyya	An Egyptian Sunni Islamist movement, it is considered a terrorist organization by the United States and the European Union. The group was dedicated to the overthrow of the Egyptian government and replacing it with an Islamic state. In 2003, the imprisoned leadership of the group renounced bloodshed; a series of high-ranking members have since been released by Egyptian authorities, and the group has been allowed to resume semilegal peaceful activities. Following the Egyptian Revolution of 2011, the movement formed a political party, the Building and Development Party, which gained 13 seats in the 2011–2012 elections to the lower house of the Egyptian Parliament.
Al-Jihad (Egyptian Islamist Group)	An Egyptian Islamist group active since the late 1970s. It is under worldwide embargo by the United Nations as an affiliate of al-Qaeda. It is also banned by several individual governments including that of the Russian Federation. The organization's original primary goal was to overthrow the Egyptian government and replace it with an Islamic state. Later it broadened its aims to include attacking United States and Israeli interests in Egypt and abroad.
Al-Qaeda	A global militant Islamist organization founded by Osama bin Laden in Peshawar, Pakistan, at some point between August 1988 and late 1989, with its origins being traceable to the Soviet War in Afghanistan. It operates as a network comprising both a multinational, stateless army and a radical Sunni Muslim movement calling for global Jihad and a strict interpretation of sharia law.

(Continued)

Organization	Description
Armed Islamic Group (GIA)	An Islamist organization that wants to overthrow the Algerian government and replace it with an Islamic state. The GIA adopted violent tactics in 1992 after the military government voided the victory of the Islamic Salvation Front, the largest Islamic opposition party, in the first round of legislative elections held in December 1991. During their 1994 hijack of Air France Flight 8969, the GIA announced, "We are the Soldiers of Mercy."
Basque Fatherland and Liberty—Euskadi Ta Askatasuna (ETA)	An armed Basque nationalist and separatist organization. The group was founded in 1959 and has since evolved from a group promoting traditional Basque culture to a paramilitary group with the goal of gaining independence for the Greater Basque Country. ETA is the main organization of the Basque National Liberation Movement and is the most important participant in the Basque conflict. ETA declared ceasefires in 1989, 1996, 1998, and 2006, and subsequently broke them. On September 5, 2010, ETA declared a new ceasefire that is still in force, and on October 20, 2011, ETA announced a "definitive cessation of its armed activity." On November 24, 2012, it was reported that the group was ready to negotiate a "definitive end" to its operations and disband completely.
Hamas (Islamic Resistance Movement)	The Palestinian Sunni Islamic or Islamist organization, with an associated military wing, the Izz ad-Din al-Qassam Brigades, located in the Palestinian territories. Since June 2007, Hamas has governed the Gaza Strip, after it won a majority of seats in the Palestinian Parliament in the January 2006 Palestinian parliamentary elections and then defeated the Fatah political organization in a series of violent clashes. Israel, the United States, Canada, the European Union, and Japan classify Hamas as a terrorist organization, while Iran, Russia, Turkey, and the Arab nations do not. Based on the principles of Islamic fundamentalism gaining momentum throughout the Arab world in the 1980s, Hamas was founded in 1987 (during the First Intifada) as an offshoot of the Egyptian Muslim Brotherhood. Cofounder Sheik Ahmed Yassin stated in 1987 and the Hamas Charter affirmed in 1988 that Hamas was founded to liberate Palestine from Israeli occupation and to establish an Islamic state in the area that is now Israel, the West Bank, and the Gaza Strip.
Hezbollah (Party of God)	A Shi'a Islamic militant group and political party based in Lebanon. Its paramilitary wing is regarded as a resistance movement throughout much of the Arab and Muslim worlds, and is considered more powerful than the Lebanese Army. It has taken the side of the government in the Syrian civil war and in May–June 2013 it successfully assisted in the recapture of the strategic town of Al-Qusayr. Hezbollah was conceived by Muslim clerics and funded by Iran following the Israeli invasion of Lebanon, and was primarily formed to offer resistance to the Israeli occupation. Hezbollah, which started with only a small militia, has grown to an organization with seats in the Lebanese government, a radio and a satellite television station, and programs for social development. Hezbollah receives military training, weapons, and financial support

(Continued)

Organization	Description
	from Iran, and political support from Syria. Despite a June 2008 certification by the United Nations that Israel had withdrawn from all Lebanese territory, in August, Lebanon's new cabinet unanimously approved a draft policy statement that secures Hezbollah's existence as an armed organization and guarantees its right to "liberate or recover occupied lands."
Liberation Tigers of Tamil Eelam (LTTE)	This organization was a separatist militant organization that was based in northern Sri Lanka. Founded in May 1976 by Velupillai Prabhakaran, it waged a violent secessionist and nationalist campaign to create an independent state in the north and east of Sri Lanka for Tamil people. This campaign evolved into the Sri Lankan Civil War, which ran from 1983 until 2009, when the LTTE was defeated by the Sri Lankan Military.
PIRA (Provisional Irish Republican Party)	An Irish republican paramilitary organization whose aim is to remove Northern Ireland from the United Kingdom and bring about a socialist republic within a united Ireland by force of arms and political persuasion. It emerged out of the December 1969 split of the Irish Republican Army over differences of ideology and how to respond to violence against the nationalist community. This violence had followed the community's demands for civil rights in 1968 and 1969, which met with resistance from some of the unionist community and from the authorities, and culminated in the 1969 Northern Ireland riots.
Revolutionary Armed Forces of Columbia (FRAC)	A Colombian Marxist–Leninist revolutionary guerrilla organization involved in the continuing Colombian armed conflict since 1964. The FARC is considered a terrorist organization by the Government of Colombia. The FARC–EP claim to be a peasant army with a political platform of agrarianism and anti-imperialism inspired by Bolivarianism. The FARC say they represent the poor people of rural Colombia.
Revolutionary People's Liberation Front	A Marxist–Leninist party in Turkey. It was founded in 1978 as Revolutionary Left and was renamed in 1994 after factional infighting. Having carried out a number of assassinations and suicide bombings, it is considered a terrorist group by Turkey, the United States, and the European Union. This group opposes Turkey's pro-Western stance and its membership in NATO.
Sendero Luminoso (Shining Path)	A terrorist organization in Peru. When it first launched the internal conflict in Peru in 1980, its stated goal was to replace what it saw as bourgeois democracy with "New Democracy." The Shining Path believed that by imposing a dictatorship of the proletariat, inducing cultural revolution, and eventually sparking world revolution, they could arrive at pure communism. Their representatives said that existing socialist countries were revisionist, and claimed to be the vanguard of the world communist movement. The Shining Path's ideology and tactics have been influential on other Maoist insurgent groups, notably the Communist Party of Nepal (Maoist) and other Revolutionary Internationalist Movement-affiliated organizations.

Bibliography

Abrahms, M. (2008). "What terrorists really want: Terrorist motives and counterterrorism strategy?" *International Security* 32(4): 86–89.

Adams, C. (2007). "The straight dope." *The Straight Dope—Fighting Ignorance Since 1973*, October 12.

Adger, W. N., Brooks, N., Bentham, G., Agnew, M., and Eriksen, S. (2004). "New indicators of vulnerability and adaptive capacity." Tyndall Centre Technical Report 7, Tyndall Centre for Climate Change Research, Norwich.

Agnew, R. (1992). "Foundation for a general strain theory of crime and delinquency." *Criminology* 30: 47–88.

Akers, R. L. (1979). "Social learning and deviant behavior: A specific test of a general theory." *American Sociological Review* 44(4): 636–655.

Akers, R. L. (1991). "Self-control as a general theory of crime." *Journal of Quantitative Criminology* 7(2): 201–211.

Akers, R. L. (1996). "Is differential association/social learning cultural deviance theory?" *Criminology* 34(2): 229–247.

Albanese, J. S. (2000). "The causes of organized crime: Do criminals organize around opportunities for crime or do criminal opportunities create new offenders?" *Journal of Contemporary Criminal Justice* 16(4): 409–423.

Albanese, J. S. (2008). "Risk assessment in organized crime developing a market and product-based model to determine threat levels." *Journal of Contemporary Criminal Justice* 24(3): 263–273.

Albini, J. L. (1971). *The American Mafia: Genesis of a Legend.* New York: Appleton-Century-Crofts.

Albini, J. L. (1988). "Donald Cressey's contributions to the study of organized crime: An evaluation." *Crime & Delinquency* 34(3): 338–354.

Albini, J. L. (1995). "Russian organized crime: Its history, structure and function." *Journal of Contemporary Criminal Justice* 11(4): 213–243.

Alexeev, M., Janeba, E., and Osborne, S. (2004). "Taxation and evasion in the presence of extortion by organized crime." *Journal of Comparative Economics* 32(3): 375–387.

Alic, J. (2012). "Is oil smuggling and organized crime the cause of Greece's economic crisis?" April 10. www.oilprice.com.

Allmer, T. (2012). *Towards a Critical Theory of Surveillance in Informational Capitalism.* Frankfurt am Main, Germany: Peter Lang.

Alsott, J. D. (1991). "The search for honor: An inquiry into the factors that influence the ethics of federal acquisition." In *Institutionalizing Organizational Ethics Programs: Contemporary Perspectives*, Petrick, J. A., Claunch, W. M., and Scherer, R. F. (eds.). Dayton, OH: Wright State University, pp. 182–194.

American Management Association and the ePolicy Institute. (2008). Electronic Monitoring and Surveillance 2007 Survey.

American National Standards Institute. The Volunteer Protection Act. www.ansi.org/public/news/1998jan/vpa_9.html.

American Water Works Association. (2001). *M-19 Emergency Planning for Water Utilities.* Denver, CO: American Water Works Association.

Andersen, D. (1999). "The aggregate burden of crime." *The Journal of Law & Economics* 42(2): 611–642.

Aniskiewicz, R. (1994). "Metatheoretical issues in the study of organized crime." *Journal of Contemporary Criminal Justice* 10(4): 314–324.

Apps, M. J., Price, D. T., and Wisniewski, J. (1995). *Boreal Forests and Global Change.* Dordrecht, the Netherlands: Kluwer Academic Publishers.

Apps, P. (2013). "Have hired guns finally scuppered Somali pirates?" *Reuters*, February 10. www.reuters.com.

Arnold, K. R. (2011). *Anti-Immigration in the United States: A Historical Encyclopedia.* Santa Barbara, CA: Greenwood Press.

Associated Press. (2008). "Pirates seize French yacht." *CNN, Associated Press.* April 4.

Associated Press. (2009). "Pirates seize Belgian ship; NATO frees 20 hostages." *Associated Press*, April 18.

ATAB (American Anti-Terrorism Accreditation Board). (2008). "Aviation security." In *Emergency Response Manual: CAS PowerPoint Training.* American Anti-Terrorism Accreditation Board. 8th edition. www.goatab.org.

ATAB. (2008). "Avoiding an attack." In *Emergency Response Manual: CAS PowerPoint Training.* American Anti-Terrorism Accreditation Board. 8th edition. www.goatab.org.

ATAB. (2008). "Crisis negotiations." In *Emergency Response Manual: CAS PowerPoint Training.* American Anti-Terrorism Accreditation Board. 8th edition. www.goatab.org.

ATAB. (2008). "Cyber terrorism." In *Emergency Response Manual: CAS PowerPoint Training.* American Anti-Terrorism Accreditation Board. 8th edition. www.goatab.org.

ATAB. (2008). "Force protection." In *Emergency Response Manual: CAS PowerPoint Training.* American Anti-Terrorism Accreditation Board. 8th edition. www.goatab.org.

ATAB. (2008). "Hostage survival." In *Emergency Response Manual: CAS PowerPoint Training.* American Anti-Terrorism Accreditation Board. 8th edition. www.goatab.org.

ATAB. (2008). "Incidents and indicators." In *Emergency Response Manual: CAS PowerPoint Training.* American Anti-Terrorism Accreditation Board. 8th edition. www.goatab.org.

ATAB. (2008). "Managing a TSCM." In *Emergency Response Manual: CAS PowerPoint Training.* American Anti-Terrorism Accreditation Board. 8th edition. www.goatab.org.

ATAB. (2008). "Managing an executive protection detail." In *Emergency Response Manual: CAS PowerPoint Training.* American Anti-Terrorism Accreditation Board. 8th edition. www.goatab.org.

ATAB. (2008). "Notification and coordination." In *Emergency Response Manual: CAS PowerPoint Training.* American Anti-Terrorism Accreditation Board. 8th edition. www.goatab.org.

ATAB. (2008). "Principles of executive protection." In *Emergency Response Manual: CAS PowerPoint Training.* American Anti-Terrorism Accreditation Board. 8th edition. www .goatab.org.

ATAB. (2008). "Responding to CBRNE Events." In *Emergency Response Manual: CAS PowerPoint Training.* American Anti-Terrorism Accreditation Board. 8th edition. www .goatab.org.

ATAB. (2008). "Responding to Terrorism." In *Emergency Response Manual: CAS PowerPoint Training.* American Anti-Terrorism Accreditation Board. 8th edition. www.goatab.org.

ATAB. (2008). "Scene control." In *Emergency Response Manual: CAS PowerPoint Training.* American Anti-Terrorism Accreditation Board. 8th edition. www.goatab.org.

ATAB. (2008). "Security energy facilities." In *Emergency Response Manual: CAS PowerPoint Training.* American Anti-Terrorism Accreditation Board. 8th edition. www.goatab.org.

ATAB. (2008). "Surveillance detection." In *Emergency Response Manual: CAS PowerPoint Training.* American Anti-Terrorism Accreditation Board. 8th edition. www.goatab.org.

ATAB. (2008). "Suspicious activity." In *Emergency Response Manual: CAS PowerPoint Training.* American Anti-Terrorism Accreditation Board. 8th edition. www.goatab.org.

ATAB. (2008). "Technical surveillance and counter measures." In *Emergency Response Manual: CAS PowerPoint Training.* American Anti-Terrorism Accreditation Board. 8th edition. www.goatab.org.

ATAB. (2008). "Travel security." In *Emergency Response Manual: CAS PowerPoint Training*. American Anti-Terrorism Accreditation Board. 8th edition. www.goatab.org.

ATAB. (2008). "Violence in the workplace." In *Emergency Response Manual: CAS PowerPoint Training*. American Anti-Terrorism Accreditation Board. 8th edition. www.goatab.org.

Atwood, D. J. (1990). "Living up to the public trust." *Defense Issues* 5: 1.

Aysan, Y. (1999). "Putting floors under the vulnerable: Disaster reduction as a strategy to reduce poverty." Presented to the World Bank *Consultation Group for Global Disaster Reduction Meeting*. June 1–2, Paris, France.

Bacastow, T. S. (2010). The Learner's Guide to Geospatial Analysis. Dutton eEducation Institute, Penn State University, PA.

Bailey, R. (2009). "Earth liberation front terrorist gets 22 years in prison for anti-biotech arson." *Reason Magazine*, February 6.

Bales, K. (1999). *Disposable People: New Slavery in the Global Economy*. Berkeley, CA: University of California Press.

Ballesteros, L. F. (2010). "Who's getting the worst of natural disasters?" *54 Pesos*, May. 54pesos.org/2008/10/04/who%e2%80%99s-getting-the-worst-of-natural-disasters.

Bankoff, G. (2002). *Cultures of Disaster: Society and Natural Hazards in the Philippines*. London: Routledge.

Baumer, E. P. and Gustafson, R. (2007). "Social organization and instrumental crime: Assessing the empirical validity of classic and contemporary anomie theories." *Criminology* 45(3): 617–663.

BBC News. (2000). "China executes pirates." news.bbc.co.uk/2/hi/asia-pacific/622435.stm, accessed January 10, 2012.

BBC News. (2005). "1983: Car bomb in South Africa kills 16." news.bbc.co.uk/onthisday/hi/dates/stories/may/20/newsid_4326000/4326975.stm, accessed January 10, 2012.

BBC News. (2008). "Cruise ship evades pirate attack." news.bbc.co.uk/2/hi/africa/7760216.stm, accessed January 10, 2012.

BBC News. (2008). "France raid ship after crew freed." news.bbc.co.uk/2/hi/africa/7342292.stm, accessed January 10, 2012.

BBC News. (2008). "Hackers warn high street chains." news.bbc.co.uk/2/hi/7366995.stm, accessed January 11, 2012.

BBC News. (2009). "Pirates attack second US vessel." www.news.bbc.co.uk/2/hi/africa/7999350.stm, accessed January 11, 2012.

BBC News. (2009). "US captain held by pirates freed." news.bbc.co.uk/2/hi/africa/7996087.stm, accessed January 11, 2012.

BBC News. (2010). "Pirates attack Russian oil tanker off Somalia coast." news.bbc.co.uk/2/hi/8661816.stm, accessed January 12, 2012.

BBC News. (2011). "Pakistan: A failed state or a clever gambler?" www.bbc.co.uk/news/world-south-asia-13318673, accessed January 12, 2012.

Belasco, J. A. and Stayer, R. C. (1994). *Flight of the Buffalo: Soaring to Excellence, Learning to Let Employees Lead*. New York: Warner Books.

Bento, L. (2011). "Toward an international law of piracy sui generis: How the dual nature of maritime piracy law enables piracy to flourish." *Berkeley Journal of International Law* 29(2): 399.

Black, H. C., Nolan, J. R., and Connolly, M. J. (1979). *Black's Law Dictionary with Pronunciations*. 5th edition. St. Paul, MN: West Publishing

Blaikie, P., Cannon, T., Davis, I., and Wisner, B. (1994). *At Risk: Natural Hazards, People's Vulnerability, and Disasters*. London: Routledge.

Blanchard, K. H. and Bowles, S. M. (1997). *Gung Ho! Turn on the People in any Organization*. New York: William Morrow & Co.

Blanchard, K. H., Carlos, J. C., and Randolph, A. (1999). *The 3 Keys to Empowerment: Release the Power Within People for Astonishing Results*. San Fransico, CA: Berrett-Koehler Publishers.

Bockstette, C. (2008). "Jihadist terrorist use of strategic communication management techniques." *George C. Marshall Center Occasional Paper Series*, 20.

Bonanos, C. "Did pirates really say 'arrrr'?" *Slate Magazine*. www.slate.com/articles/news_ and politics/explainer/2007/06/did_pirates_really_say_arrrr.html.

Bonner, R. (1998). "Getting attention: A scholar's historical and political survey of terrorism finds that it works." *The New York Times*, November 1.

Borum, R., Fein, R., Vossekuil, B., and Berglund, J. (1999). "Threat assessment: Defining an approach for evaluating risk of targeted violence." *Behavioral Sciences & the Law* 17(3): 323–337.

Bousquet, E. (2001). "Ocean warriors confront lucian fishermen." Government of Saint Lucia Web site, July 23. www.archive.stlucia.gov.lc/pr2001/ocean_warriors_confront_lucian_ fishermen.htm.

Bovenkerk, F., Siegel, D., and Zaitch, D. (2003). "Organized crime and ethnic reputation manipulation." *Crime, Law and Social Change* 39(1): 23–38.

Braithwaite, J. (1989). "Criminological theory and organizational crime." *Justice Quarterly* 6(3): 333–358.

British Standards Institution. (2006). *Business Continuity Management. Part 1: Code of Practice*. London: British Standards Institution.

British Standards Institution. (2012). *Societal Security—Business Continuity Management Systems: Requirements*. London: British Standards Institution.

Buergenthal, T. and Murphy, S. D. (2002). *Public International Law in a Nutshell*. 3rd edition. St. Paul, MN: West Group.

Burgess, E. W. (1928). "Residential segregation in American cities." *Annals of the American Academy of Political and Social Science* 140: 105–115.

Burrows, W. D. and Renner, S. E. (1999). "Biological warfare agents as threats to potable water." *Environmental Health Perspectives* 107(12): 975–984.

Burton, I., White, G. F., and Kates, R. W. (1978). *The Environment as Hazard*. New York: Oxford University Press.

The Business Times Singapore. (2006). "Piracy is still troubling the shipping industry: Report, industry fears revival of attacks though current situation has improved." August 14.

Byrnes, R. (2009). "Mexican drug traffickers now 'greatest organized crime threat' to U.S." *CNS News*. January 21.

Cabinet Office. (2004). "Overview of the act." In *Civil Contingencies Act 2004: Consultation on the Draft Regulations and Guidance*. London: Civil Contingencies Secretariat.

Campbell, K. (2001). "When is 'terrorist' a subjective term?" *Christian Science Monitor*, September 27.

Cannon, T., Twigg, J., and Rowell, J. (2005). "Social vulnerability, sustainable livelihoods, and disasters." Report to DFID Conflict and Humanitarian Assistance Department (CHAD) and Sustainable Livelihoods Support Office. London: Department for International Development.

Carolina, N. (2012). "Nature vs disaster." Canadian geographic. January. www.canadiangeographic .ca/magazine/jf12/lessen_impact_of_severe_storms.asp.

Carstairs, V. D. L. and Morris, R. (1991). *Deprivation and Health in Scotland*. Aberdeen: Aberdeen University Press.

Carter, D. L. (1994). "International organized crime: emerging trends in entrepreneurial crime." *Journal of Contemporary Criminal Justice* 10(4): 239–266.

Casciani, D. (2011). "Record £1bn worth of criminal assets seized by police." *BBC News*, July 28.

Chaliand, G., Blin, A., Schneider, E. D., Pulver, K., and Browner, J. (2007). *The History of Terrorism: From Antiquity to Al Qaeda*. Berkeley, CA: University of California Press.

Choo, K.-K. R. (2008). "Organized crime groups in cyberspace: A typology." *Trends in Organized Crime* 11(3): 270–295.

The CIA World Fact Book. (2013). www.cia.gov/library/publications/the-world-factbook/.

Cialdini, R. B. (2000). *Influence: Science and Practice*. New York: Allyn & Bacon.

Clark, R. M. and Deininger, R. A. (2000). "Protecting the nation's critical infrastructure: The vulnerability of US water supply systems." *Journal of Contingencies and Crisis Management* 8(2): 73–80.

Clark, R. M., Geldreich, E. E., Fox, K. R., Rice, E. W., Johnson, C. H., Goodrich, J. A., Barnick, J. A., and Abdesaken, F. (1996). "Tracking a *Salmonella* serovar *Typhimurium* outbreak in Gideon, Missouri: Role of contaminant propogation modelling." *Journal of Water Supply: Research and Technology* 45(4): 171–183.

Connors, S. (2002). "How terrorism prevented smallpox from being wiped off the face of the earth?" *The Independent*, January 1.

Cottino, A. (1999). "Sicilian cultures of violence: The interconnections between organized crime and local society." *Crime, Law and Social Change* 32(2): 103–113.

Couttie, B. (2009). "Maersk Alabama 'followed best practice'." *Maritime Accident Casebook*, November 20.

Craun, G. F., Swerdlow, D., Tauxe, R., Clark, R., Fox, K., Geldreich, E., Reasoner, D., and Rice, E. (1991). "Prevention of waterborne Cholera in the United States." *Journal of the American Water Works Association* 83(11): 43.

Crawford, S. J. III. (1990). "Wind and well-learned lessons." *Defense* 90: 15.

Cressey, D. R. and Finckenauer, J. O. (2008). *Theft of the Nation: The Structure and Operations of Organised Crime in America*. Piscataway, NJ: Transaction Publishers.

Cressey, D. R. and Ward, D. A. (1969). *Delinquency, Crime, and Social Process*. New York: Harper & Row.

Cutter, S. L., Boruff, B. J., and Shirley, W. L. (2003). "Social vulnerability to environmental hazards." *Social Science Quarterly* 84(1): 242–261.

Decker, S. H., Bynum, T., and Weisel, D. (1998). "A tale of two cities: Gangs as organized crime groups." *Justice Quarterly* 15(3): 395–425.

Decker, S. H. and Van Winkle, B. (1996). *Life in the Gang: Family, Friends and Violence*. Cambridge: Cambridge University Press.

Deen, T. (2005). "POLITICS: UN member states struggle to define terrorism." *Inter Press Service*, July 25.

Deininger, R. A. (2000). "Constituents of concern: The threat of chemical and biological agents to public water supply systems." In *Pipeline Net User's Guide (Appendix F)*, McLean, VA: SAIC.

Deininger, R. A. and Meier, P. G. (2000). "Sabotage of public water supply systems." In *Security of Public Water Supplies*, Deininger, R. A., Literathy, P., and Bartram, J. (eds.). Dordrecht, the Netherlands: Kluwer Academic Publishers.

DeLuca, J. R. (1999). *Political Savvy: Systematic Approaches to Leadership Behind the Scenes*. Berwyn, IL: Evergreen Business Group.

Denileon, G. P. (2001). "The who, what, why, and how of counterterrorism issues." *Journal of the American Water Works Association* 93(5): 78.

Desai, R. and Eckstein, H. (1990). "Insurgency: The transformation of peasant rebellion." *World Politics* 42(4): 441–465.

Diaz-Paniagua, C. F. (2008). "Negotiating terrorism: The negotiation dynamics of four UN counter-terrorism treaties, 1997–2005." PhD Thesis, City University of New York, New York. p. 47.

Doole, C. (2006). "Europe's new fears." *The Bridge Magazine*, Winter.

Dwyer, A., Zoppou, C., Nielson, O., Day, S., and Roberts, S. (2004). "Quantifying social vulnerability: A methodology for identifying those at risk to natural hazards." *Geoscience Australia Record* 14.

Elliot, D., Swartz, E., and Herbane, B. (1999). "Just waiting for the next big bang: Business continuity planning in the UK finance sector." *Journal of Applied Management Studies* 8(1): 43–60.

Emergency Measures Ontario. (2012). "Emergency management and civil protection act." www.e-laws.gov.on.ca/html/statutes/english/elaws_statutes_90e09_e.htm.

England, A., Wright, R., and Sevastopulo, D. (2008). "Pirates seize another ship in Gulf of Aden." *FT*, November 17. www.ft.com.

Environment Canada. (2001). "The science of climate change." April 24. www.msc.ec.gc.ca/saib/climate/Climatechange/CC_presentation_e.PDF.

Environment Canada. (2002). "CO2/Climate Report." Environment Canada technical report.

Etkin, D. A. (1995). "Beyond the year 2000, more tornadoes in western Canada? Implications from the historical record." *Natural Hazards* 12(1): 19–27.

Etymonline.com. (1979). "Online etymology dictionary." Etymonline.com. October 20.

Executive Order 12731. (1990). "Principles of ethical conduct for government officers and employees." *Federal Register* 55: 203.

Federation of American Scientists. (2000). *Remarks by the President in photo opportunity with leaders of high-tech industry and experts on computer security*, February 15. www.fas.org/irp/news/2000/02/000215-secure-wh1.htm.

Feige, E. L. and Cebula, R. (2011). "America's underground economy: Measuring the size, growth, and determinants of income tax evasion in the US." *Munich Personal RePEc Archive*. January.

Fein, R. A. and Vossekuil, B. (1999). "Assassination in the United States: An operational study of recent assassins, attackers, and near-lethal approaches." *Journal of Forensic Sciences* 44(2): 321–333.

Fein, R. A., Vossekuil, B., and Holden, G. A. (1995). "Threat assessment: An approach to prevent targeted violence." *NCJ 155000. Research in Action*, September, US Department of Justice, Washington, DC.

Fein, R. A., Vossekuil, B., Pollack, W. S., Borum, R., Reddy, M., and Modzeleski, W. (2002). "Threat assessment in schools: A guide to managing threatening situations and creating safe school climates." *NCJ 195290*, May, US Department of Education and US Secret Service, Washington, DC.

Feldman, J. (2011). *Manufacturing Hysteria: A History of Scapegoating, Surveillance, and Secrecy in Modern America*. New York: Pantheon Books.

FEMA. National Hazard Loss Estimating Methodology. www.fema.gov/hazus.

FEMA. Understanding your risks: Identifying hazards and estimating losses (FEMA 386-2). Step-by-step guidance on how to accomplish a risk assessment. www.fema.gov/mit/planning_toc3.htm.

FEMA. (2012). Flood hazard mapping hazard analysis tools (list of commercially available tools). www.fema.gov/national-flood-insurance-program-flood-hazard-mapping.

Ferris-Rotman, A. (2010). "Russian warship frees hijacked tanker, no one hurt." *Reuters*, May 6.

Fijnaut, C. and Paoli, L. (2004). *Organized Crime in Europe: Concepts, Patterns, and Control Policies in the European Union and Beyond*. The Netherlands: Springer.

Financial Action Task Force. (2010). "Global Money Laundering and Terrorist Financing Threat Assessment." July 27. www.fatf.gafi.org.

Finkenauer, J. O. (2005). "Problems of definition: What is organized crime?" *Trends in Organized Crime* 8(3): 63–83.

Flax, L. K., Jackson, R. W., and Stein, D. N. (2002). "Community vulnerability assessment tool methodology." *Natural Hazards Review* 3(4): 163–176.

Forrest, R. and Kearns, A. (2001). "Social cohesion, social capital and the neighborhood." *Urban Studies* 38(12): 2125–2143.

Fox News. (2010). "911 Tape released in Mexican pirate attack on US couple." *Fox News*, October 3.

Fox, K. R. and Lytle, D. A. (1996). "Milwaukee's crypto outbreak: Investigation and recommendations." *Journal of the American Water Works Association* 88(9): 87–94.

Fruin, J. (1971). *Pedestrian Planning and Design*. Metropolitan Association of Urban Designers and Environmental Planners, University of Michigan, MI.

Fuchs, C., Boersma, K., Albrechtslund, A., and Sandoval, M. (2012). *Internet and Surveillance: The Challenges of Web 2.0 and Social Media*. New York: Routledge.

Fugazza, M. and Jacques, J.-F. (2004). "Labor market institutions, taxation, and the underground economy." *Journal of Public Economics* 88(1–2): 395–418.

Galemba, R. G. (2008). "Informal and illicit entrepreneurs: Fighting for a place in the neoliberal economic order." *Anthropology of Work Review* 29(2): 19–25.

Gambetta, D. (1996). *The Sicilian Mafia: The Business of Private Protection*. Cambridge, MA: Harvard University Press.

Garfinkel, S. (2000). *Database Nation: The Death of Privacy in the 21st Century*. Sebastopol, CA: O'Reilly & Associates, Inc.

Geis, G. (1966). "Violence and organized crime." *Annals of the American Academy of Political and Social Science* 364(1): 86–95.

George Mason University Institute for Conflict Analysis and Resolution. (2003). *Terrorism: Concepts, Causes, and Conflict Resolution*. Fort Belvoir, VA: Defense Threat Reduction Agency.

Gilliom, J. (2001). *Overseers of the Poor: Surveillance, Resistance, and the Limits of Privacy*. London: University of Chicago Press.

Government Ethics Center of the Joseph and Edna Josephson Institute of Ethics. (1991). "Ethics at the IRS: A Quest for the Highest Standards." Internal Revenue Service Management Training Program: Workshop and Resource Materials. Marina del Rey, CA.

Grabowsky, P. (2007). "The Internet, technology, and organized crime." *Asian Journal of Criminology* 2(2): 145–161.

Grayman, W. M., Deininger, R. A., and Males, R. M. (2002). *Design of Early Warning and Predictive Source-Water Monitoring Systems*. Denver, CO: AWWA Research Foundataion.

Grayman, W. M. and Males, R. M. (2001). "Risk-based modeling of early warning systems for pollution accidents." *Proceedings of the IWA 2nd World Water Congress*. Berlin, Germany. October 15–19. London: IWA.

Grayman, W. M., Rossman, L. A., Arnold, C., Deininger, R. A., Smith, C., Smith, J. F., and Schnipke, R. (2000). *Water Quality Modeling of Distribution System Storage Facilities*. Denver, CO: AWWA Research Foundation.

Green, C. (2004). "The evaluation of vulnerability to flooding." *Disaster Prevention and Management* 13(4): 323–329.

The Guardian. (2009). "Somali pirates beaten off in second attack on Maersk Alabama." November 18.

The Guardian. (2011). "Brazil creating anti-pirate force after spate of attacks on Amazon riverboats." June 17.

Haestad Methods, Inc. (2002). *Water Security Summit Proceedings*. Waterbury, CT: Haestad Press.

Hagan, F. E. (1983). "The organized crime continuum: A further specification of a new conceptual model." *Criminal Justice Review* 8: 52–57.

Hagan, J. L. and McCarthy, B. (1997). "Anomie, social capital and street criminology." In *The Future of Anomie*, Passas, N. and Agnew, R. (eds.). Boston, MA: Northeastern University Press, pp. 124–141.

Hall, C. and Abdollah, T. (2008). "Pellicano found guilty of racketeering." *Los Angeles Times*, May 15.

Haller, M. H. (1990). "Illegal enterprise: A theoretical and historical interpretation." *Criminology* 28: 207–236.

Haller, M. H. (1992). "Bureaucracy and the mafia: An alternative view." *Journal of Contemporary Criminal Justice* 8(1): 1–10.

Hampson, R. (2009). "Statue of Liberty gets her view back." *USA Today*, July 6.

Harding, B. L. and Walski, T. M. (2000). "Long time-series simulation of water quality in distribution systems." *Journal of Water Resources Planning and Management* 126(4): 1199.

Hargesheimer, E. (2002). *Online Monitoring for Drinking Water Utilities.* Denver, CO: AWWA Research Foundation.

Harr, J. (1995). *A Civil Action.* New York: Vintage Books.

Harris, S. (2011). *The Watchers: The Rise of America's Surveillance State.* London: Penguin Books Ltd.

Hewitt, K. (1997). *Regions of Risk: A Geographical Introduction to Disasters.* Essex: Longman.

Hickman, D. C. (1999). *A Chemical and Biological Warfare Threat: USAF Water Systems at Risk.* Maxwell Air Force Base, AL: USAF Counterproliferation Center.

Hier, S. P. and Greenberg, J. (2009). *Surveillance: Power, Problems, and Politics.* Vancouver, BC: UBC Press.

Hilhorst, D. and Bankoff, G. (2004). "Introduction: Mapping vulnerability." In *Mapping Vulnerability: Disasters, Development & People*, Bankoff, G., Frerks, G., and Hilhorst, D. (eds.). Sterling, VA: Earth Scan Publications, pp. 1–9.

Hoffman, B. (2003). "The Logic of Suicide Terrorism." *Atlantic Monthly* 291(5): 40–47.

Hoffman, B. (2006). *Inside Terrorism.* 2nd edition. New York: Columbia University Press.

Hoover, J. E. (1941). "Water supply facilities and national defense." *Journal of the American Water Works Association* 33(11): 1861–1865.

Howard, D. (2013). "Tom Cruise accused of wiretap conspiracy with convicted criminal Anthony Pellicano during Nicole Kidman divorce (exclusive)." July 6, www.celebuzz .com/2012-07-06/tom-cruise-accused-of-wiretap-conspiracy-with-convicted-criminal-anthony-pellicano-during-nicole-kidman-divorce-exclusive/.

Hudson, R. A. (2002). *Who Becomes a Terrorist and Why: The 1999 Government Report on Profiling Terrorists.* Guilford, CT: The Lyons Press.

Humphreys, A. (2006). "One official's 'refugee' is another's 'terrorist'." *National Post*, January 17.

IFRC. (2009). "The epidemic divide." Health and Care Department, International Federation of Red Cross and Red Crescent Society, Geneva, Switzerland, July.

The Independent. (2010). "Russian special forces arrest Somali pirates." May 6.

Intergovernmental Panel on Climate Change. (2007). *Climate Change 2007, Fourth Assessment Report.* New York: Cambridge University Press.

International Federation of Red Cross and Red Crescent Societies. Plan 2010–2011—*Disaster Management and Risk Reduction: Strategy and Coordination.* Executive Summary Report. www.ifrc.org/docs/appeals/annual10/MAA0002910p.pdf.

International Federation of Red Cross and Red Crescent Societies. (2013). "Types of disasters: Definition of a hazard." January 1. www.ifrc.org/en/what-we-do/disaster-management/about-disasters/definition-of-hazard.

Institute for Management Excellence. Traits of managers and leaders, leadership trends. www .itstime.com.

The Irish Times. (2009). "Captain freed unhurt, pirates killed." April 12.

Jablin, F. M. and Putnam, L. (2001). *The New Handbook of Organizational Communication: Advances in Theory, Research, and Methods.* Thousand Oaks, CA: Sage Publications.

Jacobs, J. B. and Peters, E. (2003). "Labor racketeering: The mafia and the unions." *Crime and Justice* 30: 229–282.

James, B., Burton, I., and Egener, M. (1999). "Disaster mitigation and preparedness in a changing climate." May. Environment Canada.

Jenkins, P. (2003). *Advanced Surveillance: The Complete Manual of Surveillance Training.* Keighley: Intel Publishing.

Jensen, D. (2006). *Endgame: Resistance.* New York: Seven Stories Press.

Jensen, D. and Draffan, G. (2004). *Welcome to the Machine: Science, Surveillance, and the Culture of Control.* White River Junction, VT: Chelsea Green Publishing Company.

Josephson, M. (1993). *Making Ethical Decisions.* Marina del Rey, CA: The Josephson Institute of Ethics.

Juergensmeyer, M. (2000). *Terror in the Mind of God.* Berkeley, CA: University of California Press.

Karp, H. B. and Abramms, B. (1992). "Doing the right thing." *Training and Development,* August, pp. 37–41.

Kates, R. W., Hohenemser, C., and Kasperson, J. X. (1985). *Perilous Progress: Managing the Hazards of Technology.* Boulder, CO: Westview Press.

Khan, A. (1987). "A theory of international terrorism." *Connecticut Law Review* 19: 945.

Kissinger, H. (2001). "The pitfalls of universal jurisdiction." *Foreign Affairs,* July/August.

Klein, M. W., Kerner, H.-J., Maxson, C., and Weitekamp, E. (2001). *The Eurogang Paradox: Street Gangs and Youth Groups in the US and Europe.* Boston, MA: Springer.

Klein, M. W., Weerman, F. M., and Thornberry, T. P. (2006). "Street gang violence in Europe." *European Journal of Criminology* 3(4): 413–437.

Konstam, A. (2008). *Piracy: The Complete History.* Oxford: Osprey Publishing.

Kovacs, P. and Kunreuthler, H. (2001). "Managing catastrophic risk: Lessons from Canada." *Paper presented at the ICLR/IBC Earthquake Conference,* March 23, Simon Fraser University, Vancouver, BC.

Krane, J. (2006). "US Navy warships exchange gunfire with suspected pirates off Somali coast." *The Seattle Times,* March 19.

Krug, E. G., Dahlberg, L. L., Mercy, J. A., Zwi, A. B., and Lozano, R. (2002). "World report on violence and health." World Health Organization, Geneva, Switzerland.

Laidler, K. (2008). *Surveillance Unlimited: How We've Become the Most Watched People on Earth.* Cambridge: Icon Books Ltd.

The Land League. (2009). "The 'no rent' manifesto: Text of the document." *The New York Times.* August 2.

Landler, M. (2013). "President Obama calls the Boston marathon bombings 'an act of terror'." *Daily News,* April 17.

Law Lords Department. (1997). "Semco Salvage & Marine Pte. Ltd. v. Lancer Navigation." The Stationery Office Ltd. February 6, p. 1.

Lee, S.-H. and Drew, K. (2011). "South Korea rescues crew and ship from pirates." *The New York Times,* January 21.

Leeson, P. T. (2009). "Want to prevent piracy? Privatize the ocean." *National Review,* April 13.

Lekuthai, A. and Vongvisessomjai, S. (2001). "Intangible flood damage quantification." *Water Resources Management* 15: 343–362.

Lewis, M. (2010). "USS Nicholas captures suspected pirates." *United States Department of Defense American Forces Press Service,* April 1.

Li, X. and Correa, C. M. (2009). *Intellectual Property Enforcement: International Perspectives.* Northampton, MA: Edward Elgar Publishing.

The Library of Congress. (1999). "The sociology and psychology of terrorism." A report prepared under an interagency agreement by the Federal Research Division, Library of Congress, Washington, DC.

Lind, M. (2005). "The legal debate is over: Terrorism is a war crime." *The New America foundation,* May 5. www.Newamerica.net.

Lloyd, J. (2001). "Building political savvy in a good way." *The Business Journal.* milwaukee .bcentral.com/milwaukee/stories/2001/07/23/smallb2.html.

Logan, T. (2005). "Robotic vessels against pirates." *BBC World Service,* December 14.

Lyman, M. D. and Potter, G. W. (2011). "Organized crime and the drug trade (From Drugs and Society: Causes, Concepts and Control)." *Criminal Justice Policy Review* 22(2): 213–227.

Lynch, W. S. and Phillips, J. W. (1971). "Organized crime-violence and corruption." *Journal of Public Law* 20: 59.

Lynn, W. J. III. (2010). "Defending a new domain: The pentagon's cyberstrategy." *Foreign Affairs*, September/October, pp. 97–108.

Lyon, D. (2001). *Surveillance Society: Monitoring in Everyday Life*. Philadelphia, PA: Open University Press.

Lyon, D. (2006). *Theorizing Surveillance: The Panopticon and Beyond*. Cullompton: Willan Publishing.

Lyon, D. (2007). *Surveillance Studies: An Overview*. Cambridge: Polity Press.

Macionis, J. J. and Gerber, L. M. (2010). *Sociology*. 7th Canadian Edition. Toronto, ON: Pearson Canada Inc.

Mackey, R. (2009). "Can soldiers be victims of terrorism?" *The New York Times*, November 20.

Maltz, M. D. (1976). "On defining 'organized crime': The development of a definition and a typology." *Crime & Delinquency* 22(3): 338–346.

Maltz, M. D. (1990). *Measuring the Effectiveness of Organized Crime Control Efforts*. Chicago, IL: University of Illinois.

Martyn, A. (2002). "The right of self-defence under international law–the response to the terrorist attacks of 11 September." Australian Law and Bills Digest Group, Parliament of Australia Web site, February 12. lr.law.qut.edu.au/article/download/200/194.

Masciandaro, D. (2000). "The illegal sector, money laundering and the legal economy: A macro-economic analysis." *Journal of Financial Crime* 8(2): 103–112.

Matsueda, R. L. (1982). "Testing control theory and differential association: A causal modeling approach." *American Sociological Review* 47(4): 489–504.

Matsueda, R. L. (1988). "The current state of differential association theory." *Crime & Delinquency* 34(3): 277–306.

Matsueda, R. L. and Heimer, K. (1987). "Race, family structure, and delinquency: A test of differential association and social control theories." *American Sociological Review* 52(6): 826–840.

Matteralt, A. (2010). *The Globalization of Surveillance*. Cambridge: Polity Press.

Maxwell, J. C. (1999). *The 21 Indispensable Qualities of a Leader: Becoming the Person Others Will Want to Follow*. Nashville, TN: Thomas Nelson Publishers.

Mays, L. W. (2004). *Water Supply Systems Security*. New York: McGraw-Hill Books.

McBean, E. A., Gorrie, J., Fortin, M., Ding, J., and Moulton, R. (1988). "Flood depth-damage curves by interview survey." *Journal of Water Resources Planning and Management* 114(6): 613–633.

Mendelsohn, B. (2005). "Sovereignty under attack: The international society meets the Al Qaeda network." *Review of International Studies* 31(1): 45–68.

Merton, R. (1938). "Social structure and Anomie." *American Sociological Review* 3(5): 672–682.

Messner, F. and Meyer, V. (2005). "Flood damage, vulnerability and risk perception— Challenges for flood damage research." UFZ Discussion Papers, Department of Economics 13/2005.

Miller, J. W. (2009). "Loaded: Freighters ready to shoot across pirate bow." *Wall Street Journal*. January 5, p. A9.

Miller, W. B. (1992). *Crime by Youth Gangs and Groups in the United States*. Washington, DC: US Department of Justice, Office of Justice Programs, Office of Juvenile Justice and Delinquency Prevention.

Mills, E. (2005). "Insurance in a climate of change." *Science* 309: 1040–1044.

Morrison, S. (2002). "Approaching organized crime: Where are we now and where are we going?" Canberra, ACT: Australian Institute of Criminology.

Mukundan, P. (2009). "Pirates attack tanker, NATO frees 20 fishermen." *Associated Press*, April 18.

Murphy, P. J. (1986). *Water Distribution in Woburn, Massachusetts*. Amherst, MA: University of Massachusetts.

Nagin, D. S. and Pogarsky, G. (2001). "Integrating celerity, impulsivity, and extralegal sanction threats into a model of general deterrence: Theory and evidence*." *Criminology* 39(4): 865–892.

National Advisory Committee on Criminal Justice Standards and Goals. (1976). *Disorders and Terrorism*. Washington, DC.

Nelson, D. (2009). "Pakistani president Asif Zardari admits creating terrorist groups." *The Daily Telegraph*, July 8.

The Netherlands Red Cross. (2006). *Preparedness for Climate Change Report*. The Netherlands: Red Cross/Red Crescent Climate Centre.

Nightingale, A. and Bockmann, M. W. (2012). "Somalia piracy falls to six-year low as guards defend ships." *Bloomberg News*, October 22.

Non-profit risk management centre. Understanding the volunteer protection act and a summary of state volunteer protection laws. www.nonprofitrisk.org/library/articles/employee_or_volunteer.shtml.

North American Emergency Response Guidebook. (2012). Secretariat of Transport and Communications, Transport Canada, US Department of Transportation, Pipeline and Hazardous Material Safety Administration, Ottawa, ON, Canada.

Nunberg, G. (2001). "Head games/It all started with Robespierre/'terrorism': The history of a very frightening word." *San Francisco Chronicle*, October 28.

O'Neil, M. (2006). "Rebels for the system? Virus writers, general intellect, cyberpunk and criminal capitalism." *Continuum: Journal of Media & Culture Studies* 20(2): 225–241.

Paoli, L. (2002). "The paradoxes of organized crime." *Crime, Law and Social Change* 37(1): 51–97.

Pape, R. A. (2003). "The strategic logic of suicide terrorism." *American Political Science Review* 97(3): 1–19.

Parenti, C. (2003). *The Soft Cage: Surveillance in America from Slavery to the War on Terror.* New York: Basic Books.

Passas, N. (1990). "Anomie and corporate deviance." *Crime, Law and Social Change* 14(2): 157–178.

Pastor, J. F. (2009). *Terrorism & Public Safety Policing: Implications of the Obama Presidency*. New York: Taylor & Francis.

Pelisek, C. (2011) "Anthony Pellicano: The Hollywood phone hacker breaks his silence." *Newsweek*, August 7.

Pelling, M. (1999). "The political ecology of flood hazard in urban Guyana." *Geoforum* 30: 249–261.

Powell, E. (1966). "Crime as a function of anomie." *The Journal of Criminal Law, Criminology, and Police Science* 57(2): 161–171.

Press TV. (2009). "French forces seize pirates mother ship." *Press TV*, April 16.

Priest, D. and Arkin, W. (2010). "A hidden world, growing beyond control." *The Washington Post*, July 19.

Public Safety Canada. (2002). " Threat analysis: Geomagnetic storms—Reducing the threat to critical infrastructure in Canada." Communication Division, Ottawa, ON, April.

Public Safety Canada. (2003). *Threat Analysis: Threats to Canada's Critical Infrastructure*. Ottawa, ON: Office of Critical Infrastructure Protection and Emergency Preparedness.

Public Safety Canada. (2009). *National Strategy for Critical Infrastructure*. Ottawa, ON: Government of Canada.

Public Safety Canada. Canadian Disaster Database. www.publicsafety.gc.ca/cnt/rsrcs/cndn-dsstr-dtbs/index-eng.aspx.

Pugliese, D. (2002). *Canada's Secret Commandos: The Unauthorized Story of Joint Task Force II*. Ottawa, ON: Espirit de Corps Books.

Purpura, P. P. (2007). "Terrorism and homeland security: An introduction with applications." Boston, MA: Butterworth-Heinemann.

Quarantelli, E. L. (1998). "Where we have been and where we might go?" In *What Is A Disaster?* London: Routledge, pp. 146–159.

Radinsky, M. (1994). "Retaliation: The genesis of a law and the evolution toward international cooperation: An application of game theory to modern international conflicts." *George Mason Law Review (Student Edition)* 2(53): 52–75.

Reddy, M., Borum, R., Vossekuil, B., Fein, R. A., Berglund, J., and Modzeleski, W. (2001). "Evaluating risk for targeted violence in schools: Comparing risk assessment, threat assessment, and other approaches." *Psychology in the Schools* 38(2): 157–172.

Renfroe, N. A. and Smith, J. L. (2011). "Threat/vulnerability assessments and risk analysis." *World Building Design Guide.* www.wbdg.org.

Rex, J. (1968). *The Sociology of a Zone of Transition.* Oxford: Oxford University Press.

Reynolds, P. (2005). "UN staggers on road to reform." *BBC News,* September 14.

Richardson, L. (2006). *What Terrorists Want: Understanding the Terrorist Threat?* London: John Murray.

Richardson, P. J. and Thomas, D. (1999). *Archbold Criminal Pleading, Evidence and Practice.* London: Sweet & Maxwell.

Ricolfi, L. (2005). "Palestinians 1981–2003." In *Making Sense of Suicide Missions,* Gambetta, D. (ed.). 1st edition. Oxford: Oxford University Press, pp. 76–130.

Roberts, R. and Konrad, J. (2009). "Mariner details life aboard Maersk Alabama lifeboat." *NPR,* April 11.

Rodin, D. (2006). "Terrorism." In *Routledge Encyclopedia of Philosophy,* Craig, E. (ed.). London: Routledge.

Romero, S. (2009). "Shining path." *The New York Times,* March 18.

Roots, R. (2005). "Mafia brotherhoods: Organized crime, Italian style." *Contemporary Sociology: A Journal of Reviews* 34(1): 67–68.

Rose, P. (2003). "Disciples of religious terrorism share one faith." *Christian Science Monitor,* August 28.

Ruby, C. L. (2002). "The definition of terrorism." *Analyses of Social Issues and Public Policy* 2(1): 9–14.

Rygel, L., O'Sullivan, D., and Yarnal, B. (2005). "A method for constructing a social vulnerability index." *Mitigation and Adaptation Strategies for Global Change,* January 21.

Sageman, M. (2004). *Understanding Terror Networks.* Philadelphia, PA: University of Pennsylvania Press.

Salter, F. (2002). *Risky Transactions: Trust, Kinship, and Ethnicity.* New York: Berghahn Books.

Sampson, R. J. and Groves W. B. (1989). "Community structure and crime: Testing social-disorganization theory." *American Journal of Sociology* 99(4): 774–802.

Sanchez-Jankowski, M. (1991). "Gangs and social change." *Theoretical Criminology* 7(2): 192–216.

Sanchez-Jankowski, M. (1991). *Islands in the Street: Gangs and American Urban Society.* Berkeley, CA: University of California Press.

Sanchez-Jankowski, M. (1995). *Ethnography, Inequality, and Crime in the Low-Income Community.* Stanford, CA: Stanford University Press.

Savona, E. (1997). "Globalization of crime: The organizational variable." *15th International Symposium on Economic Crime.* September 14–20. Cambridge.

Savona, E. U. and Vettori, B. (2009). "Evaluating the cost of organized crime from a comparative perspective." *European Journal on Criminal Policy and Research* 15(4): 379–393.

Scheider, B. R. and Davis, J. A. (2006). *Avoiding the Abyss: Progress, Shortfalls, and the Way Ahead in Combating the WMD threat.* Westport, CT: Greenwood Publishing Group.

Schloenhart, A. (1999). "Organized crime and the business of migrant trafficking." *Crime, Law and Social Change* 32(3): 203–233.

Schorn, D. (2006). "Ed Bradley reports on extremists now deemed biggest domestic terror threat." *60 Minutes*, June 18. www.cbsnews.com/stories/2005/11/10/60minutes/main1036067.shtml.

Schuster, C. J., Murray, S., and McBean, E. A. (2007). "Vulnerability characterization, mapping, and assessment: A study of flooding scenarios for the credit river watershed." Report Submitted to Credit Valley Conservation. University of Guelph, Guelph, ON.

Scott, W. (1992). *Organizations. Rational, Natural, and Open Systems*. Englewood Cliffs, NJ: Prentice Hall.

Serviceleader.org. Volunteer Management/Service Leader Resources. www.serviceleader.org/manage/.

Shabad, G. and Ramo, F. J. L. (1995). "Political violence in a democratic state: Basque terrorism in Spain." In *Terrorism in Context*, Crenshaw, E. (ed.). University Park, PA: Pennsylvania State University, p. 467.

Siegel, D. and Nelen, J. M. (2008). *Organized Crime: Culture, Markets and Policies*. New York: Springer.

Slackman, M. (2009). "New status in Africa empowers an ever-eccentric Gaddafi." *The New York Times*, March 22.

Smilowitz, R. (2011). "Designing buildings to resist explosive threats." *World Building Design Guide*, October 19. www.wbdg.org.

Smith, D. C. (1976). "Mafia: The prototypical alien conspiracy." *Annals of the American Academy of Political and Social Science* 423(1): 75–88.

Smith, D. C. (1980). "Paragons, pariahs, and pirates: A spectrum-based theory of enterprise." *Crime & Delinquency* 26(3): 358–386.

Smith, D. C. (1991). "Wickersham to Sutherland to Katzenbach: Evolving an 'official' definition for organized crime." *Crime, Law and Social Change* 16(2): 135–154.

Smith, K. (2000). *Environmental Hazards: Assessing Risk and Reducing Disaster*. London: Routledge.

Smith, W. E. (2001). Terror Aboard Flight 847. *TIME Magazine*. June 24.

Sobsey, M. D. (1999). "Methods to identify and detect contaminants in drinking water." In *Identifying Future Drinking Water Contaminates*. Washington, DC: National Academy Press.

Sparks, R. and Hope, T. (2000). *Crime, Risk and Insecurity: Law and Order in Everyday Life and Political Discourse*. New York: Routledge.

Sprenger, P. (1999). "Sun on Privacy: 'Get Over It'." *Wired Magazine*. January 26.

St. Bernard G. (2003). "Towards the construction of a social vulnerability index—Theoretical and methodological considerations." *Journal of Social and Economic Studies* 53(2): 1–29.

Staples, W. G. (2000). *Everyday Surveillance: Vigilance and Visibility in Post-Modern Life*. Lanham, MD: Rowman & Littlefield Publishers.

Steinfels, P. (2003). "Beliefs: The just-war tradition, its last-resort criterion and the debate on an invasion of Iraq." *The New York Times*. March 1.

Stembach, G. (2002). "History of anthrax." *Journal of Emergency Medicine* 24(4): 463–467.

Stephens, B. (2008). "Why don't we hang pirates anymore?" *The Wall Street Journal*, November 25.

Stille, A. (2003). "Historians trace an unholy alliance; Religion as the root of nationalist feeling." *The New York Times*, May 31.

Stohl, M. (1986). "The Superpowers and International Terror." In *Government Violence and Repression: An Agenda for Research*, Stohl, M. and Lopez, G. A. (eds.). New York: Greenwood Press, 1986.

Stohl, M. and Lopez, G. A. (1988). *Terrible beyond Endurance? The Foreign Policy of State Terrorism*. New York: Greenwood Press, 1988.

Stossel, J. and Kirell, A. (2009). "Could profit motive put an end to piracy?" *ABC News*, May 8.

Style. (2009). "Number of terrorist attacks, fatalities." *The Washington Post*, June 12.

The Sydney Morning Herald. (2007). "Whaling acid attack terrorist act: Japan." February 9.

Tapsell, S. M., Penning-Roswell, E. C., Tunstall, S. M., and Wilson, T. L. (2002). "Vulnerability to flooding: Health and social dimensions." *Philosophical Transactions of the Royal Society* 360: 1511–1525.

Tapsell, S. M. and Tunstall, S. M. (2001). "The health and social effects of the June 2000 flooding in the northeast region." *Report to the Environment Agency*. Flood Hazard Research Centre, Middlesex University, Enfield.

"Terrorism." (1979). www.dictionary.com.

"Terrorism." (2012). "Cambridge international dictionary of english." Dictionary.cambridge.org.

Thompson, B. (2005). "Hollywood on crusade." *The Washington Post*, May 1.

Thompson, L. (2002). *The Counter-Insurgency Manual: Tactics of the Anti-Guerrilla Professionals*. London: Greenhill Books.

Thompson, L. (2005). *Bodyguard Manual*. London: Greenhill Books.

Thompson, L. (2005). *Hostage Rescue Manual*. London: Greenhill Books.

Thrasher, F. (1927). *The Gang: A Study of 1313 Gangs in Chicago*. Chicago, IL: University of Chicago Press.

Tilly, C. (1985). "State formation as organized crime." In *Bringing the State Back In*, Evans, P., Rueschemeyer, D., and Skocpol, T. (eds.). Cambridge: Cambridge University Press.

Time Magazine. (2009). "Basque terrorist group marks 50th anniversary with new attacks." July 31.

Toby, J. (1957). "Social disorganization and stake in conformity: Complementary factors in the predatory behavior of hoodlums." *Journal of Criminal Law, Criminology & Police Science* 48(12): 12–17.

Townsend, P., Phillimore, P., and Beattie, A. (1988). *Health and Deprivation: Inequality and the North*. London: Croom Helm.

Turner, B. L. II, Kasperson, R. E., Matson, P. A., McCarthy, J. J., Corell, R. W., Christensen, L., Eckley, N. et al. (2003). "A framework for vulnerability analysis in sustainability science." *Proceedings of the National Academy of Sciences of the United States of America* 100(14): 8074–8079.

United Nations. (1982). "United Nations Convention on the Law of the Sea (UNCLOS) of 10 December 1982, Part VII: High Seas."

United Nations. (1994). United Nations declaration on measures to eliminate international terrorism annex to UN general assembly resolution 49/60, "Measures to Eliminate International Terrorism." December 9, UN Doc. A/Res/60/49. www.un.org/documents/ga/res/49/a49r060.htm.

United Nations. (1998). "Report of the special rapporteur on systemic rape." *The United Nations Commission on Human Rights*, June 22. www.un.org/unifeed/script.asp?scriptId=73.

United Nations. (2002). "Press conference with Kofi Annan and foreign minister Kamal Kharrazi." www.un.org. January 26.

United Nations. (2005). "UN Reform." March 21.

United Nations. (2008). "UN maritime agency welcomes security council action on Somalia piracy." *UN News Centre*, June 5.

United Nations Development Programme. "United Nations development goals." www.undp.org/content/undp/en/home/mdgoverview/.

United Nations Millennium Declaration. www.un.org/millennium/declaration/ares552e.htm.

UNODC. (2006). "Trafficking in persons: Global patterns." Anti-Human Trafficking Unit report, UNODC, New York.

US Department of Transportation. (2012). "Emergency Response Guide No. 112." In *Emergency Response Guidebook—A Guidebook for First Responders during the Initial Phase of a Dangerous Goods/Hazardous Materials Transportation Incident*. Department of Transportation.

US Department of Transportation. (2012). "Emergency Response Guide No. 114." In *Emergency Response Guidebook—A Guidebook for First Responders during the Initial Phase of a Dangerous Goods/Hazardous Materials Transportation Incident*. Department of Transportation.

US Department of Transportation. (2012). "Emergency Response Guide No. 117." In *Emergency Response Guidebook—A Guidebook for First Responders during the Initial Phase of a Dangerous Goods/Hazardous Materials Transportation Incident*. Department of Transportation.

US Department of Transportation. (2012). "Emergency Response Guide No. 119." In *Emergency Response Guidebook—A Guidebook for First Responders during the Initial Phase of a Dangerous Goods/Hazardous Materials Transportation Incident*. Department of Transportation.

US Department of Transportation. (2012). "Emergency Response Guide No. 124." In *Emergency Response Guidebook—A Guidebook for First Responders during the Initial Phase of a Dangerous Goods/Hazardous Materials Transportation Incident*. Department of Transportation.

US Department of Transportation. (2012). "Emergency Response Guide No. 125." In *Emergency Response Guidebook—A Guidebook for First Responders during the Initial Phase of a Dangerous Goods/Hazardous Materials Transportation Incident*. Department of Transportation.

US Department of Transportation. (2012). "Emergency Response Guide No. 153." In *Emergency Response Guidebook—A Guidebook for First Responders during the Initial Phase of a Dangerous Goods/Hazardous Materials Transportation Incident*. Department of Transportation.

US Department of Transportation. (2012). "Emergency Response Guide No. 159." In *Emergency Response Guidebook—A Guidebook for First Responders during the Initial Phase of a Dangerous Goods/Hazardous Materials Transportation Incident*. Department of Transportation.

US Department of Transportation. (2012). "Emergency Response Guide No. 163." In *Emergency Response Guidebook—A Guidebook for First Responders during the Initial Phase of a Dangerous Goods/Hazardous Materials Transportation Incident*. Department of Transportation.

US Department of Transportation. (2012). "Emergency Response Guide No. 164." In *Emergency Response Guidebook—A Guidebook for First Responders during the Initial Phase of a Dangerous Goods/Hazardous Materials Transportation Incident*. Department of Transportation.

US District Court. (2009). *Hepting v. AT&T*, U.S. District Court for the Northern District of California, San Francisco, CA, June 3.

US Fire Association. (2013). "Dealing with the media—book/manual/guide." www.usfa.fema .gov/fserd/pro_med1_list.htm.

USA Freedom Corps. www.usafreedomcorps.gov/.

USA Today. (2007). "Navy: US ship fired at pirates off Somalia." June 6.

USA Today. (2008). "Shooting reported on pirate ship surrounded by US destroyer Doug Stanglin." September 30.

Van Duyne, P. (1997). "Organized crime, corruption and power." *Crime, Law and Social Change* 26(3): 201–238.

Verjee, Z. and Starr, B. (2009). "Captain jumps overboard, SEALs shoot pirates, official says." *CNN*, April 12.

Volunteer Canada. (2012). "The Canadian code for volunteer involvement: Values, guiding principles and standards of practice." Volunteer Canada. www.volunteer.ca.

Volunteerism in the United States. usinfo.state.gov/usa/volunteer/.

von Lampe, K. (2003). "The use of models in the study of organized crime." *Paper presented at the 2003 conference of the European Consortium for Political Research*, Marburg, Germany, September 19.

Vossekuil, B., Borum, R., Fein, R. A., and Reddy, M. (2001). "Preventing targeted violence against judicial officials and courts." *Annals of the American Academy of Political and Social Science* 576: 78–90.

Wang, P. (2013). "The rise of the Red Mafia in China: A case study of organised crime and corruption in Chongqing." *Trends in Organized Crime* 16(1): 49–73.

Warrick, J. (2007). "Domestic Use of Spy Satellites To Widen." *Washington Post*. August 16, p. A01.

Washington Military Department. (2001). "Washington state comprehensive emergency management planning guide." Washington Military Department, Emergency Management Division. www.emd.wa.gov/plans/documents/CompleteCEMP.pdf.

The Washington Post. (2005). "Pirates open fire on cruise ship off Somalia." November 5.

The Washington Post. (2013). "The economics of Somali piracy." March 3.

WBDG Secure/Safe Committee. (2013). "Fire protection." *World Building Design Guide*, August 29. www.wbdg.org.

WBDG Secure/Safe Committee. (2013). "Natural hazards and security." *World Building Design Guide*, August 29. www.wbdg.org.

WBDG Secure/Safe Committee. (2013). "Occupant safety and health." *World Building Design Guide*, August 29. www.wbdg.org.

WBDG Secure/Safe Committee. (2013). "Security for building occupants and assets." *World Building Design Guide*, August 30. www.wbdg.org.

Weick, K. E. (1987). "Organizational culture as a source of high reliability." *California Management Review* 29(2): 112–127.

Welland Canal. (2010). "Moscow University." *Welland Canal*. Retrieved May 5, www.wellandcanal.ca/salties/m/moscowuniversity/university.htm.

White, R. and Etkin, D. (1997). "Climate change and the Canadian insurance industry." *Natural Hazards* 16: 135–163.

White, T. (1997). *Fighting Skills of the SAS and Special Forces*. London: Robinson.

Whyte, W. (1943). "Social organization in the slums." *American Sociological Review* 8(1): 34–39.

Wilkinson, P. (1997). "The media and terrorism: A reassessment." *Terrorism and Political Violence* 9(2): 51–64.

Williams. (2001). "Organized crime and cybercrime: Synergies, trends, and responses." *Global Issues* 6(2): 22–26.

Wisner, B., Blaikie, P., Cannon, T., and Davis, I. (2004). *At Risk: Natural Hazards, People's Vulnerability and Disasters*. 2nd edition. London: Routledge.

Woolf, M. (2008). "Pirates can claim UK asylum." *The Sunday Times*, April 13.

World Bank. "Disaster risk management." web.worldbank.org/WBSITE/EXTERNAL/TOPICS/EXTURBANDEVELOPMENT/EXTDISMGMT/0,,menuPK:341021~pagePK:149018~piPK:149093~theSitePK:341015,00.html.

World Health Organization. (2004). "Mortality and burden of disease estimates for WHO member states in 2002." A report by WHO Press, World Health Organization, Geneva, Switzerland.

World Health Organization. (2008). "Global burden of disease." A report by WHO Press, World Health Organization, Geneva, Switzerland.

Yager, J. (2010). "Former intel chief: Homegrown terrorism is a devil of a problem." July 25, thehill.com.

Young, J. (1999). *The Exclusive Society: Social Exclusion, Crime and Difference in Late Modernity*. Thousand Oaks, CA: Sage Publications.

Young, R. (2007). "PBS Frontline: 'Spying on the Home Front'." *PBS: Frontline*, May 16.

Zuckerman, M. J. and Fratianno, J. (1987). *Vengeance is Mine: Jimmy "the Weasel" Fratianno Tells How He Brought the Kiss of Death to the Mafia*. New York: Macmillan.

Index

Note: Locators "*f*" and "*t*" denote figures and tables in the text

A

Abu Sayyaf Group, 303
Advanced GEOINT, 157
Advanced life support (ALS), 273
Air Force Special Operations Command, 302
Al-Gama'a al-Islamiyya, 303
Alien conspiracy theory, 122
Al-Jihad (Egyptian Islamist Group), 303
All-hazard design, 66
Al-Qaeda, 303
American Society of Civil Engineers (ASCEs), 66
Arc fault interruption (AFI), 56
Armed Islamic Group (GIA), 303
Arming Pilots against Terrorism Act, 97
Armstrong's mixture, 268
Army Commando Coys, 299
Automatic identification systems (AISs), 126
Avalanche, 263

B

Bacillus anthracis, 37
Basic event, 81
Basic life support (BLS), 273
Basque Fatherland and Liberty—ETA, 304
Bell's theory, 122
Biological agents, explosives
    EMS, 273–274
    fire department, 272–273
    HazMat group, 275–276
    law enforcement, 274–275
    preblast operations, 270–272
    unexploded device, 270
Bioterrorism, 265
Blizzard, 263
Bodyguards, 178
Bootlegging, 121
Building security, requirements
    construction, 52
    egress, 53
    elements, 52
    fire detection/protection, 54
Building, threats
    hand-carried weapon, 166–168
    moving vehicle attack, 165

protection
    objectives, 164
    strategies, 170
stationary vehicle
    parking garage, 165
    secured perimeter line, 165
Business continuity plan (BCP), 220, 223–224
phases, developing
    analysis, 258–259
    implementation, 260
    maintenance, 261–262
    solution design, 259–260
    testing, 260–261
recovery, 220
Business impact analysis (BIA), 258
Bust-out, 121
Buzo Tactico, 299

C

Canadian Food Inspection Agency (CFIA), 94
Canadian Special Operations Forces Command (CANSOFCOM), 300
Canadian Special Operations Regiment (CSOR), 300
Captain of the port (COTP), 100
Cardiopulmonary resuscitation (CPR), 26
Casualty collection point (CCP), 272, 274
Center for Excellence (COE), 295
Centers for Disease Control and Prevention (CDC), 38, 95
Central Intelligence Agency (CIA), 113, 116, 131
Certified information security systems professional (CISSP), 175
Chemical, biological, or radiological (CBR), 66, 70, 73, 127
Chemical, biological, radiological, nuclear, and explosive (CBRNE) events, 265
Chief Administrative Officer (CAO), 5
Chief executive officer (CEO), 7, 133
Civil Contingencies Act (CCA), 293
Civil Defence Emergency Management Groups (CDEMGs), 292
Civil unrest, 265
Clean-desk policy, 142
Close circuit television (CCTV), 138
Coercion, 116

323

Combat Zones That See, 147
Communications Assistance for Law
    Enforcement Act (CALEA), 145
Community-based organizations (CBOs), 27
Computational fluid dynamics (CFDs), 86
Computer Emergency Response Team
    (CERT), 177
Computer Incident Alert Center (CIAC), 177
Contingency plans
    BCP, phases
        analysis, 258–259
        implementation, 260
        maintenance, 261–262
        solution design, 259–260
        testing, 260–261
    force protection
        measures, 254
        operations, 254
        outline, 253
        routine security operations, 254
        threat, estimation, 253
        vulnerabilities, assessment, 254
    security, developing, 255–258
        approvals, 255
        communications/consultations, 255–256
        context, 256
        executive summary, 255
        implementation, 257–258
        risk assessment/treatment, 257
Crime Prevention through Environmental Design
    (CPTED), 67, 165
Crisis management plan (CMP), 221
    components, 221
        action procedures, 222
        appendix, 223
        considerations, 221–222
        document introduction, 221
        exercising, 223
        facility, 222
        notification procedures, 222
        postcrisis analysis, 222
        scenarios/situations, 221
        team, 222
    information technology, 221
Critical infrastructure protection (CIP), 47–48
    aviation security, 95–99
        threat areas, 95
        typical terminal layout, 98*f*
    building security, 48–71
        Air France crash, 59*f*
        ample lighting, 65*f*
        attack, types, 66
        design elements, facility, 53
        earthquake damage, 60*f*
        emergency power, 54
        fiberglass, usage, 58

        fire department hose outlets, 54*f*
        flooding, 61*f*
        intense storm, 62*f*
        interior exit, 52*f*
        issues, safe working, 55
        occupant emergency plans, 71
        plan layout, 53
        recommendations, 55–58
        requirements, *see* Building security,
            requirements
        threats, *see* Threats, building security
        tornado damage, 60*f*
        wildfire, 63*f*
    cybersecurity, 107–111
        encryption, 109–110
        requirements, 107
        tools, 109
    energy facilities, security, 87–93
        distributed, 93
        electricity, 91
        key elements, 88
        mobile networks, 88
        natural gas, 90*f*
        oil/gas pipelines, 90–91
        power plant, attack, 92
        protection, level, 89
    food/agricultural security, 93–95
        contamination, 95
        nation, attack, 95
        threats, types, 94–95
    land transportation, 103–106
        terrorist attacks, 105
        vulnerabilities, 104
    maritime security and asset protection,
        99–103
        Coast Guard, laws, 99
        facility, 99
        port, 99–100
        vessel, 99–102
    water supply systems security, 71–87
        challenges, 72
        contamination, 73, 87
        detecting spills, 77
        distribution model, 82, 85
        early warning, 76
        events, types, 72
        key elements, 72–73
        online monitoring, 77
        poisons, toxicity, 76*t*
        potential threat, 74*t*–75*t*
        removal, contamination, 78*t*
        risk assessment, 80*f*, 81*f*, 83*f*
        spill models, 86
        vulnerability, 71, 72*f*
Cults, 227
Cyberterrorism, 2

D

Data
  mining, 149
  profiling, 149–150
Defense Advanced Research Projects Agency
    (DARPA), 147
Defense diplomacy and military assistance
    (DDMA), 300
Department of Homeland Security (DHS),
    294–295
Department of Transportation (DOT), 128
Department of Transportation
    Emergency Response Guide No. 158,
    see DOT-ERG #158
Digital Subscriber Line (DSL), 108, 213
Disaster management, 1
  all-hazards approach, 12–15
    community profile, 12, 13*t*, 14*t*
    damage, prediction, 15*t*
  definition, 1
  man-made hazards, 2–3
  migration, 3
  natural hazards, 2–3
  technical task, 1, 40
  types, 2–3
Disaster(s), 1, 40
  management, *see* Disaster management
  recovery, 220
  since 1900, 297
  stress management, 285–288
    control, sense, 287
    responses, normal, 286
    time, length, 286–287
Distributed denial of service (DDoS) attack, 38
DOT-ERG #158, 266
DOT-ERG #163, 267–268
DOT-ERG #164, 267

E

Early warning systems, 30, 76–77, 87, 95
  components, 76–77
  mechanisms, 81
Earthquake, 263–264
Emergency Control Ministry (EMERCOM), 293
Emergency management
  disasters, *see* Disaster management
  planning, *see* Emergency planning
  skills
    communication, 23
    decision making, 15–18
    influence, 18–22
    leadership, 18–22
    problem solving, 15–18
    volunteers, developing/managing, 24–28

Emergency Management and Civil Protection
    Act, 4
Emergency Measures Organization
    (EMO), 291
Emergency Medical Services (EMSs), 5, 204,
    265, 269, 279, 281
Emergency operations plan (EOP), 4–11, 16
Emergency planning, 6
  HazMat, 5
  process, 4–5
  team leader, 6
Emergency public information (EPI), 9
Emergency Response Units (ERUs), 290
Environmentally sensitive areas (ESAs), 40
Escorting, formations
  destination, arrival, 185*f*
  elevator, entry
    phase I, 188*f*
    phase II, 188*f*
    phase III, 189*f*
    phase IV, 189*f*
  exit elevator
    phase I, 190*f*
    phase II, 190*f*
    phase III, 191*f*
  foot, 186*f*
  hallway, 185*f*
  speaking engagement, 187*f*
  stage/banquet deployment, 187*f*
Espionage Act of 1917, 99
European Program for Critical Infrastructure
    Protection (EPCIP), 47
European Union (EU), 289
Euskadi Ta Askatasuna (ETA), 304
Executive and close personal protection,
    178–194
  assignment, 193
  dedication, 184
  foot formations, 183*f*, 184*f*
  prevention steps, 192
  risk assessment, 191
  role, 179
  rules, 193–194
  stress, stages, 187
Explosions, 144
Explosive ordnance disposal (EOD), 275
Explosive trace detection (ETD), 97
Extended-period simulation (EPS),
    82, 84–85

F

Facial thermographs, 149
Facility security, 100
  OCS, 100
  regulations, 100

Farabundo Marti National Liberation Front
(FMLN), 225
Federal Aviation Authority (FAA), 36
Federal Bureau of Investigation (FBI), 146
Federal Flight Deck Officer (FFDO) program, 97
Field Assessment and Coordination Team
(FACT), 290
Fire department connections (FDCs), 51
Flash floods, 264
Flight Guard, 96
Floating production storage offloading (FPSO)
units, 100
Food and Drug Administration (FDA), 94
Force multiplier effect, 36
Forensic modeling, 84
Freezing rain, 264

G

GEDAPER, 202
review process, 205
steps, 202
GEOINT, 158
Geospatial information and service (GI&S), 157
Geospatial intelligence (GEOINT), 157–159
analysis, 159
data, types, 158
Glass fail first criteria, 171
Global Facility for Disaster Reduction and
Recovery (GFDRR), 290
Global positioning system (GPS), 103, 106, 125,
152, 158
Golden Shield Project, 147
Go-to-guy, 211
Grenzschutzgruppe 9 (GSG-9), 300
Gross domestic product (GDP), 1, 34
Gross national product (GNP), 41
Groupement d'Intervention de la Gendarmerie
Nationale (GIGN), 300
Groupthink, 17

H

Hamas (Islamic Resistance Movement), 304
Hancock, 148
Hazardous material (HazMat), 5
Hazardous materials spill, 265
Health and Human Service (HHS), 94
Heating, ventilation, and air conditioning
(HVAC) system, 266
confirmed agent, 267
dry agent, threat, 266–267
Heat wave, 264
Heterogeneous Aerial Reconnaissance Team
(HART), 149
Hezbollah (Party of God), 304

High-efficiency particulate air filter (HEPA), 265
Homeland Security Act of 2002, 97
Homeland Security Presidential Directive 7
(HSPD-7), 93
Homicide, 241–242
Human Identification at a Distance, 148
Human intelligence (HUMINT), 145, 150, 152
Hurricanes, 264
Katrina, 281
Hyogo Framework for Action (HFA), 289

I

Improvised explosive devices (IEDs), 96,
143, 179
Incident commander (IC), 201, 204
Incident Command System (ICS), 201
Industrial control system (ICS), 107–110, 201,
271–272
Inference, ladder, 19
Information system centers, 142
Integrated pest management (IPM), 57–58
Integrated Threat Assessment Centre (ITAC), 88
Intergovernmental Panel on Climate Change
(IPCC), 2
International Association of Emergency
Managers (IAEMs), 289
International Civil Aviation Organisation
(ICAO), 300
International Federation of Red Cross and Red
Crescent Societies (IFRCs), 290
International Maritime Bureau (IMB), 124
International Recovery Platform (IRP), 289
The International Ship and Port Facility Security
(ISPS) Code, 2002, 99
Irish Republican Army (IRA), 105, 118, 268

J

Jurisdiction, universal, 125

L

Landslides, 264
Law enforcement agency (LEA), 294
Leadership in Energy and Environmental Design
(LEED), 49
Lesak, David M., 202
Liberation Tigers of Tamil Eelam (LTTE), 305
Lighting strike, 264
Limited political terrorism, 119
Limnic eruption, 264
Line of sight (LOS), 138, 140
Long range acoustic device (LRAD), 125
Long-range reconnaissance patrols (LRRPs), 299
Low-level threats, 87

M

Magnuson Act, 1950, 99
Management strategies
    bomb threat, 206–210
        after-action report, 208
        control publicity, 208
        detonation/damage control, 207–208
        device, locating, 207
        disposal, 207
        evacuations, 206–207
        operations control, 206
        reduce, actions, 209
        telephone recording, 208–209
    close protection detail
        preoperational plans, 218–219
        steps, 217
    crisis/incident
        first responders, 201–206
        mass casualty, 203
        situation, 201–202
    crisis negotiations, 237–241
        course, 238
        field commanders, 238–241
        teams, 238–240
    data center, disasters, 219–225
    event/crowd
        facility, 247–252
        type, 247
    hostage survival, 225–237
        behavior, rules, 228
        denial, 232
        identification, 232–233
        political extremists, 226
        regression, 232
        religious fanatics, 226–227
    TSCMs
        firm, selecting, 214
        statistics, security, 210
        sweep checklists, samples, 216–217
    workplace, violence, 241–246
        avoid, 246
        coworker, hostility, 245
        deal, actions, 244–245
        precautionary measures, 245
        resolution, 246
        scenarios, 243t
        shouting, 246
        Type I—criminal intent, 242
        Type II—customer/client, 242
        Type III—worker-on-worker, 243, 244t
        Type IV—personal relationship, 243, 245t
        weapon, threatening, 246
        zero-tolerance policy, 246
Maritime Transportation Security Act of 2002
        (MTSA), 99

Mass casualty incident, 203
Maximum tolerable period of disruption
        (MTPD), 258
Millennium Development Goals (MDGs), 3
Ministry of Civil Defence & Emergency
        Management (MCDEM), 292
Mitigation, 30, 39, 46, 49–50
Mob, 120
    work, 121
Mobility, queer ladder, 122–123
Monetary vulnerability, 41
Monitoring and Information Center (MIC), 289
Monte Carlo simulation, 81
Multibarrier approach, 137
Multilateration, 146

N

National Crisis Centre (NCC), 292
National Crisis Management Centre
        (NCMC), 292
National Electronic Database Surveillance
        System (NEDSS), 95
National Fire Academy (NFA), 202
National Incident Management System
        (NIMS), 294
National Institute of Standards and Technology
        (NIST), 107
National Response Framework (NRF), 294
National Security Agency (NSA), 146
National Voluntary Organizations Active in
        Disaster (NVOAD), 27
Naval Special Warfare Command, 301
Negative pressure wave, 144
North Atlantic Treaty Organization
        (NATO), 126
$N-1$ standard, 87
Nuclear/radiation accidents, 265

O

Occupational Safety and Health Administration
        (OSHA), 56
Operational technology (OT), 107–108
Operation security (OPSEC), 143, 157, 166
Operation Virtual Shield, 147
Opportunity costs, 30
Ordinary threats, 87
Organized crime, 121
Outer continental shelf (OCS), 100

P

Parliamentary mafiocracy, 120
Patient staging area (PSA), 274
Personal identification number (PIN), 141

Personal protective equipment (PPE), 127,
     202–203, 279
Pinwale, 146
Pirates, 123
Poor devil syndrome, 233
Ports and Waterways Safety Act, 1972, 99
Power failure, 265
Pressure and release (PAR), 42
     components, 42
     risk, 42
Preventative maintenance (PM), 58
Programmable logic controllers (PLCs), 110
Protectee, 133, 178–182, 189–193
Protection strategies
     countersurveillance, 154
     cyberattacks, response, 175–178
          Information Security Risk Management
               Program, 178
          verification, 176
     executive/close, 178–194
          assignment, 193
          dedication, 184
          foot formations, 183*f*, 184*f*,
               see also Escorting, formations
          prevention steps, 192
          risk assessment, 191
          role, 179
          rules, 193–194
          stress, stages, 187
     explosives/blast effects, against, 162–175
          building, *see* Building, threats
          design recommendations, 163
          direct air-blast, 166, 167*f*
          loads, seismic and, 163, 171–175
          progressive collapse, 167–169, 168*f*
          SDOF approach, 169
          standoff distance, 171
     GEOINT
          analysis, 159
          definition, 157–158
     IED awareness, 143–145
          fragmentation effect, 144
          motivating factors, 144
          vehicle bomb data, 145*t*
     physical security, 137–143
          access controls, 140–141
          alarm systems, 139–140
          clean-desk policy, 142
          construction specifications, 139
          facilities, 137–138
          goal, 137
          requirements, types, 137
          restricted area, 140
          standoff zone, 137–138, 138*f*
     site security survey, 154–157
          preparation/planning, 155–156

     review, *see* Security survey process,
          review
surveillance, 145–154
     aerial, 149
     biometric, 148
     corporate, 150–151
     crowd/traffic control, 149
     participatory, 148
travel security, 194–199
     documentation, 197
     fire safety measures, 196
     first-aid kit, 198
     kit, emergency, 198
     preventive measures, 195–197
TSCM program
     activities, 159
     prevention, 162
     threats, report, 160, 161*f*
Provisional Irish Republican Party (PIRA), 305
Pump and dump, 121

Q

Queues, 251
     bulk, 251
     linear, 251

R

Racketeer Influenced and Corrupt Organization
     (RICO) Act, 122
Racketeering, 121
Radio frequency identification (RFID), 152
Recovery point objective (RPO), 258
Recovery time objective (RTO), 258
Reflected pressure, 166
Remote terminal units (RTUs), 110
Response/recovery operations
     biological agents, 265–276
          chemical, 268–269
          confirmed, 267
          dry, threat, 266–267
          explosives, *see* Biological agents,
               explosives
          incendiary devices, 268
          nuclear/radiological, 267–268
          wet/dry, 266
     disasters
          components, 282
          holistic approach, 281
          human-induced, 265
          natural, 263–264
          processes, 283
          stress management, *see* Disaster(s), stress
               management
     special response teams, 283–285

terrorist event, 276–281
    command, 277
    consideration, *see* Terrorist(s), event
        considerations
    defensive operations, 280
    need, responder(s), 278
    rescue operations, 279–280
    resources, need, 277
    undertake, actions, 276
    victims, symptoms, 278
Retrospective modeling, 84
Revolutionary Armed Forces of Colombia
        (FARC), 225, 305
Revolutionary People's Liberation Front, 305
RFID tags, 152
Risk, 30, 42
    assessment, 45–46
    management, 30
        approach, 30–31
        stages, 30
Risk–hazard (RH), 42

S

Safe Food for Canadians Act, 94
Salmonellosis, 84
Sapphire Marketing, 212
Seafarers, 124
Sea Shepherd, 124
Security survey process, review
    access controls, 156
    existing systems, 156–157
    facility(s)
        function/integration, 156
        perimeter, 156
    internal assets, 157
    policies/procedures, 157
    surrounding areas, 156
Self-contained breathing apparatus (SCBA),
        202–203, 266–268
Sendero Luminoso (Shining Path), 305
Sensitive compartment and information facility
        (SCIF), 139
Sensitive compartmented information
        (SCI), 139
Sicilian Mafia, 122–123
Single-degree-of-freedom (SDOF), 169
Sites, 129
Sky marshals, 97
SLUDGEM, 268, 278
Slums, 3
Social Flood Vulnerability Index (SFVI), 43
Social vulnerability, 41
Solitary confinement, 236
Somali Disaster Management Agency
        (SDMA), 293

Source-water protection program, 73
Special-agent-in-charge (SAC), 218
Special mission units (SMUs), 284, 301
Special weapons and tactics (SWAT), 208, 237,
        239–240, 284
Spontaneous Unaffiliated Volunteers
        (SUVs), 295
Standard operating guidelines (SOGs), 204
Standard operating procedures (SOPs), 7, 10, 16,
        64, 97, 201, 204
Standoff distance, 68
Statesmen, 118
Stockholm syndrome, 231–232
    responses, 234
    victims, experience, 233
Stress, 231
    control, ways, 231
    debriefings, 241
    psychological, 232
Structural terrorism, 119
Supervisory Control and Data Acquisition
        (SCADA), 66, 81, 85, 107–109

T

Targeted killings, 113
Targeted violence, 113
Task force
    Cooper, H. H. A., 118
    terrorism, classification
        civil disorder, 118
        limited political, 119
        nonpolitical, 119
        official/state, 119
        political, 118
        quasi, 119
Technical surveillance and countermeasures
        (TSCMs), 159–162, 214
    firm, selecting, 214
    statistics, security, 210
    sweep checklists, samples, 216–217
        electronic analysis, 216–217
        phones/telephone lines, 216
        physical examination notes, 217
Technisches Hilfswerk, 291
Tension leg platforms (TLPs), 100
Terrorism
    definition, 116–117, 122
    structural, 119
Terrorist(s), 118
    acts, 118
    event considerations
        command, 277
        defensive operations, 280
        first responders, 276–277
        life safety, 279

Threats
  assessment
    anthrax, 37
    biological attack, 38
    cyberattacks, 38
    epidemic, 32
    force multiplier effect, 36
    forest fires, 35
    multipronged approach, 39
  building security
    ballistic, 69–70
    cyber/information, 71
    explosive, 68–69
    insider, 68
    unauthorized entry, 67–68
    WMD: CBR, 70
Top event, 80
Tornado, 264
Toronto Stock Exchange (TSE), 35
Transportation Security Administration
  (TSA), 97
Travel security, 194–199
  documentation, 197
  fire safety measures, 196
  first-aid kit, 198
  kit, emergency, 198
  preventive measures, 195–197
Triage, 110
Trichloroethylene (TCE), 84
Tsunami, 264

U

Undeveloped events, 81
UN Disaster Assessment and Coordination
  (UNDAC) team, 290
United Kingdom Special Forces (UKSF), 301
United Nations (UN), 290
United States Coast Guard, 99
  laws, 99
  maritime security activities, 99
United States Computer Emergency Readiness
  Team (US-CERT), 108
Universality principle, 125
Unmanned aerial vehicles (UAVs), 125–126,
  149, 154
Unmanned surveillance vehicles (USVs), 126
UN Office for the Coordination of Humanitarian
  Affair (UN-OCHA), 290
Urban fires, 265
U.S. Army Special Operations Command, 301
U.S. Department of Agriculture (USDA), 94
U.S. Department of Defense (DOD), 141, 164,
  228–229
U.S. Department of Homeland Security
  (DHS), 147

U.S. Environmental Protection Agency
  (EPA), 80
U.S. Federal Communication Commission
  (FCC), 216
U.S. Federal Emergency Management Agency
  (FEMA), 292, 294

V

van Eck phreaking, 146
Vehicle-borne IED (VBIED), 144
Violence, 242
  spectrum, 243
  targeted/behavior, *see* Violence,
    targeted/behavior
  types, 242
Violence, targeted/behavior
  attack, avoiding, 133–136
  definition, 133
  incidents/indicators, 126–129
  maritime piracy, 123–126
    plane hijacker, 125
    robbery, 123
  methods, 114–116
  mind-set, terrorist, 114–116
  organized crime, 120–123
  suspicious activity, 129–133
    security forces, 129
    terror network, 130
  terrorism, 116–120
    activities, 120
    communications, 120
    conflict, types, 119
    definition, 116
    Martyn, Angus, 117
    media exposure, 120
    symbolism, 116
    Walzer, Michael, 117
Violent nonstate actors, 115
Virtual private network (VPN), 109
Volatile organic compounds (VOCs), 55, 57
Volcanic eruption, 264
Voluntary agency (VOLAG), 15, 24, 27
Vulnerability, 39–40
  assessment
    criticality, 40
    health, 43
    index, 39–40
  monetary, 41
  social, 41
  types, 41
Vulnerability index (VI), 39, 43–46
  aspects, 39–40
  Monte Carlo simulation, 44–45
    indices, 45
    weighting factors, 44–45

W

Weapons of mass destruction (WMD), 64, 66, 70,
    100–103, 105, 283
Wenchuan earthquake, 282
Wild fires, 264

World Bank, 290
World Building Design Guide (WBDG), 50
World Conference on Disaster Reduction
    (WCDR), 289
World Health Organization (WHO), 37
World Trade Center (WTC), 36

Printed and bound by CPI Group (UK) Ltd, Croydon, CR0 4YY

18/10/2024

01776266-0004